NUCLEOPHILIC AROMATIC SUBSTITUTION OF HYDROGEN

NUCLEOPHILIC AROMATIC SUBSTITUTION OF HYDROGEN

Oleg N. Chupakhin
Department of Organic Chemistry
Urals State Technical University
Ekaterinburg, Russia

Valery N. Charushin
Department of Organic Chemistry
Urals State Technical University
Ekaterinburg, Russia

Henk C. van der Plas
Department of Organic Chemistry
Agricultural University
Wageningen, The Netherlands

Academic Press
San Diego   New York   Boston   London   Sydney   Tokyo   Toronto

This book is printed on acid-free paper. ∞

Copyright © 1994 by ACADEMIC PRESS, INC.

All Rights Reserved.
No part of this publication may be reproduced or transmitted in any form or by any means, electronic or mechanical, including photocopy, recording, or any information storage and retrieval system, without permission in writing from the publisher.

Academic Press, Inc.
A Division of Harcourt Brace & Company
525 B Street, Suite 1900, San Diego, California 92101-4495

United Kingdom Edition published by
Academic Press Limited
24-28 Oval Road, London NW1 7DX

Library of Congress Cataloging-in-Publication Data

Chupakhin, O. N. (Oleg Nikolaevich)
  Nucleophilic aromatic substitution of hydrogen / by Oleg N. Chupakhin, Valery N. Charushin, Henk C. van der Plas.
    p.   cm.
  Includes bibliographical references and index.
  ISBN 0-12-174640-2
  1. Substitution reactions. 2. Aromatic compounds. I. Charushin, Valery N. II. Plas, H. C. van der. III. Title.
  QD281.S6C48 1994
  547'.604593--dc20   94-20812
                                                                    CIP

PRINTED IN THE UNITED STATES OF AMERICA
94 95 96 97 98 99 BB 9 8 7 6 5 4 3 2 1

"How far it was and how cold..."
Dedicated to our wives Lyubov, Olga, and Mien

CONTENTS

PREFACE		ix
1.	INTRODUCTION	1
2.	NUCLEOPHILIC SUBSTITUTION OF HYDROGEN IN ARENES	16
	I. Unactivated arenes	16
	II. Arene-metal complexes	18
	III. Nitroarenes	31
	A. Reactions with C-nucleophiles	33
	B. Reactions with O-nucleophiles	52
	C. Reactions with N-nucleophiles	59
	D. Reactions with P-nucleophiles	66

IV. Electrochemical S_N^H reactions ... 72

V. Tropylium salts and other non-benzenoid compounds ... 76

VI. Intramolecular S_N^H reactions and rearrangements accompanied by S_N^H substitution ... 81

    A. The Smiles and Smiles-Truce rearrangements ... 81

    B. The Sommelet-Hauser rearrangment and related reactions ... 82

    C. The von Richter rearrangement ... 84

    D. Other examples of intramolecular S_N^H reactions ... 86

3. NUCLEOPHILIC SUBSTITUTION OF HYDROGEN IN HETEROAROMATICS ... 89

    I. Azines ... 89

        A. Reactions with uncharged nucleophiles ... 89

        B. Reactions with anionic nucleophiles ... 108

        C. Reactions with organometallic compounds ... 119

    II. Azaaromatic substrates containing quaternary nitrogen ... 127

        A. NH-Azinium salts ... 127

        B. N-Alkyl and N-arylazinium salts ... 142

        C. N-Acylazinium salts ... 161

        D. Azine N-oxides, N-alkoxy and N-acyloxyazinium salts ... 176

        E. N-Fluoroazinium salts ... 200

        F. N-Aminoazinium salts ... 216

    III. Pyrylium and thiopyrylium cations ... 218

    IV. Five-membered heterocycles ... 227

    V. Electrochemical S_N^H reactions ... 236

    VI. Intramolecular S_N^H reactions in heteroaromatics ... 241

4. REACTIVITY OF ARENES AND HETARENES AND MECHANISMS OF THE S_N^H REACTIONS — 246

   I. The Addition-Elimination S_N^H (AE) mechanism — 246

      A. The addition stage — 246

      B. The aromatization stage — 270

      C. Kinetic studies — 274

   II. The Elimination-Addition S_N^H (EA) mechanism — 280

   III. The S_N^H (ANRORC) mechanism — 283

5. REFERENCES — 287

SUBJECT INDEX — 356

PREFACE

Nucleophilic aromatic substitution in carbo- and heteroaromatic systems has been a subject of considerable interest to many chemists for many decades. In particular, the nucleophilic displacement reactions of halogens and other groups with nucleofugal properties have attracted much attention. An overwhelming amount of information related to product formation, kinetics, and mechanisms, supported by labeling techniques, UV, ESR, and NMR spectroscopic studies, has been collected. In this respect, it is remarkable that textbook treatment of the subject of nucleophilic aromatic substitution is sometimes very limited, in contrast to the similarly important field of electrophilic substitution of aromatics.

Whereas investigations on the substitution of hydrogen in aromatics by use of electrophiles have a long history, broad studies on nucleophilic displacement reactions of hydrogen (S_N^H) in π-deficient carboaromatics and heteroaromatics have a much shorter history. Early examples of hydrogen substitution in π-deficient systems (the Chichibabin amination and the hydroxylation reaction) showed that these nucleophilic displacements required very drastic reaction conditions and oxidizing agents, a situation that did not stimulate chemists in earlier days to enter this field of S_N^H substitutions. Today, this picture has completely changed. Many new methodologies are being developed, making it possible to perform the replacement reaction of hydrogen in π-deficient systems under mild conditions leading to C_{arom}—C, C_{arom}—Hal, C_{arom}—N, C_{arom}—O, C_{arom}—S, and C_{arom}—P bond formation.

It is the intention of the authors of this book to review the material on S_N^H substitution reactions published in the literature up to 1993, to show the potential synthetic applications of S_N^H substitutions, and to discuss the mechanisms involved in these reactions.

The authors will show that nucleophilic aromatic substitution of hydrogen is an advanced field of chemistry. Electrophilic and nucleophilic substitutions of hydrogen in aromatics have the common feature that, in the first step of the reaction, the aromatic system is destroyed by addition of the attacking electrophile or nucleophile, respectively. However, whereas in an electrophilic substitution the aromaticity is restored by loss of a proton, a nucleophilic substitution requires an oxidizing agent, or an *auto*-aromatization to remove the formally "hybride" ion. In this way, nucleophilic aromatic substitution is complementary to electrophilic substitution of hydrogen in aromatics, both processes being of fundamental value in the chemistry of (hetero)aromatic compounds.

PREFACE

This book was written through the close cooperation of three chemists: two from Russia (Chupakhin and Charushin) and one from The Netherlands (van der Plas). This cooperation made it possible to discuss the Russian literature, which is usually difficult to access. The authors thank those colleagues who provided references, reviews, and other valuable information. These colleagues also helped the authors in the selection of the material published and encouraged them to prepare the manuscript.

O. N. Chupakhin
V. N. Charushin
H. C. van der Plas

Ekaterinburg–Moscow–Wageningen
January 1994

Introduction

Nucleophilic aromatic substitution of hydrogen comprises a large group of reactions which have in common that through interaction of a π-deficient arene or hetarene with a nucleophile the hydrogen in the aromatic ring is displaced (Scheme 1).

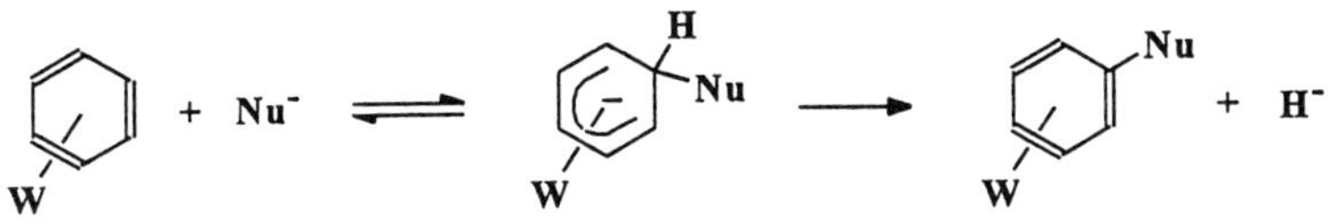

W - electron-withdrawing group

Scheme 1

The first examples of the displacement of hydrogen in the benzene ring by action of nucleophiles were reported in the literature about one hundred years ago: reactions of nitroaromatic compounds with the hydroxide anion in the presence of air oxygen, yielding a mixture of *ortho*- and *para*-nitrophenol (Scheme 2) [1882LA344; 1899CB3486].

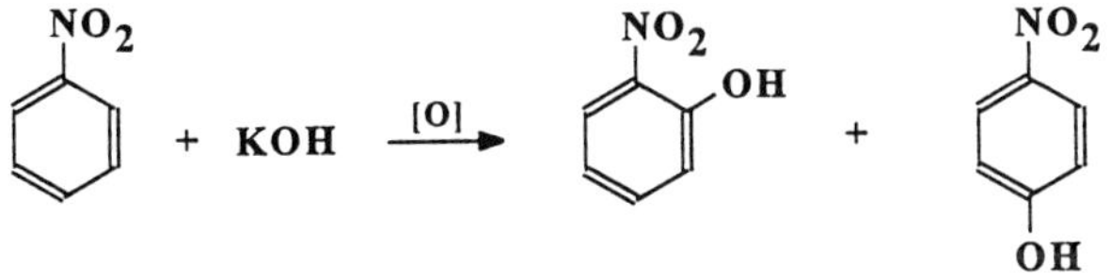

Scheme 2

A decisive role of an electron-withdrawing group activating the aromatic ring for a nucleophilic displacement of hydrogen has been established by numerous studies. An aromatic compound undergoing this kind of nucleophilic reaction usually needs activation by either an electron-withdrawing substituent or heteroatom in the ring; it can also be activated by annelation with the benzene ring or by forming an organometallic complex (Scheme 3).

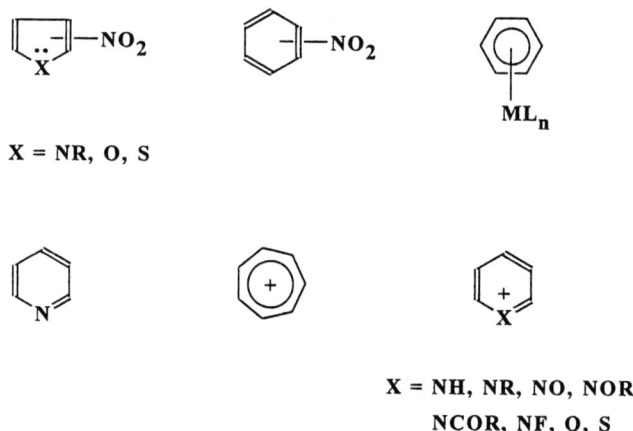

Scheme 3. Activated aromatic and heteroaromatic compounds undergoing nucleophilic displacement of hydrogen.

The reactions we are going to discuss in this book differ substantially from the classical displacements of nucleofugal groups: by their reaction pathways, by the specific mechanisms for departure of the hydride ion, by the conditions they do require, by the nature of by-products, and so on. In order to underline the specific nature of the reactions described formally in Scheme 1, the symbols $S_N^H(AE)$ [76RCR454; 88T1], NASH (Nucleophilic Aromatic Substitution of Hydrogen), and $S_N(Ar)^H$ [91MI2] were suggested. We prefer to leave the symbol Ar and to use the abbreviation S_N^H [76RCR454; 88T1]. To indicate by the symbol Ar that the substitution takes place in an aromatic system is not necessary, since nucleophilic substitution of hydrogen is a privilege of compounds of aromatic nature.

Reactions of electrophilic aromatic substitution of hydrogen (S_EAr) have been extensively studied and formed a solid basis for the development of theoretical concepts, as well as for industrial use of aromatic raw materials (see, for example, textbooks on organic chemistry [85MI2; 88MI2; 92MI1; 92MI2]). However, treatment in textbooks of the subject nucleophilic aromatic substitution attracted less interest, although it is evident that nucleophilic aromatic substitution of halogen or other nucleofugal groups capable of forming stable anions (Hal$^-$, RO$^-$, RS$^-$, NO$_2^-$, etc.) is of great synthetic importance for both laboratory use and wide industrial applications [68MI1; 76MI2; 90MI6]. Even modern textbooks on organic chemistry [88MI2; 92MI1] limit discussions on these reactions to only a few pages, although one should realize that about 30% of the organic compounds used in industry are of aromatic nature [88MI3; 92MI3].

The monograph "Aromatic Nucleophilic Substitution" written by J. Miller [68MI1] was the first and for a long time the only classical book on this subject till the beginning of the nineties, when an excellent monograph of F. Terrier "Nucleophilic Aromatic Displacement: The Influence of the Nitro Group" appeared, dealing with nucleophilic aromatic substitution reactions in the series of nitroarenes [91MI2].

S_N^H Reactions have been unnoticed for a long period of time, since the σ-adducts derived from the addition of nucleophiles to an unsubstituted ring carbon (σ^H-adducts) cannot easily lose the hydride anion. Although a few examples of the S_N^H reactions were reported at the end of the last century it was not until 1976 when the first review on nucleophilic substitution of hydrogen appeared [76RCR454]. For many organic chemists S_N^H reactions have long been associated with the Chichibabin amination, a reaction which is known to require rather severe conditions [71MI4; 88AHC2]. Reviews appeared dedicated to particular S_N^H reactions such as nucleophilic substitution of hydrogen in azines [88T1], and the so-called vicarious nucleophilic substitution [83MI2; 87ACR282; 89RCR747; 90MI4; 92PJC3]. Also, some aspects of the S_N^H reactions are reflected in review articles: on nucleophilic substitution of nucleofugal groups [51CRV273; 60AG294; 76MI2; 90MI5], on the formation and structural elucidation of the anionic σ-adducts [84MI1; 82CRV77; 82CRV427], on the chemistry of nitroarenes [69MI1; 91MI1], and on reactions of electron-deficient heteroaromatics [67MI1; 71MI3; 88H2659; 88AHC2; 88AHC199; 90AHC117]. In the book of F. Terrier a whole chapter of over 60 pages is dedicated to substitution of hydrogen in nitroarenes [91MI2]. A similar chapter is presented in the book by Russian authors V.L. Rusinov and O.N. Chupakhin on the chemistry of nitroazines [91MI1]. The reactions of nucleophilic aromatic substitution of hydrogen have been discussed briefly by J. March in his book "Advanced Organic Chemistry" containing a chapter on nucleophilic aromatic substitution with a subsection "Hydrogen as Leaving Group" [92MI2]. In accordance with common terminology J. March suggested to phrase these reactions by first naming the substituent which is introduced followed by ending with "dehydrogenation reaction." For instance, the displacement of hydrogen by the amino group is defined as the "amino-dehydrogenation" reaction [92MI2].

Since the displacement of hydrogen proceeds with various aromatic compounds and through different reaction pathways, they are considered in different sections of chemical literature with a variety of terminology, which, of course, does not stimulate progress in this field. Sometimes the S_N^H reactions are defined by mistake as addition reactions, even if the intermediate adduct has not been registered at all (for instance, see [74TL1361]). In other cases the S_N^H reactions are rationalized as carbenic [71JOC2907], electrophilic [73UK1416], or radical ones [86MI1; 89MI1; 90T2525; 92TL3057]. In the chemistry of arene-metal complexes they are simply interpreted as functionalization of an aromatic ligand [76MI1; 77MI2]. In a number of papers the S_N^H processes are described as "unexpected," "abnormal," and "unusual" reactions [73UK1416; 64JCS2806; 64JOC3381; 90TL3217].

The huge amount of data on direct nucleophilic amination, hydroxylation, halogenation, alkylation, arylation, and heteroarylation of π-deficient aromatic compounds being accumulated in the literature during the last two decades prompted us to undertake a systematic analysis of the vast range of reactions which can formally be regarded as nucleophilic substitution of aromatic hydrogen (S_N^H). We have tried to provide a comprehensive and coherent concept of both theoretical and practical aspects of nucleophilic displacement of hydrogen occurring in different types of aromatics (nitroarenes, arene–metal complexes), and heteroaromatic substrates (pyridines, their aza and benzo analogs, as well as quaternary N-alkyl, N-acyl, N-alkoxy, N-amino, and N-fluoroazinium salts, N-oxides, pyrylium cations, five-membered heterocycles, etc.). It is our purpose to pay in this book considerable attention to synthetic aspects of the S_N^H reactions and the scope of their applications, to discuss mechanisms for the displacement of hydrogen, to define terminology, and finally to show the common features of S_N^H reactions, which at first glance seem to be very different. At the same time, the book does not pretend to be an exhaustive compilation of all data on S_N^H reactions. The emphasis lies on reactions of arenes and hetarenes with typical C-, N-, O-, S-, and P-nucleophiles; less attention is paid to nonbenzenoid systems and other heteroatomic nucleophiles. Quinones are omitted as not being aromatic systems, although quinones react in S_N^H reactions in a manner very similar to that of nitroarenes. Also photochemical reactions are not discussed in the book due to their specific nature.

One of the aims of this book is also to show the common character of S_N^H reactions in the series of arenes, hetarenes, arene–metal complexes, carbo and heteroaromatic cations, and other π-deficient aromatics (Scheme 3), and to stress their features and specific traits. In this book we do not discuss in detail the aryne $S_N(EA)$ mechanism because it has extensively been reviewed (see Chapter 4,II) and the S_N^H(ANRORC) mechanism [78ACR462], since it is rather seldom operative in the nucleophilic displacement of hydrogen (see Chapter 4,III). Therefore we concentrate on the overwhelming majority of the S_N^H reactions which proceed via the two-step "Addition-Elimination" pathway [S_N^H(AE)], being very similar to that established for nucleophilic substitution of nucleofugal groups [S_NAr(AE) (Scheme 4)]. It is evident that the S_N^H(AE) and S_NAr reactions are formally very similar to each other, but differ in the nature of the leaving group. An important feature of S_NAr reactions with aromatic substrates containing nucleofugal groups (Le= Hal, OR, SR, SAr, SO_2R, CN, and NO_2) is that the adducts with hydrogen at the sp^3-carbon are formed faster than the σ^{Le}-adducts (Scheme 4). Due to poor leaving ability of the hydride ion the intermediate σ^H-adducts are more stable than the σ^{Le}-adducts [92PJC3]. On the other hand, the stability of σ^H-adducts greatly varies – from very unstable short-lived species to rather stable anionic complexes or dihydro compounds (see Chapter 4,I,A). It affects greatly their further conversions, which also depend on the presence of oxidizing agents, auxiliary groups in reactants, reaction conditions, and so on (see further discussion in this chapter).

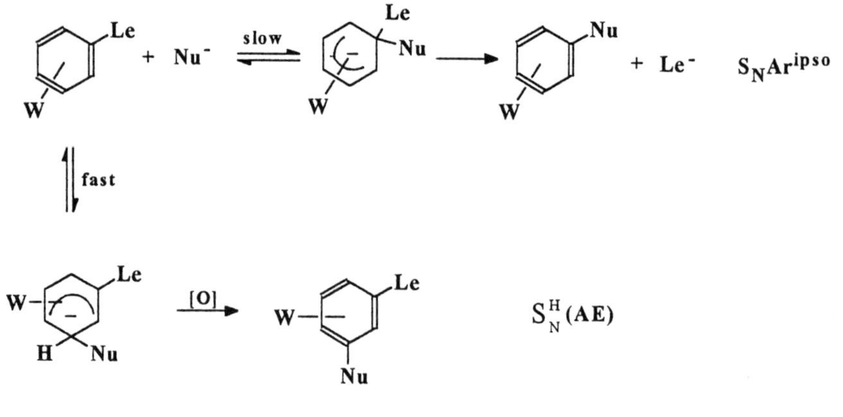

Scheme 4

Although in S_N^H reactions the leaving group is formally the hydride ion, its existence as a kinetically independent particle is rejected in the literature [74MI2], and its ability to depart as H⁻ is considered to be too doubtful [84KG1299]. On the other hand, there are examples of S_N^H reactions mentioned in the literature, in which lithium hydride was suggested to be formed (Scheme 5) [30LA123; 31LA174]. Some evidence for the departure of hydrogen with a pair of electrons has also been obtained from kinetic studies (for review of the mechanism of elimination of the hydride ion see [84KG1299]).

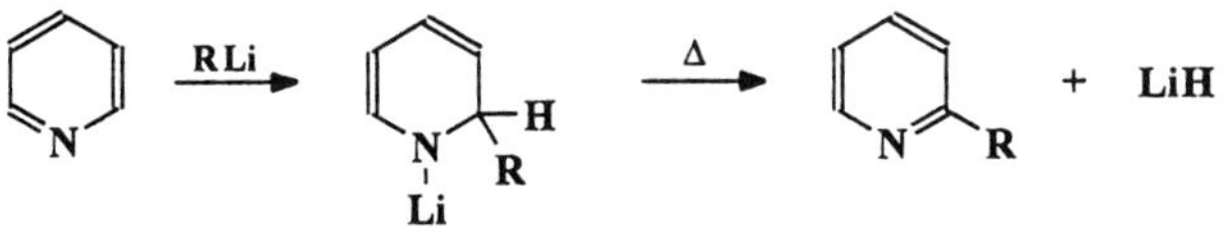

Scheme 5

Being not capable of solvation the hydride ion is not stable; the existence of hydrogen as H⁺ or H• is thermodynamically more favored. Since the tendency of σ^H-adducts toward aromatization contradicts the nature of the group which has to depart as H⁻, aromatization of the σ^H-adducts is not performed directly with elimination of the hydride ion, but by means of either an oxidative pathway or through the so-called *auto*-aromatization process.

Oxidative Aromatization. This process involves the successive transfer of two electrons and proton (or one electron and a hydrogen atom) from the intermediary π–excessive σ^H-adducts in which hydrogen is attached to the sp³-carbon. Use of air oxygen as oxidant is an old and well-known procedure to perform S_N^H reactions, as illustrated by the early synthesis of nitrophenols (Scheme 2) or alizarin (1,2-dihydroxyanthraquinone) [1899CB3486; 01CB2442]. A great variety of aromatic and heteroaromatic compounds (Scheme 3) were successfully subjected to direct nucleophilic amination, hydroxylation, halogenation, alkylation, arylation, and other S_N^H reactions using as oxidants both inorganic (halogens, hypohalites, permanganate, sulphur, oxygen, etc.) and organic compounds (tropylium and triphenylmethyl cations, chloranil, dichlorodicyanobenzoquinone, N-bromosuccinimide, etc.) [69MI1; 73ZOR2354; 74ZOR133; 86JOC1704; 88T1; 91MI2].

In the absence of an oxidant the π-deficient substrate itself can act as electron receiver and can perform the dehydrogenation of σ^H-adducts into the final products. The starting material is then partly consumed forming by-products with a different stage of reduction (see Scheme 6 [62JCS367]). This type of dehydrogenation reaction has been defined as a "spontaneous" aromatization [91MI2].

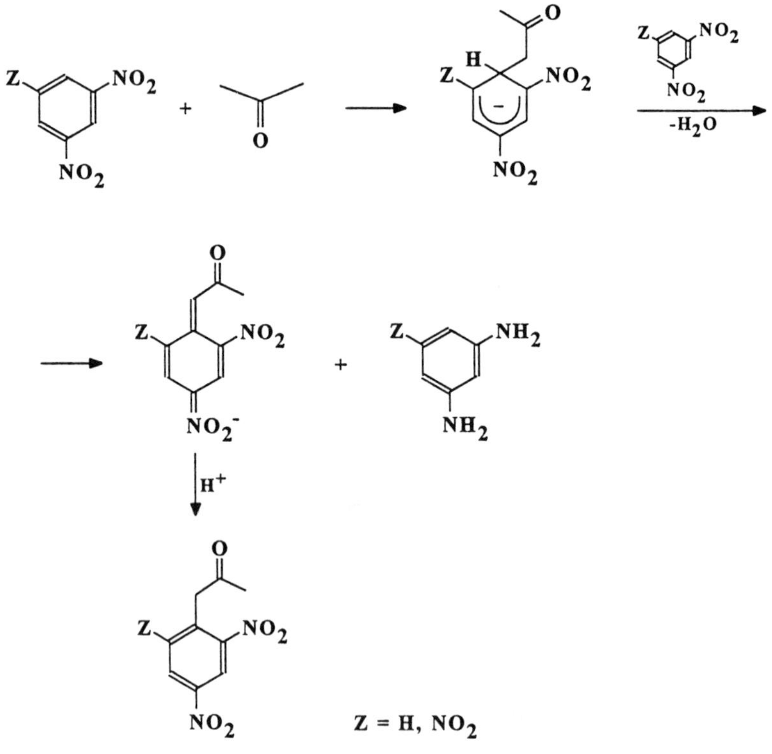

Z = H, NO_2

Scheme 6

Auto-Aromatization. In *auto*-aromatization processes an important role is played by the presence of an auxiliary group. If the intermediate S_N^H adduct contains an auxiliary nucleofugal group, redistribution of the electron density in the adduct can occur in such a way that the abstraction of this auxiliary group as an anion as well as the hydrogen as proton is facilitated. Amination of nitrobenzenes by hydroxylamine is a long-known example of the S_N^H reaction involving *auto*-aromatization (Scheme 7) [06CB2533].

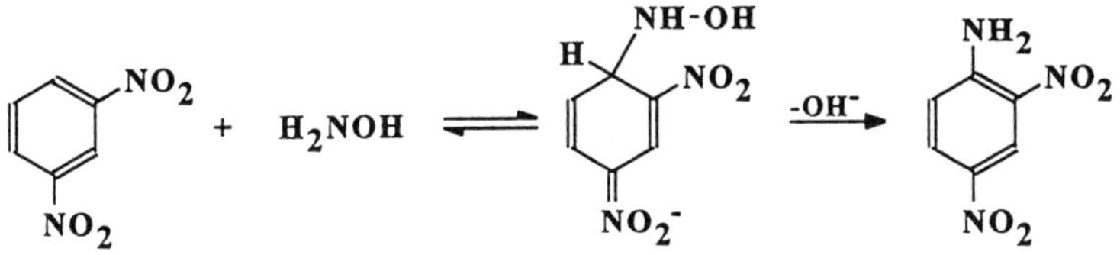

Scheme 7

Depending on mutual positions of an auxiliary group A and the hydrogen attached to the sp³-carbon in the adduct, several plausible ways of *auto*-aromatization can occur. A particular case is shown in Scheme 8, where an auxiliary group A is present at the *ortho*-position of the sp³-carbon in the ring. The product obtained after *auto*-aromatization is formed according to a mechanism, known as an $S_N(AE)^{cine}$ substitution reaction, but this process can formally be regarded as a nucleophilic aromatic substitution of hydrogen (S_N^H) assisted by the departure of a nucleofugal (auxiliary) group from the *ortho*-position.

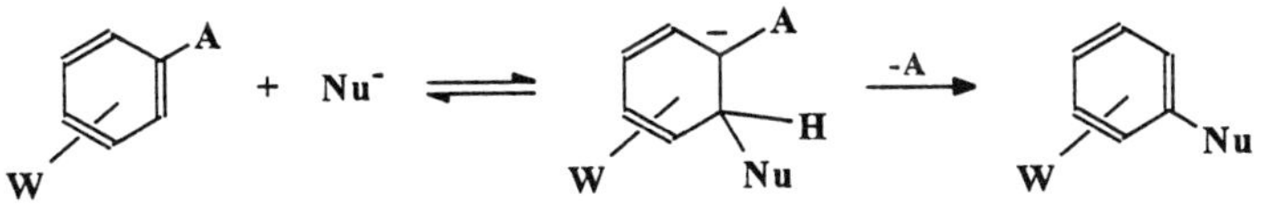

Scheme 8

Other mutual positions of an auxiliary group and hydrogen at the sp^3-carbon may result in the formation of S_N^H products which are usually considered as derived from *tele-* and *vicarious* nucleophilic substitution reactions. *Tele*-substitutions are defined as S_N reactions in which a nucleophile adds to a ring carbon separated from the leaving (auxiliary) group by several (odd and even, at least two) atoms. This concept can be exemplified by the $S_N(AE)^{tele}$ amination reaction of 8-halogeno-1,7-naphthyridines, yielding 2-amino-1,7-naphthyridine (Scheme 9) [67TL2087; 77RTC151]. It is evident that this reaction is formally a S_N^H reaction, which is facilitated by the presence of the *tele*-nucleofugal group at position 8 (Scheme 9).

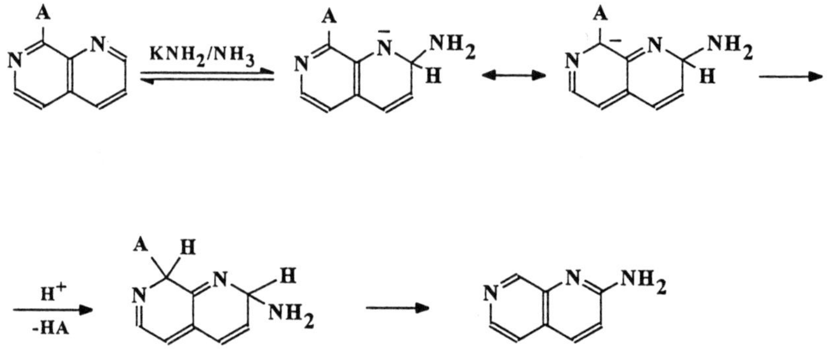

Scheme 9

Tele-substitutions are also observed in aromatic compounds which contain an auxiliary group not being present in the ring, but in the α-position of a side-chain, as illustrated in Scheme 10 [92LA19].

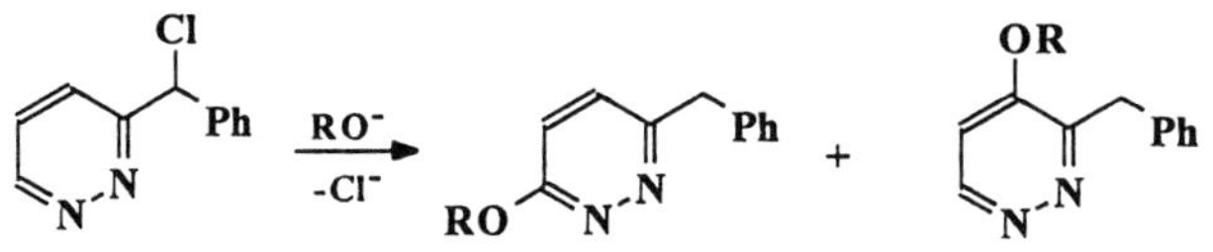

Scheme 10

Also the so-called deoxygenative reaction of N-oxides and their cations (see Chapter 3,II,D), when they react with nucleophiles and result in the displacement of hydrogen, involves the *auto*-aromatization step, in which the N-oxide, N-alkoxy, or N-acyloxy functions are auxiliary groups facilitating elimination of hydrogen from the intermediate adducts (Scheme11) [67MI1; 71MI3; 86CCA89].

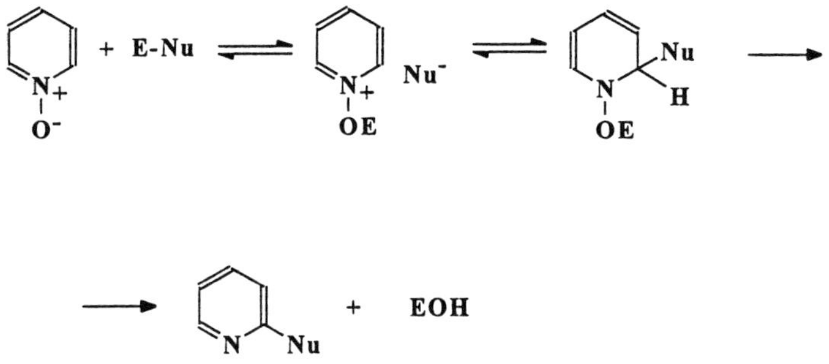

Scheme 11

A special and well-studied case of *auto*-aromatization is the reaction in which the attacking nucleophile carries a nucleofugal group at its α-position so that after addition to the arene, a σ-adduct is formed, in which the auxiliary (vicarious) group facilitates aromatization (Scheme 12). This approach for performing S_N^H reactions has been extensively investigated by M. Makosza [83MI2; 87ACR282; 89RCR747; 91MI2; 92PJC3]. It has been suggested to call these S_N^H reactions "vicarious nucleophilic substitution" (VNS) [83MI2; 87ACR282], later on abbreviated as VS_NAr^H in accordance with a common terminology accepted for the nucleophilic aromatic substitution [91MI2]. The VS_NAr^H reactions are of great synthetic value, since they are characterized by high yields, mild conditions, and, being very important, they enable functionalization of an arene (hetarene) containing a nucleofugal group in the ring (Scheme 12).

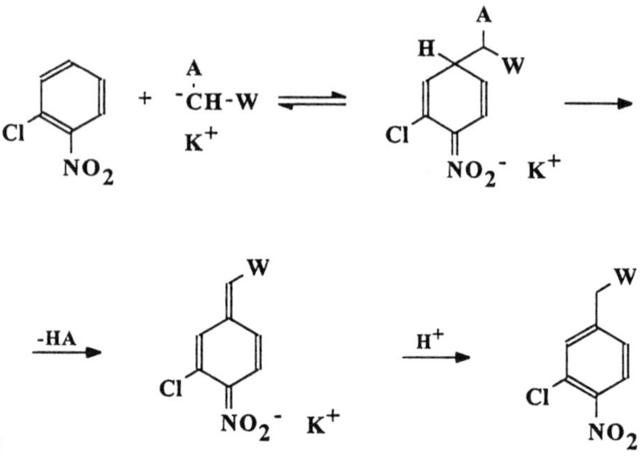

Scheme 12

When considering the different types of S_N^H reactions, in spite of the great variety of aromatic substrates and nucleophilic reagents used, they share common features, regardless of which kind of dehydrogenation, oxidative or *auto*-aromatization, is involved in the aromatization of the intermediary σ-adducts.

In a simplified form the oxidative pathway involves abstraction of two electrons from an intermediate σ-adduct and rejection of hydrogen as proton from the cation (Scheme 13).

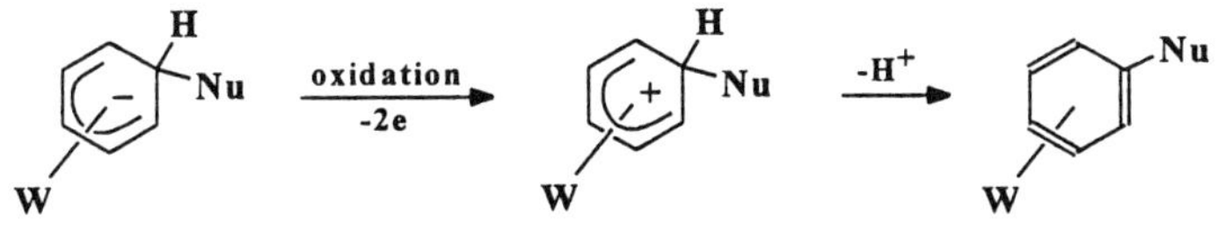

Scheme 13

All *auto*-aromatization reactions have in common that the auxiliary group, when departing as anion takes off two electrons leaving a vacant sp^2- or sp^3-orbital, which then accepts the two electrons, becoming available when the hydrogen leaves as a proton. Whether these processes are concerted or occur in two steps is a matter of discussion (see Chapter 4). The auxiliary group may be attached to the ring (Scheme 14) or be present at the α-position of a side-chain substituent (Scheme 15).

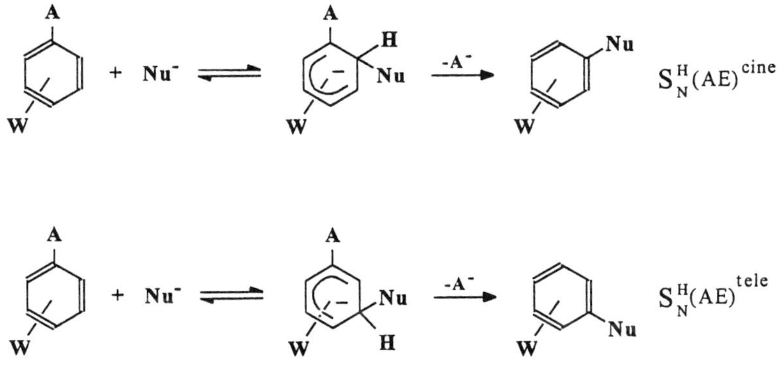

Scheme 14. Nucleophilic substitution of hydrogen facilitating by an auxiliary group present in an aromatic ring.

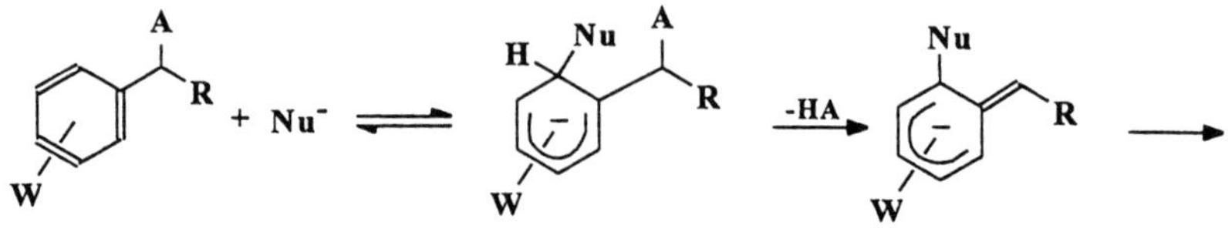

An auxiliary group A
is present in substrate

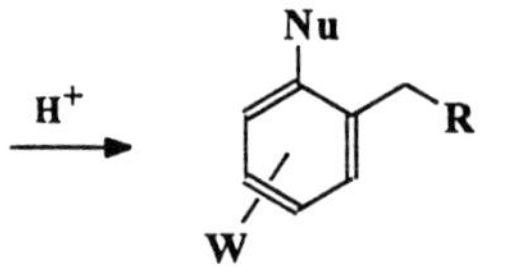

$S_N^H(AE)^{tele}$

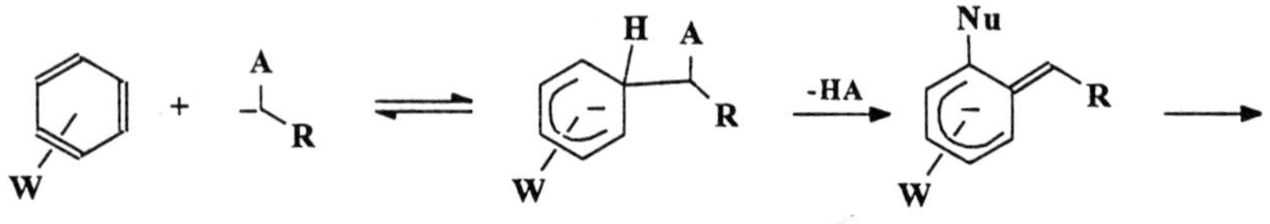

An auxiliary group A
is present in nucleophile

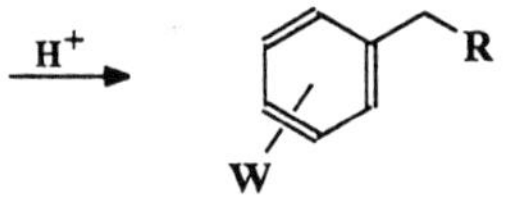

$VS_N^H(AE)$

Scheme 15. Nucleophilic substitution of hydrogen facilitating by an auxiliary group present at the α-position of a side-chain substituent.

The *auto*-aromatization reactions can occur regardless of the position of the auxiliary group. It does not matter whether this group is available in the starting aromatic substrate, like in the case of *tele*-substitution reactions (see the examples, mentioned in

Schemes 9, 10, 11, 14, and 15) or is present in the nucleophile in the case of vicarious S_N^H reactions (Schemes 12 and 15). We consider all S_N^H reactions involving the *auto*-aromatization step (*cine*- and *tele*-substitutions, deoxygenative substitutions in azine N-oxides, etc.) as "vicarious" when they require an auxiliary group capable of elimination as anion. More detailed analysis of the reaction mechanisms is given in Chapter 4 of this book.

It is worth noting that the synthetic value of S_N^H reactions is constantly growing. Whereas the electrophilic substitution in aromatic compounds has been thoroughly studied and has led to many industrial applications, functionalization of an aromatic ring can now also be achieved via nucleophilic alkylation, arylation, halogenation, acylation, nitration, and other kinds of nucleophilic reactions. They can be considered as complementary to reactions which occur by action of electrophilic agents. Where appropriate we will reflect in the book on the differences in nucleophilic and electrophilic behavior of carbo- and heteroaromatics. For π-deficient aromatic substrates, such as nitroarenes, azines, and azinium cations, the methods for introduction of substituents into the ring via the direct nucleophilic displacement of hydrogen are, of course, more favored, if not the only possible method. Besides functionalization, the use of the S_N^H methodology for the synthesis of carbo- and heterocyclic systems will also be discussed.

*　　*　　*

We found it useful to present at the end of each chapter a number of typical original experimental procedures in order to show how the different methodologies can be used. We hope that these experimental procedures might stimulate synthetic chemists in both industrial and university laboratories to consider S_N^H methodologies in their practical work.

Nucleophilic Substitution of Hydrogen in Arenes

I. UNACTIVATED ARENES

Benzene and its benzo analogs are usually very reluctant to undergo nucleophilic substitution reactions, since the addition of nucleophiles to such systems is associated with the loss of relatively high aromatic delocalization energies, whereas no electron-withdrawing substituents or heteroatoms stabilizing the formation of anionic σ-adducts are present in the ring. There are, however, a few S_N^H reactions reported in the literature of unactivated arenes: the nucleophilic alkylation of anthracene and phenanthrene by action of dimethyl sulfoxide in the presence of a strong base (sodium hydride or potassium t-butoxide) (Scheme 16) [66JOC248; 68TL4625], the alkylation of benzene and naphthalene with alkyllithiums [63JA1356; 64TL613; 68JA1606]) and Grignard reagents [64TL3295], and the Chichibabin amination of naphthalene in a melt with sodium amide (Scheme 17) [06CB3006].

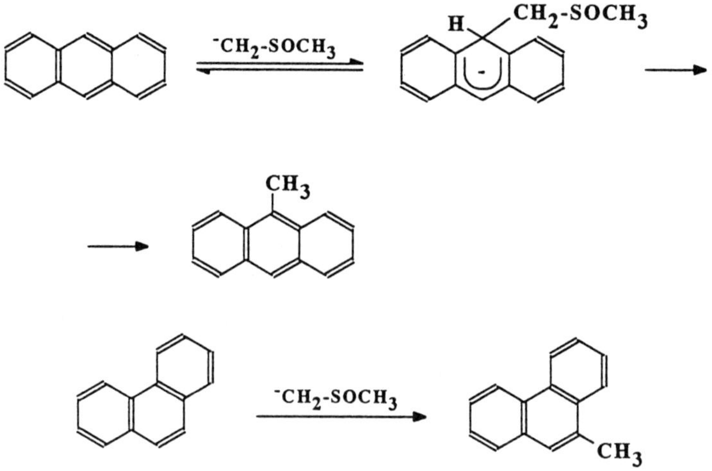

Scheme 16

It is remarkable that even benzene can undergo nucleophilic monoalkylation reaction; treatment by *t*-butyllithium in decalin at 165°C gives *t*-butylbenzene, although in a poor yield (15%) [63JA1356]. The reaction of naphthalene with *t*-butyllithium under the same reaction conditions results in a mixture of mono- and dialkyl naphthalenes; the monoalkylation occurs predominantly in the 1-position (at least 95%), while further alkylation occurs at the 3-, 6-, or 7-positions (Scheme 17) [63JA1356].

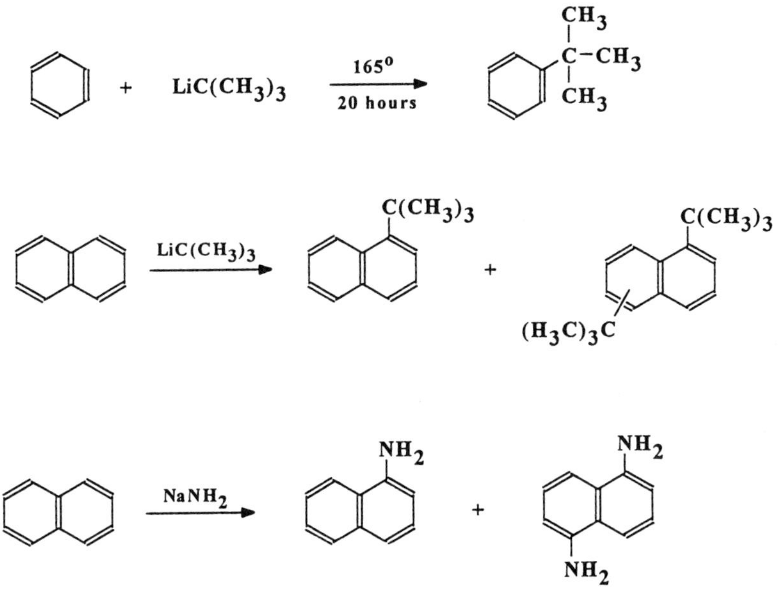

Scheme 17

Experimental Procedures

Example 1. Methylation of phenanthrene, preparation of 9-methylphenanthrene

To a solution of 2.64 g of sodium hydride (110 mmol) in 100 ml of DMSO at 70°C was added 2.00 g (11 mmol) of phenanthrene in 50 ml of DMSO. The solution was stirred for 4 hours at 70°C followed by the addition of 100 ml of a 1:1 mixture of concentrated hydrochloric acid and water. The resulting precipitate was filtered off, dissolved in chloroform; the chloroform solution was washed with water and dried with magnesium sulfate. The chloroform solution was estimated by GLPC to contain 93% of 9-methylphenanthrene. Removal of the chloroform gave a solid that was twice recrystallized from methanol to yield 1.64 g (76%) of 9-methylphenanthrene, melting point 92-93 °C [66JOC248; Copyright permission 1994 American Chemical Society].

Example 2. *t*-Butylation of naphthalene

Naphthalene (13.0 g, 0.11 mol) was added to a decalin solution of *t*-butyllithium (7.0 g, 0.11 mol) which was prepared by evaporating the pentane from a commercial solution and replacing it with decalin. The homogeneous mixture obtained was heated at 165°C for 41 hours. After hydrolysis, the organic products (oils) were separated by fractional distillation and identified (IR, NMR, and the gas-liquid chromatography analysis) as 1-*t*-butylnaphthalene (5.69 g, 30%) and a mixture of di-*t*-butylnaphthalenes (6.81 g, 50%) [63JA1356; 68JA1606;Copyright permission 1994 American Chemical Society].

II. ARENE–METAL COMPLEXES

It is well known that coordination of arenes with a transition metal lowers the electron density of the arene ring and enhances the susceptibility of the benzene ring toward nucleophilic attack [76MI1; 77MI2; 78MI1; 80JOM207; 81T3957; 82MI2; 83JCS(P1)3065; 84CB152; 84CB161; 85JCS(CC)417; 85JCS(CC)1181; 85MI3; 85TL3989; 87MI4].

Indeed, arene–metal complexes **1** are reactive with carbanions R⁻ to form rather stable adducts **2** in which substituent R usually occupies the *exo*-position, that is, on that side of the cyclohexadienyl ring being opposite to the coordinated metal (Scheme 18) [76MI1; 77MI2; 8 0JOM207; 81PAC2379; 84CB152; 84CB161; 87MI4]. Attempts to cause elimination of the hydrogen ion from complexes **2** by action of the triphenylmethyl cation or other electrophiles are not always successful, since an electrophilic attack at the *endo*-oriented hydrogen in complexes **2** is sterically hindered. Many papers describe however, reactions in which elimination of this *endo*-hydrogen has been achieved, often simultaneously followed by decomposition of the arene–metal complexes **3** into R-substituted arenes **4** (Scheme 18) [67JA3744; 69JCS(C)2024; 75JOM365; 79JA3535; 84CB161; 87MI4].

The overall process can be regarded as nucleophilic aromatic substitution of hydrogen. The use of different types of arene–metal complexes discussed below shows that this methodology is actually a powerful synthetic tool for the preparation of substituted arenes.

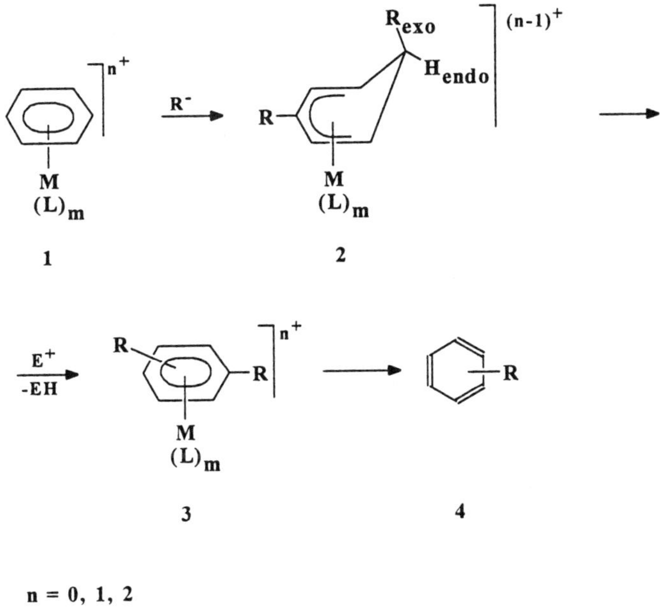

$n = 0, 1, 2$

Scheme 18

The chromium tricarbonyl unit proved to be a very suitable group for activating nucleophilic substitution of hydrogen in arenes. Since this unit can easily be attached to the benzene ring and later on be removed in the last stage of the reaction [74JA7091; 74JA7092; 75JA1247, 76MI1; 79JA3535; 79IS154; 80JOC2555; 80JOC2560; 81PAC2379; 81T3957; 87MI4], transition-metal coordination with arenes gives us a unique possibility for overall nucleophilic substitution of hydrogen in arenes, provided, however, that the reactions are carried out in the presence of a suitable oxidant. The function of this oxidant is twofold: to remove the hydrogen attached to the sp^3-carbon of the cyclohexadienyl intermediate as well as to oxidize the chromium complex for releasing the arene unit. Indeed, treatment of arenechromium tricarbonyl complexes with alkyl- or aryllithiums, and subsequently with iodine results in alkyl- or aryl-substituted arenes in good yields (Table I) [74JA7092; 75JA1247; 79JA3535; 81T3957; 87MI4]. It can be exemplified by the formation of *t*-butylbenzene **6** in 68% yield on treatment of η^6-benzenetricarbonylchromium(0) **5** with *t*-butyllithium followed by oxidation with iodine of the resulting intermediate (Scheme 19) [75JA1247; 79JA3535]. In a similar way, compound **5** reacts with dithiane lithium **7** to give 2-phenyl-1,3-dithiane **8** in excellent yield (Scheme 19) [75JA1247].

Complexing of arenes with tricarbonylchromium has been successfully applied to introduce cyanomethyl, isobutyronitril, *n*-butyl, *t*-butyl, and phenyl groups into the benzene ring (Table I). Limitations concern the alkyllithium species, such as methyllithium, ethyllithium, or *n*-propyllithium. Being too basic they are able to perform a concurrent hydrogen-lithium exchange in the benzene ligand, leading after treatment with iodine to the formation of iodobenzene (Scheme 20) [75JA1247; 77MI2; 80JOC2555]. By changing reaction conditions this hydrogen–lithium exchange can be avoided, as reported by the successful alkylation of η^6-ethylbenzenetricarbonyl-chromium(0) with *n*-butyllithium at −40°C which results in a mixture of *meta-* and *para-*

para-ethyl-substituted *n*-butylbenzenes, **9** and **10**, in 55% and 29% yields, respectively (Table I; Scheme 20) [80JOC2560].

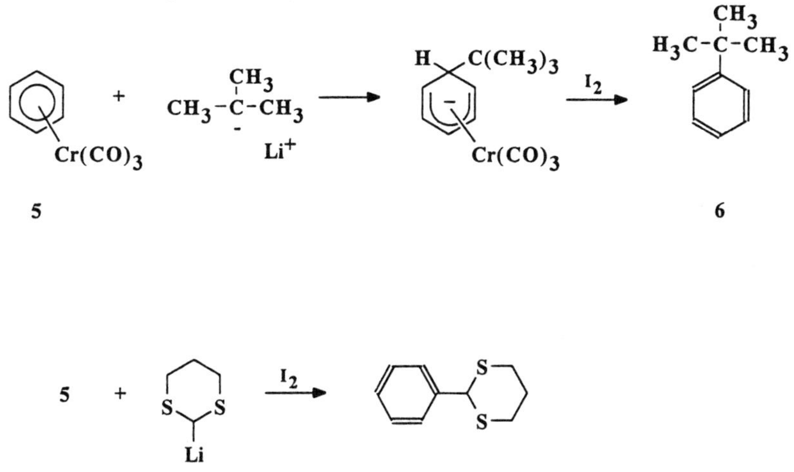

7 **8**

Scheme 19

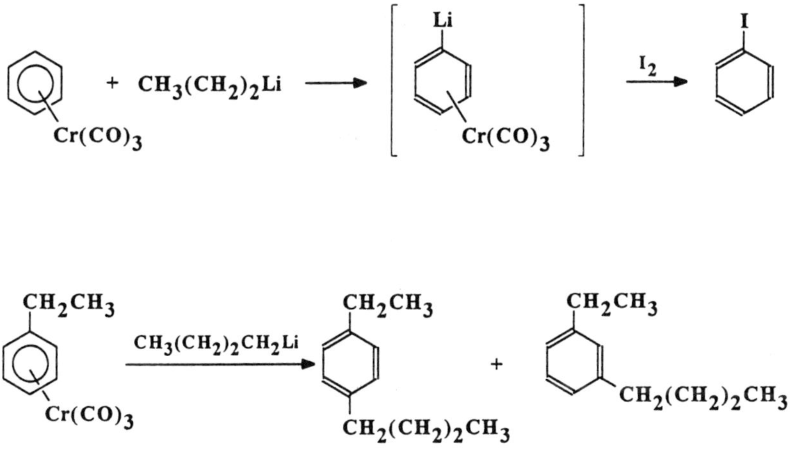

9 **10**

Scheme 20

TABLE I
Substitution of Hydrogen in η^6-Arenetricarbonylchromium
Complexes by Action of Carbanions and I_2 as Oxidant

Arene coordinated with $Cr(CO)_3$	Carbanion	S_N^H product(s)	Yield (%)	References
Benzene	$LiCH_2CN$	Benzyl cyanide	63	75JA1247
Benzene	$LiC(CH_3)_2CN$	Phenyl isobutyronitrile	94	75JA1247
Benzene	$LiC(CH_3)_3$	t-Butylbenzene	68	79JA3535
Benzene	LiPh	Diphenyl	71	75JA1247
Toluene	$LiCH_2CN$	o-Tolyl acetonitrile	30	77MI2
		m-Tolyl acetonitrile	56	77MI2
		p-Tolyl acetonitrile	2	77MI2
Toluene	$LiC(CH_3)_2CN$	o-Tolyl isobutyronitrile	1	77MI2
		m-Tolyl isobutyronitrile	92	77MI2
		p-Tolyl isobutyronitrile	2	77MI2
Ethylbenzene	n-BuLi	1-n-Butyl-3-ethylbenzene	55	80JOC2560
		1-n-Butyl-4-ethylbenzene	29	80JOC2560
Chlorobenzene	$LiC(CH_3)_2CN$	o-Chlorophenyl isobutyronitrile	2	74JA7092
		m-Chlorophenyl isobutyronitrile	56	74JA7092
Anisole	$LiCH_2CN$	o-Methoxyphenyl acetonitrile	1	77MI2
		m-Methoxyphenyl acetonitrile	37	77MI2
Anisole	$LiC(CH_3)_2CN$	o-Methoxyphenyl isobutyronitrile	3	77MI2
		m-Methoxyphenyl isobutyronitrile	79	77MI2

It is worth noting that the reactions of arene-metal complexes with carbanions often feature high regiospecificity, being generally interpreted as reaction which are kinetically controlled. This interpretation appears to be reasonable since the addition of many carbanions to arenetricarbonylchromium complexes is complete within a few minutes at −70°C [80IC3092; 80OMR426; 81JOM147; 83JA2034; 87MI4], while rearrangement processes require reaction times of many hours and temperatures of 0 to 50°C [79JA768; 79JA3535]. The distribution of products obtained from methyl, ethyl, chloro, and methoxy-substituted benzenes (Table I) suggests that the transition state leading to the η^5-cyclohexadienyl intermediate has the pentadienyl anion character. The *meta*-directing effect of the electron-donating substituents is then quite understood since they have the least destabilizing effect when located at the *meta*-position of the cyclohexadienyl system [74JA7091; 74JA7092; 75JA1247; 76MI1; 79JA3535; 80JOC2560; 81PAC2379; 81T3957; 87MI4]. However, the situation is often more complex, as illustrated by the alkylation of 1,4-dimethoxynaphthalenetricarbonyl-chromium with lithium isobutyronitrile. The reaction was found to be temperature dependent: at −60°C; it occurs preferentially at the β-carbon, while on warming to 0°C the kinetically favored β-substituted adduct rearranges rapidly, via dissociation and readdition steps, into the thermodynamically favored α-isomer (Scheme 21) [83JA6962; 85JOM183]

The cationic complexes of iron and manganese offer much stronger activation of the arene ligand to nucleophiles than the neutral arene-chromium complexes. However, in general they proved to be of less importance for organic synthetic work because the high reactivity of these complexes gives rise to several other reaction products. Therefore the yields of the S_N^H products are only moderate. Moreover, no general method for removing and separating the coordinated metals from the substituted arenes has been found [76MI1; 77MI2].

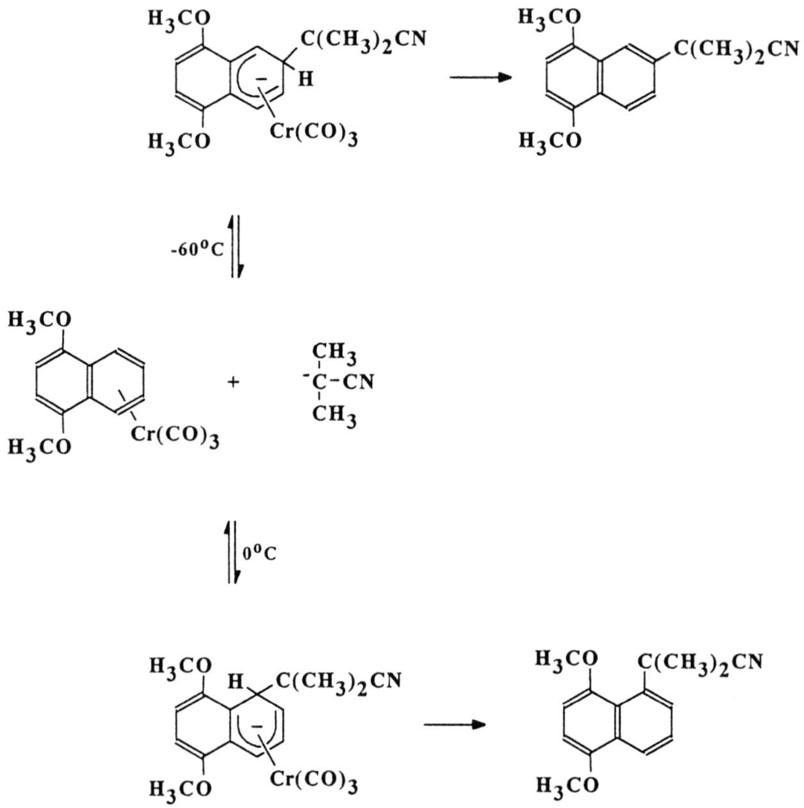

Scheme 21

Reactions of η^6-arene-η^5-cyclopentadienyliron(II) fluoroborate complexes **11** with nucleophiles under oxidative conditions follow the course outlined in Scheme 22 [60JCS989]. Oxidation of the intermediary adducts **12** produces cationic η^5-cyclopentadienyliron complexes **13** which can be further oxidized under mild conditions to give the corresponding substituted arenes **14** (Scheme 22) [75JOM365]. The course of the reaction is strongly influenced by the nature of the substituent: complex **11a** (R= Cl) adds methyllithium preferentially to the *ortho*-position ($o{:}m{:}p$ = 4:1:0) due to the inductive activating effect of the chloro substituent, while the anisole analog **11b** (R= OCH$_3$) reacts preferentially at the *meta*-position [69JCS(C)2024].

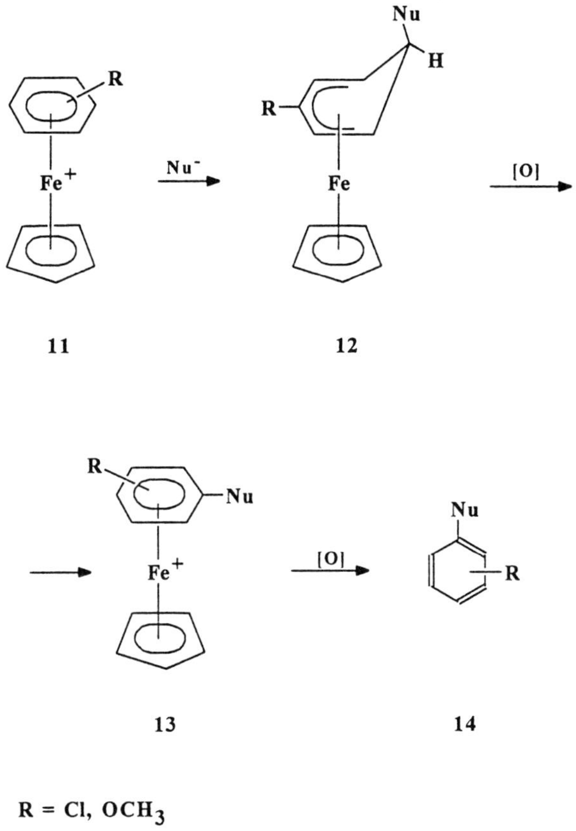

R = Cl, OCH$_3$

Scheme 22

The arene in η^6-arenetricarbonylmanganese(I) complexes **15** is also reactive toward a variety of nucleophiles, such as methyl- and phenyllithium [75JCS(D)1683], cyanide, malonate, and acetylacetonate [73ICA621]. Oxidation of the intermediary η^5-cyclohexadienyltricarbonylmanganese(I) derivatives fails in most cases. However, aqueous Ce (IV) proved to be an appropriate reagent to remove the sp^3-*endo*-hydrogen from the adduct of η^6-mesitylenetricarbonylmanganese(I) with the cyanide ion, that is, **16**. This reagent is also able to remove the metal ligand from the arene, producing cyanomesitylene **17** in 70% yield (Scheme 23) [73JCS(D)622].

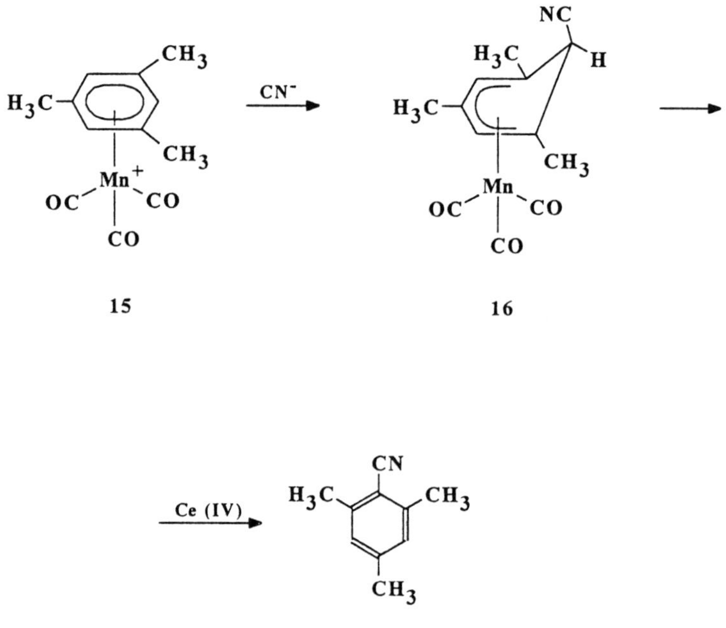

Scheme 23

Alkyllithium compounds also react with benzeneruthenium(II) and benzene-osmium(II) complexes. They give the rather stable cyclohexadienates **18** and **19** (Scheme 24) [82CB3766; 84CB152; 84CB161]. Structural elucidation of **18** and **19** by ^1H and ^{31}P NMR, and X-ray analysis for the ruthenium complex **18a** (R= n-C_4H_9) have revealed that the cyclohexadienyl ring possesses an envelope conformation, in which the *n*-butyl group occupies the *exo*-position at the sp^3-C atom [84CB152].

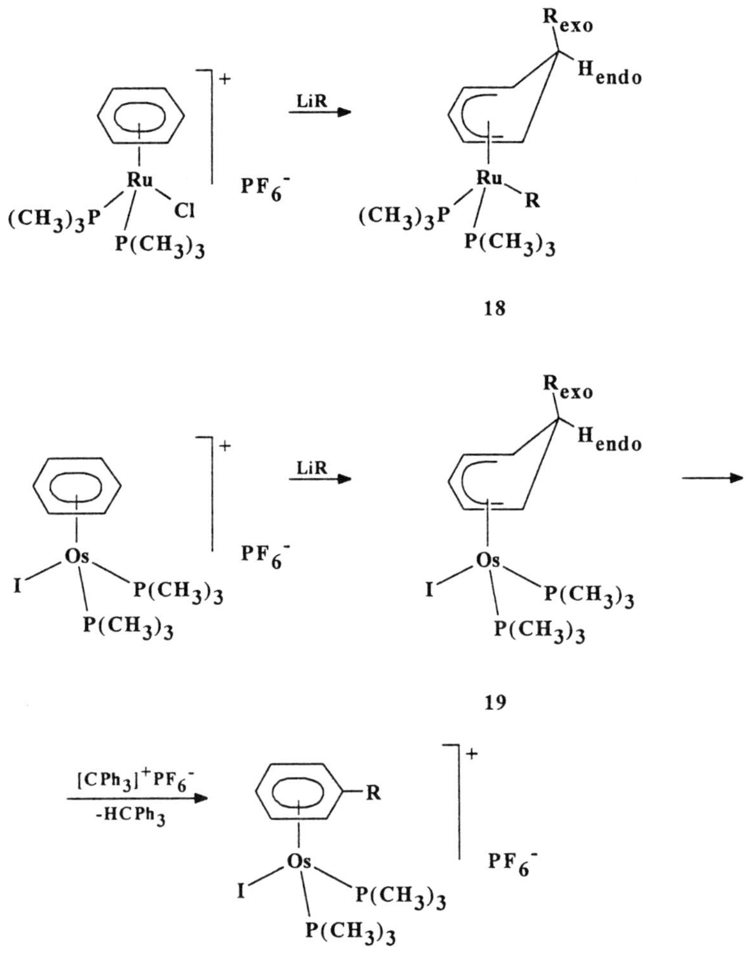

Scheme 24

Oxidation of the *exo*-R-*endo*-H osmiumcyclohexadienates **19** with triphenylmethyl hexafluorophosphate in acetone proceeds very easily at room temperature and is complete in a few minutes with the formation of yellow or orange areneosmium salts **20** in nearly quantitative yields (Scheme 24) [84CB161]. The mechanism of this oxidative reaction seems to involve the formation of isomeric cyclohexadienate complexes in which the alkyl group is no longer present at the sp^3-carbon, due to migration of the alkyl group. It follows from experiments with (*exo*-6-*t*-butyl-η^5-cyclohexadienyl)-iodo-*bis*(trimethylphosphan)osmium(II) (**19a**, R= *t*-Bu), which, being more stable than the complexes **19** bearing *n*-alkyl groups, is oxidized by tritylhexafluorophosphate in methylene chloride only at 55°C for 5 hours. NMR studies have revealed that protonation of the osmium complex **19a** with CF_3COOD in the presence of NH_4PF_6 takes place on osmium resulting in the formation of the hydridoosmium(IV) complex **21**, and that **21** undergoes an isomerization to compound **22** (Scheme 25). The isomerization process proceeds via the intermediate complex **23**, thus involving the reversible transfer of a proton from osmium to the cyclohexadienyl ring and vice versa. The cyclohexadienate complex **22** contains the CH_2 fragment in which the *exo*-hydrogen is certainly more vulnerable to oxidation [84CB161]. Also, it is worth mentioning that somewhat similar isomerizations have been observed in the series of cyclohexadienyl complexes of tricarbonylchromium, manganese, and molybdenum [72JCS(P2)1141].

Experimental Procedures

Example 1. *t*-Butylation of benzene

A solution of *t*-butyllithium (1.0 ml of a 2.1 *M* solution in *n*-pentane, 4.2 mmol) was mixed with THF (5 ml) at −78°C under argon. A solution of η^6-benzenetricarbonyl-

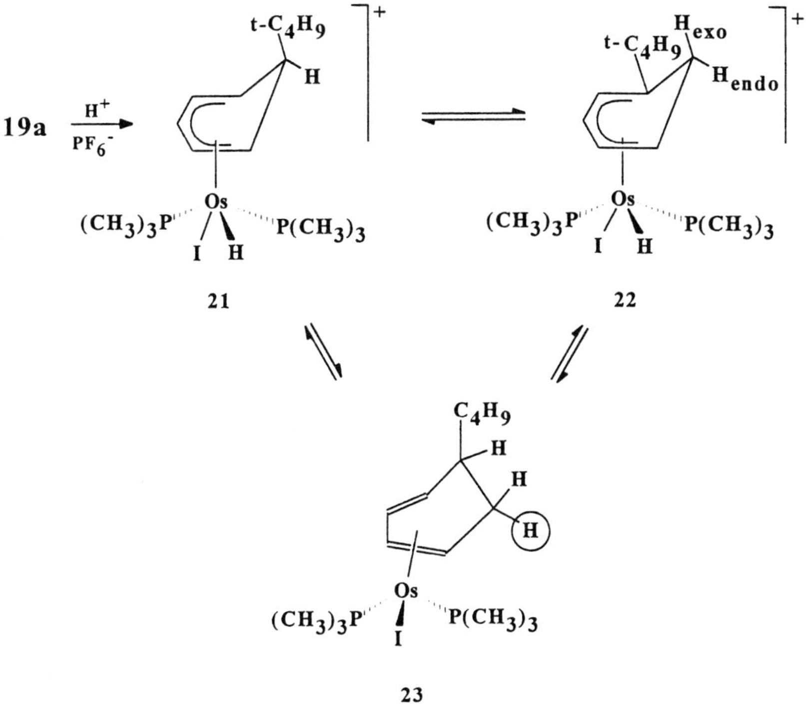

S_N^H REACTIONS IN ARENES

19a $\xrightarrow[PF_6^-]{H^+}$ **21** ⇌ **22**

↕ ↕

23

Scheme 25

chromium (434 mg, 2.03 mmol) in 5 ml of THF was added dropwise over a period of a few minutes at −78°C. After being stirred at −78°C for 1.5 hours, the solution was warmed at 0°C for 1.0 hour and then cooled to −78°C again. A solution of iodine (2.0 g, 7.9 mg-atoms) in 5 ml of THF was added rapidly via syringe. After 3 hours at 25°C, the mixture was diluted with ether, washed successively with aqueous solution of sodium bisulfite and sodium chloride, and then dried. The ether solution was concentrated and flash distilled (0.001 Torr, 25°C) to yield a colorless liquid of *t*-butyl-benzene. Yield 185 mg (68%) [79JA3535; Copyright permission 1994 American Chemical Society].

Example 2. Preparation of phenyl isobutyronitrile

Lithium diisopropylamide was prepared from *n*-butyllithium (12.8 ml of a 1.95 M solution in hexane, 25.2 mmol) and diisopropylamine (3.84 ml, 27.5 mmol) in 50 ml of THF by mixing the reagents at -78°C under argon and allowing the mixture to stir at 0°C for 15 minutes. The pale yellow solution was cooled to -78°C, and isobutyronitrile (2.20 ml, 25 mmol) was added dropwise over a period of 5 minutes. The mixture was warmed at 0°C for 15 minutes, cooled to -78°C, and a solution of η^6-benzene-tricarbonylchromium in 25 mL of THF was added dropwise over 5 minutes. The yellow solution was warmed at 0°C for 30 minutes, cooled to -78°C, and a solution of iodine (25 g, 98 mmol of I_2) in 75 ml of THF was added dropwise over a period of 15 minutes. After the resulting mixture had been stirred at 24°C for 3 hours (CO evolution) it was partitioned between ether (200 ml) and 5% aqueous sodium bisulfite solution (100 ml). The ether layer was washed successively with water and aqueous salt solution, dried over anhydrous magnesium sulfate, and concentrated to leave a yellow liquid. Flash distillation at 25°C (0.001 Torr) afforded 3.562 ml (94%) of phenyl isobutyronitrile as a colorless liquid [75JA1247; Copyright permission 1994 American Chemical Society].

Example 3. Preparation of (η^6-*n*-butylbenzene)iodo-*bis*(trimethylphosphan)osmium hexafluorophosphate (**20**, R = *n*-Bu)

The Addition Step. A solution of *n*-butyllithium (378 mg, 6.0 mmol) was added to a suspension of (η^6-benzene)iodo-*bis*(trimethylphosphan)osmium(II) hexafluoro-phosphate (692 mg, 1.0 mmol) in 4 ml of ether. When the reaction was over (the solution becomes nearly transparent), methanol (0.243 ml, 6.0 mmol) was added at -78°C to destroy an excess of *n*-butyllithium and the reaction mixture was heated up to room temperature. The precipitate obtained was filtered off, dried and suspended in 10 ml of benzene. After filtration of this suspension, the solid was washed twice with 5 ml of benzene and dried. It was then dissolved in 40 ml of hexane and filtrated. The filtrate was concentrated till 10 ml and cooled slowly till $-78^{\circ}C$ to give yellow crystals of (*exo*-butyl-η^5-cyclohexadienyl)iodo-*bis*(trimethylphosphan)osmium(II), which were collected by filtration. Yield 363 mg (60%), melting point $75^{\circ}C$ [84CB152].

The Aromatization Step. A solution of (*exo-n*-butyl-η^5-cyclohexadienyl)iodo-*bis*-(trimethylphosphan)osmium(II) (152 mg, 0.25 mmol) in 2 ml of acetone was treated at room temperature with triphenylmethyl hexafluorophosphate (194 mg, 0.50 mmol). After addition of methanol (0.04 ml, 1 mmol) and, subsequently, of ether (10 ml), the product precipitate was filtered off, washed with ether (3 x 5 ml), and dried *in vacuo*. Recrystallization from nitromethane gave (η^6-*n*-butylbenzene)iodo-*bis*(trimethylphosphan)osmium hexafluorophosphate as yellow crystals. Yield 150 mg (80%) [84CB161].

III. NITROARENES

Nucleophilic displacements of hydrogen in arenes being activated by the presence of a nitro group have been the subject of continuous investigations. Numerous publications concerning the chemistry of nitroarenes [69AG136; 69MI1; 84ACR109; 91MI1; 91MI2], the formation of anionic σ-complexes [82CRV77; 82CRV427; 84MI1], and the substitution of nucleofugal groups [60AG294; 68MI1] have appeared. Because of their great interest, the reactions of nucleophilic substitution of hydrogen in aromatic systems bearing the nitro group have extensively been reviewed [76RCR454; 80MI1; 83MI2; 87ACR282; 88T1; 89RCR747; 91MI2; 91S103; 92PJC3].

The S_N^H reactions of nitroarenes are usually considered to proceed via a "conventional" two-step "addition-elimination" (AE) mechanism [76RCR454; 88T1; 91MI2]; however, it is not always taken into account that the reactions may involve the formation of charge-transfer complexes, ion-radical salts, and other plausible elementary steps. Indeed, interaction of nitroarenes with nucleophilic reagents is often accompanied by the formation of charge-transfer complexes, although in only a few cases have convincing experimental arguments for their existence on the reaction coordinate been obtained [74MI1].

The reactions of nitroarenes with nucleophiles can either be terminated at the addition stage by forming very stable Meisenheimer-type σ-adducts, or can proceed further by aromatization into the final aromatic products. This is mainly depending on how effective nitro groups participate in delocalization of the negative charge. The differences in delocalization energies between arenes and their anionic σ-complexes are very indicative in this respect, changing from -33.5 kJ/mol for nitrobenzene to 4.2 kJ/mol for 1,3-dinitro- and 41.9 kJ/mol for 1,3,5-trinitrobenzene, respectively [69AG136].

Aromatization of anionic σ-adducts can be realized either with the aid of oxidants supplied to the reaction mixture or through an *auto*-aromatization process, as exemplified with a variety of vicarious nucleophilic substitution reactions (see Chapter 2,III,A) [83MI2; 87ACR282; 91MI2; 92PJC3].

When appropriate conditions are found, S_N^H reactions of nitroaromatic compounds may proceed smoothly enough to have preparative value and, when appropriate, to apply them industrially.

In the following sections the potential synthetic use of S_N^H reactions of nitroarenes with C-, O-, N-, and P-nucleophilic reagents will be discussed, and the general mechanisms will be put forward. A more detailed picture of the mechanism is presented in Chapter 4.

A. Reactions with C-Nucleophiles

1. *Cyanides*. A series of nitrobenzenes carrying electron-withdrawing substituents in the *ortho*- or/and *para*-positions have been found to undergo direct cyano-dehydrogenation reaction in DMSO. However, the reaction is accompanied by conversion of the nitro substituent into the hydroxy group (Scheme 1) [71JCS(CC)1120; 75JOC3746]. A Nef-type process is suggested involving isomerization of the *ortho*-cyano adduct into the spirocyclohexadienyl oxaziridine oxide followed by loss of the hyponitrite anion (Scheme 26) [75JOC3746].

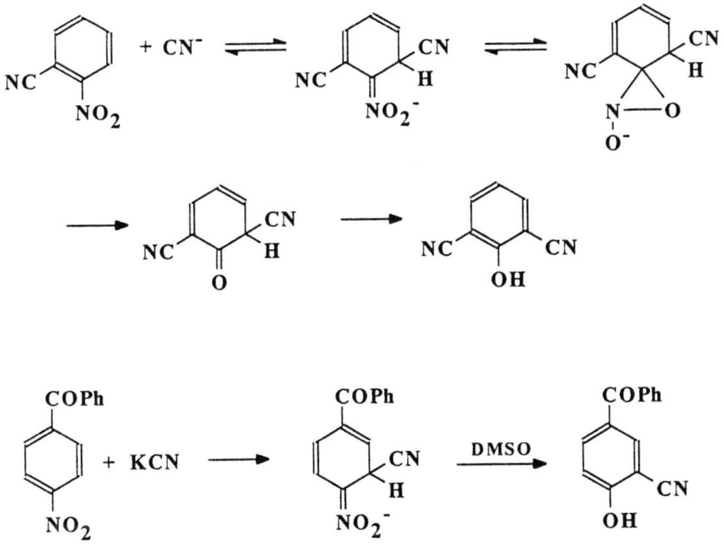

Scheme 26

In a similar manner 4-nitrobenzophenone is transformed in 60% yield into 3-cyano-4-hydroxybenzophenone on treatment with potassium cyanide in DMSO (Scheme 26) [71JCS(CC)1120].

The reaction of 9-nitroanthracene with sodium cyanide in DMF is more complex. It results in the formation of a series of products: besides 9-cyano-10-nitroanthracene, 9-cyano-10-anthranol, and 9-cyano-10-anthrone, also 9-amino-10-cyanoanthracene and 9,10-dicyanoanthracene are obtained (Scheme 27) [68JOC403].

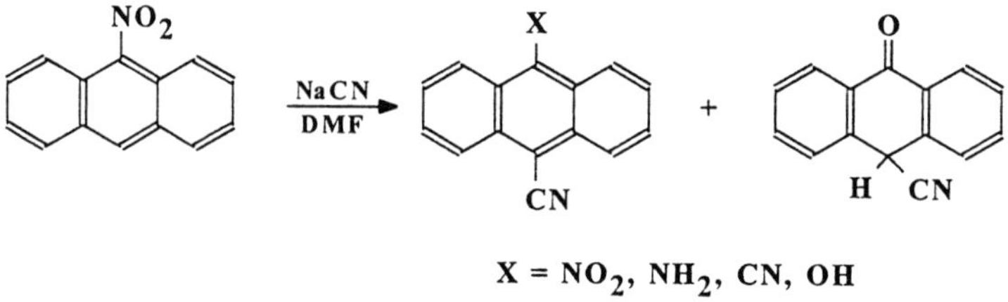

$X = NO_2, NH_2, CN, OH$

Scheme 27

2. Base-Induced Carbanions. The formation of 2-substituted 4-chloronitrobenzenes by reacting 4-chloronitrobenzene with active methylene compounds in the presence of a strong base (potassium *t*-butoxide in *t*-butylamine or potassium amide in liquid ammonia) at low temperatures can be regarded as a characteristic example of the many S_N^H reactions reported with nitroarenes (Table II) [82H177; 84H1811]. It is remarkable that under these conditions no $S_N(AE)$ substitution of the chloro atom, but only the S_N^H replacement, is observed.

Phenoxide ions have been found to act as carbon nucleophiles and not as oxygen nucleophiles in reactions with 1,3-dinitro- and 1,3,5-trinitrobenzenes in the presence of sodium hydroxide in DMSO [84CJC534; 87JOC488]. The reaction leads to biphenyl derivatives and can be regarded as a formal displacement of the hydrogen atom of the nitroaromatic ring by the carbon atom of the phenolate (Scheme 28) [85JOC3091].

TABLE II

Nucleophilic Substitution of Hydrogen in *para*-Chloronitrobenzene by Action of Base-Induced Carbanions

Nucleophile	S_N^H Product(s)	Yield (%)	References
Acetonitrile	5-Chloro-2-nitrobenzyl cyanide	22	82H177
Acetone	5-Chloro-2-nitrophenylacetone	56	82H177
Methyl-*t*-butyl ketone	5-Chloro-2-nitrobenzyl-*t*-butylketone	62	82H177
Acetophenone	(5-Chloro-2-nitrobenzyl)-phenylketone	40	82H177
2-Picoline N-oxide	(4-Chloro-1-nitrophenyl-2)-(pyridinyl-2)methane N-oxide	45	84H1811
2,6-Lutidine N-oxide	(4-Chloro-1-nitrophenyl-2)-(2-methylpyridinyl-6)methane N-oxide	39	84H1811
Quinaldine	(4-Chloro-1-nitrophenyl-2)-(quinolinyl-2)methane	23	84H1811
Quinaldine N-oxide	(4-Chloro-1-nitrophenyl-2)-(quinolinyl-2)methane N-oxide	23	84H1811

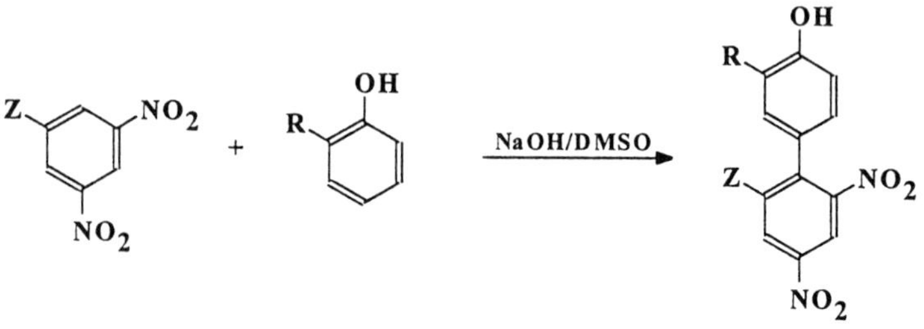

[Scheme 28 reaction diagram]

R = H, CH₃, t-C₄H₉
Z = H, NO₂

Scheme 28

3. Nucleophilic Radicals. Highly electrophilic polynitroarenes, such as 1,4-dinitrobenzene, and 1,2,4- and 1,3,5-trinitrobenzene, were demonstrated to undergo the displacement of hydrogen by action of poor nucleophilic phenyl and methyl radicals generated from the corresponding carboxylic acids [79JCS(P2)469]. For instance, when heating a mixture of 1,2,4-trinitrobenzene and benzoic acid with ammonium persulfate in the presence of AgNO₃, 2,4,5-trinitrodiphenyl is formed in 40% yield (Scheme 29) [79JCS(P2)469].

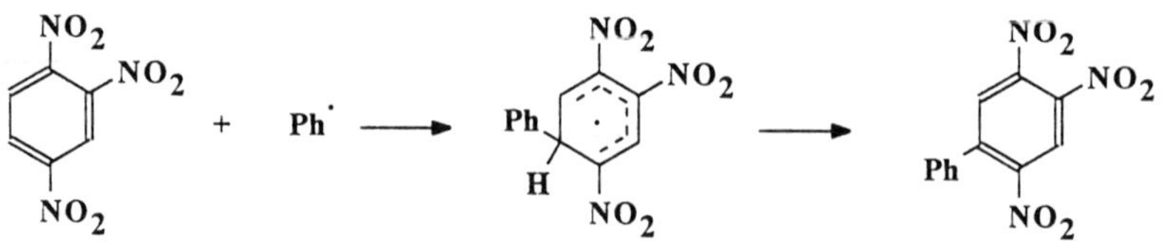

[Scheme 29 reaction diagram]

Scheme 29

It is of interest to note that the reaction of 1,2,4-trinitrobenzene with the more nucleophilic 1-adamantyl radical [76JCS(P2)662] results in adamantyl-denitration products; no displacement of hydrogen could be detected [79JCS(P2)469]. The same features in behavior of phenyl and adamantyl radicals were observed in their reactions with nitro-thiophen derivatives [80JCS(P2)1331].

4. Organometallic Compounds. Nitrobenzene reacts with alkylmagnesium halides at *ortho-* and *para-*positions, yielding the corresponding nitronate adducts. The *ortho/para* alkylation ratio is the statistically predicted one, that is, 2:1. In the reactions of *para-*Z-substituted nitrobenzenes with the alkyl Grignard reagents the addition of RMgX takes place exclusively at the *ortho-*position relative to the nitro group (Scheme 30). These nitronate σ-adducts are relatively stable, particularly at lower temperatures, due to effective participation of the nitro group in delocalization of the negative charge and a partial covalent bond O-MgX formation. They are easily decomposed by action of mineral acids into the corresponding alkyl-substituted nitrosobenzenes or can be oxidized with bromine, dichlorodicyano-*para*-benzoquinone (DDQ), or potassium permanganate into fully aromatized nitro compounds (Scheme 30) [84ACR109]. For instance, acid treatment of a mixture obtained from the reaction of nitrobenzene with *n*-butyl-magnesium bromide results in *ortho-* and *para-*butyl-substituted nitrosobenzenes in 42% and 22% yields, respectively (Scheme 30) [79JOC2087; 84ACR109]. Yields and products derived from the Grignard alkylation of some substituted nitrobenzenes are presented in Table III.

Tris(dimethylamino)sulfonium difluorotrimethylsiliconate (TASF) has been found to be an efficient compound for the fluoride-assisted addition of silyl enol ethers to aromatic nitro compounds [84JOC4571; 85JA5473]. In the presence of TASF nitroarenes are able to react with trimethylsilyl ethers of enols into the silylated σ-adducts which are substantially stabilized, probably due to O-silylation of the nitrohexadienates.

For instance, nitrobenzene **24** reacts with trimethylsilyl ketene acetal **25** in THF/ CH_3CN in the presence of one equivalent of TASF at –60°C into adduct **26**, which by an *in situ* oxidation of a rapidly cooled solution with an equivalent amount of bromine or DDQ at –78°C gave the product **27** in 79% yield (Scheme 31; Table III) [85JA5473].

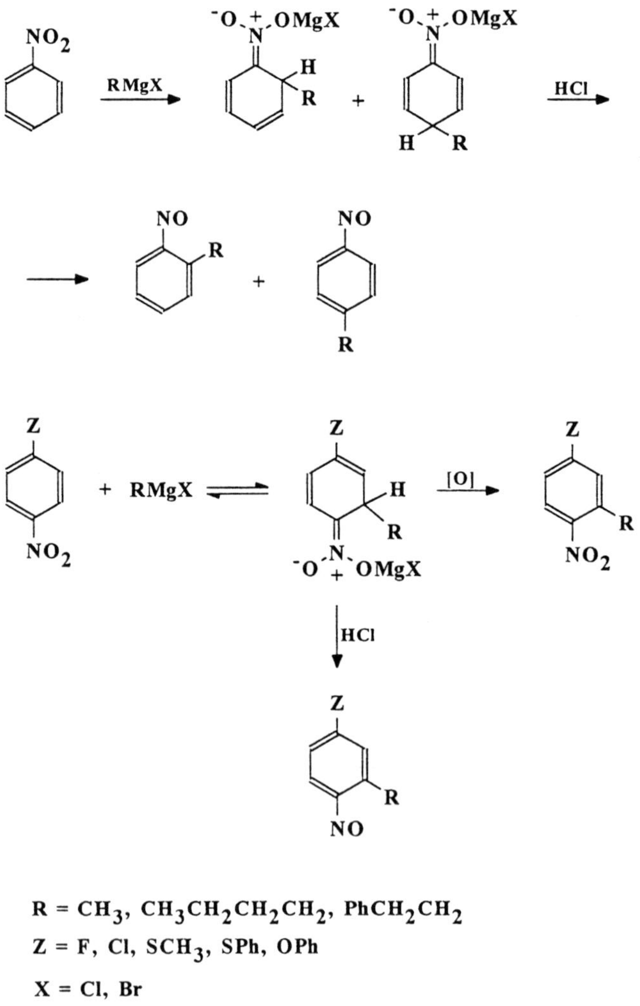
NO2
RMgX
-O + OMgX N
H
R
+
HCl
H R
NO
R
Z
+ RMgX
[O]
NO2
-O + OMgX
HCl
NO
R = CH3, CH3CH2CH2CH2, PhCH2CH2
Z = F, Cl, SCH3, SPh, OPh
X = Cl, Br

Scheme 30

TABLE III
Nucleophilic Substitution of Hydrogen in Nitrobenzenes by Action of Organometallic Compounds

Nitroarene	Reagent[1]	S_N^H Product(s)	Yield (%)	References
Nitrobenzene	CH$_3$Li	2-Nitrotoluene	32	78HCA449
		4-Nitrotoluene	20	
		4-Nitro-*m*-xylene	7	
Nitrobenzene	*n*-BuMgBr	2-Nitroso-*n*-butylbenzene	42	84ACR109
		4-Nitroso-*n*-butylbenzene	22	
p-Nitrotoluene	CH$_3$MgBr	4-Nitroso-*m*-xylene	50	84ACR109
p-Methylthio-nitrobenzene	CH$_3$MgBr	2-Nitroso-5-methylthio-toluene	73	79JOC2087
p-Phenylthio-nitrobenzene	CH$_3$MgBr	2-Nitroso-5-phenylthio-toluene	65	79JOC2087
p-Phenoxy-nitrobenzene	CH$_3$MgBr	2-Nitroso-5-phenoxy-toluene	53	79JOC2087
1-Nitro-naphthalene	*n*-BuLi	2-*n*-Butyl-1-nitro-naphthalene	25	78HCA449
		4-*n*-Butyl-1-nitro-naphthalene	48	
Nitrobenzene	A	Methyl α–methyl-α-(4-nitro-phenyl)-propionate	79	85JA5473
Nitrobenzene	B	Methyl α-(2-nitrophenyl)-propionate	36	85JA5473
		Methyl α-(4-nitrophenyl)-propionate	16	
p-Fluoronitro-benzene	B	Methyl α-(5-fluoro-2-nitrophenyl)-propionate	77	85JA5473

[1]A: With 1-methoxy-1-(trimethylsiloxy)-2-methyl-1-propene in THF in the presence of TASF followed by oxidation with bromine; B: With 1-methoxy-1-(trimethylsiloxy)-1-propene under the same conditions.

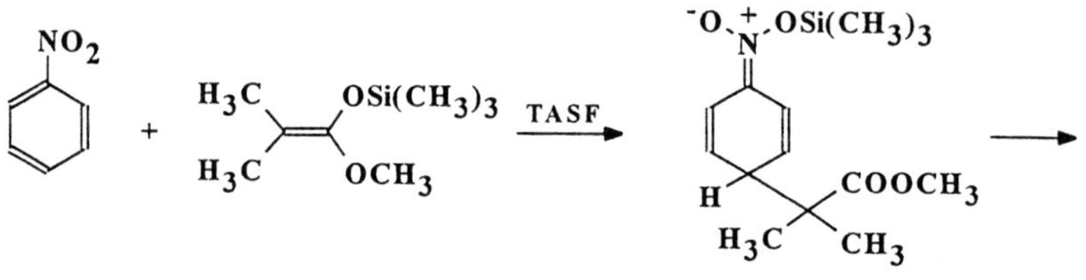

24 25 26

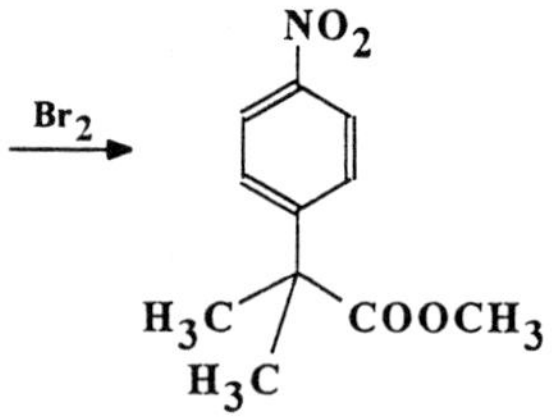

27

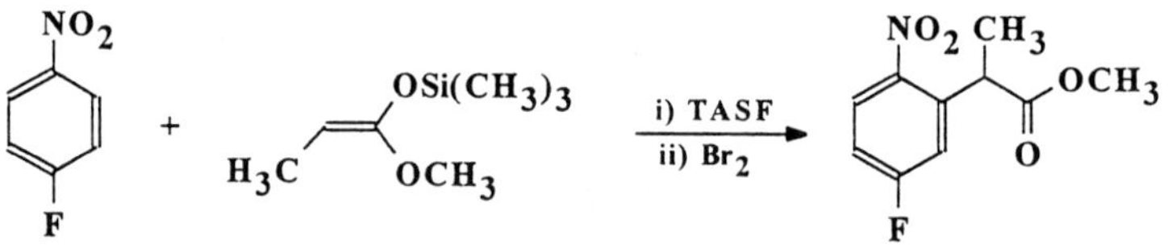

28 29 30

Scheme 31

The regiochemistry of this reaction seems to be controlled by the size of the silicon reagent, as can be deduced from the fact that reactions of nitrobenzenes with less hindered trimethylsilyl ketene acetal **29** result in a mixture of *ortho-* and *para-*substituted products with a strong preference for the *ortho*-addition (Table III) [85JA5473]. Again, it is worth noting that in all cases studied the substitution of hydrogen in the benzene ring prevailed over that of halogen. Even with the lable fluoro substituent in *para*-fluoronitrobenzene **28** no displacement of halogen was observed (Scheme 31) [85JA5473].

In a similar way, 1,3,5-trinitrobenzene was alkylated with trimethylsilanes $RSiMe_3$ in the presence of KF and a crown ether [87JOM139; 90ZOR933].

5. Vicarious Nucleophilic Substitution. The concept of vicarious nucleophilic substitution of hydrogen in arenes was introduced by M. Makosza in the late 1970s. It concerns reactions of nitroarenes or other π-deficient systems with carbanions featuring a good leaving group L at the carbanionic center. Using this kind of nucleophile creates the interesting possibility that in the intermediary σ-adduct both the anion L⁻, and the sp^3-hydrogen can be eliminated (Scheme 32). Since the L fragment acts as a vicarious leaving group, the reaction has been called "vicarious nucleophilic aromatic substitution of hydrogen" [83MI2; 87ACR282; 87MI3; 89RCR747; 91S103; 92PJC3].

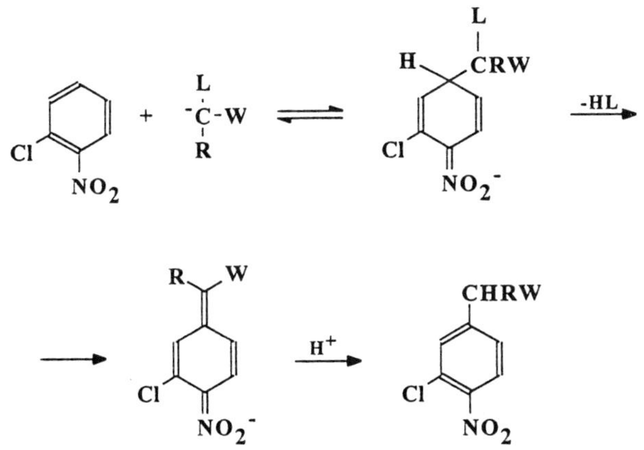

Scheme 32

In general, the course of the vicarious nucleophilic substitution is not influenced by the substituent present on nitrobenzene or nitronaphthalene [83MI2; 87ACR282;

89RCR747; 91S103]. Most substituents, such as the nucleofugal groups F, Cl, Br, I, the electron-donating OCH_3, NR_2, SCH_3, CH_3 and electron-withdrawing COOR, CF_3, NO_2 groups do not interfere with the reaction [86S50; 91S103]. Only the presence of substituents, which under the applied reaction conditions are deprotonated into anions which delocalize on the ring and the nitro group (for instance, OH and SH), inhibits the reaction [87MI3; 91S103].

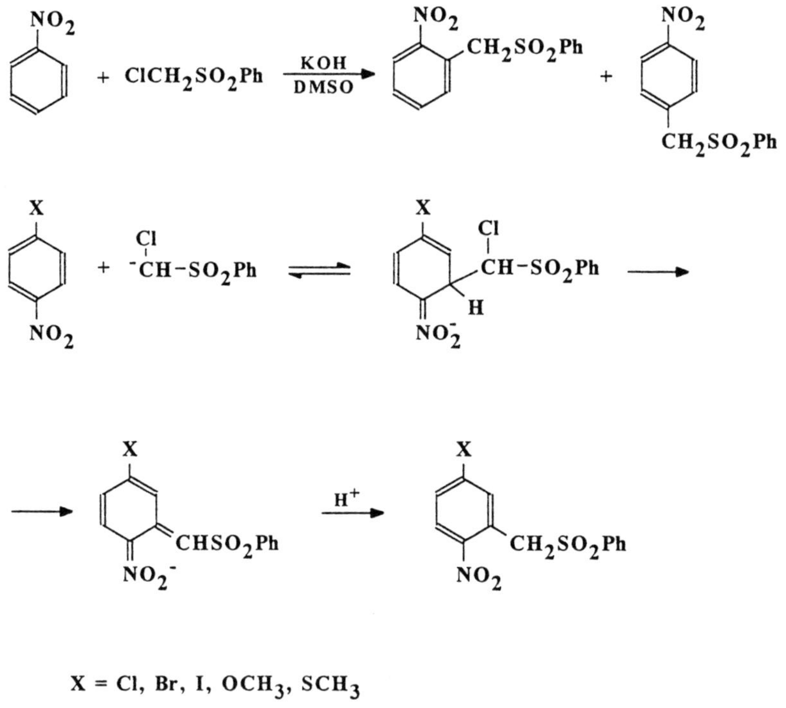

X = Cl, Br, I, OCH_3, SCH_3

Scheme 33

A great variety of carbanions with the structure R(W)-C⁻-L, where L represents a leaving group (Cl, PhO, PhS, etc.) and W is an electron-withdrawing group stabilizing the formation of carbanion (CN, COR, COOR, SO_2Ar, etc.), are able to perform the reaction, thus providing elegant synthetic methods for nucleophilic aromatic substitution of hydrogen [78TL3495; 80JOC1534; 83JOC3860; 83MI2; 83S40; 84JOC1488; 84JOC1494; 84T1843; 84TL803; 84TL4791; 87ACR282; 87MI3; 89RCR747; 91S103; 92PJC3: 92PJC2005; 92S571]. For instance, chloromethyl phenyl sulfone, when treated with a strong base, converts nitrobenzene into a mixture of 2- and 4-nitrobenzyl phenyl sulfones (Scheme 33; Table IV) [78TL3495]. An analogous reaction of nitrobenzene with α-chlorobenzyl phenyl sulfone proceeds more selectively affording the *para*-substituted nitrobenzene in 93% yield (Table IV) [84JOC1488].

One of the most important features of vicarious S_N^H nucleophilic substitution in halo-nitroarenes is that the S_N^H reaction proceeds faster than the conventional nucleophilic substitution of halogen and other nucleofugal groups [87MI3; 92PJC3]. Thus, in *para*-X-substituted nitrobenzenes vicarious carbanions substitute the *ortho*-hydrogen atom without replacing such nucleofugal groups X as Cl, Br, I, OCH_3, and SCH_3! (Scheme 33; Table IV). A similar observation has been made in the reaction of *ortho*-chloro-nitrobenzene with vicarious carbanions in which only the *para*-hydrogen is replaced [91MI2; 92PJC3].

Another promising process for nucleophilic alkylation of nitroaromatics is the reaction with alkyl trifluoromethyl sulfones (Scheme 34).

The trifluoromethyl group provides a very efficient stabilization for the formed carbanion and the trifluoromethyl sulfone group has a high nucleofugal character. Combination of these two features makes alkyl trifluoromethyl sulfone a typical vicarious reagent. Indeed, the reaction of nitroarenes with butyl trifluoromethyl sulfone in the presence of a strong base gave alkylated nitroaromatic compounds, although in moderate yields (Scheme 34; Table IV) [90OPP575].

TABLE IV
Vicarious Nucleophilic Substitution of Hydrogen in Nitroarenes with Sulfone-Stabilized Carbanions

Arene	Reagent[1]	S_N^H Product(s)	Yield (%)	References
Nitrobenzene	A	2-Nitrobenzyl phenyl sulfone	36	84JOC1488
		4-Nitrobenzyl phenyl sulfone	39	
Nitrobenzene	B	α-(4-Nitrophenyl)benzyl phenyl sulfone	93	84JOC1488
m-Dinitrobenzene	D	2,4-Dinitrophenylacetone	68	92PJC2005
m-Dinitrobenzene	F	t-Butyl (2,4-dinitrophenyl)chloroacetate	81	91LA605
m-Dinitrobenzene	G	4-(n-Butyl)-1,3-dinitrobenzene	23	90OPP1575
4-Fluoro-1,3-dinitrobenzene	F	t-Butyl (5-fluoro-2,4-dinitrophenyl)chloroacetate	44	91LA605
4-Fluoro-1,3-dinitrobenzene	C	(5-Fluoro-2,4-dinitrophenyl)chloromethyl p-tolyl sulfone	55	91LA605
p-Fluoronitrobenzene	A	5-Fluoro-2-nitrobenzyl phenyl sulfone	18	84JOC1488
		α-Chloro-4-nitrobenzyl phenyl sulfone	27	
p-Chloronitrobenzene	A	5-Chloro-2-nitrobenzyl phenyl sulfone	69	84JOC1488
p-Chloronitrobenzene	E	5-Chloro-2-nitrobenzyl cyanide	75	84JOC1494

TABLE IV (Continued)

Arene	Reagent[1]	S_N^H Product(s)	Yield (%)	References
p-Bromonitrobenzene	A	5-Bromo-2-nitrobenzyl phenyl sulfone	61	84JOC1488
p-Iodonitrobenzene	A	5-Iodo-2-nitrobenzyl phenyl sulfone	74	84JOC1488
p-Methoxynitrobenzene	A	5-Methoxy-2-nitrobenzyl phenyl sulfone	72	84JOC1488
p-Methylthionitrobenzene	A	5-Methylthio-2-nitrobenzyl phenyl sulfone	72	84JOC1488
1-Nitronaphthalene	G	2-(n-Butyl)-1-nitronaphthalene	19	90OPP575
		4-(n-Butyl)-1-nitronaphthalene	18	

[1]A: Chloromethyl phenyl sulfone; B: α-Chlorobenzyl phenyl sulfone; C: Dichloromethyl p-Tolyl Sulfone; D: α-Chloroacetone; E: Chloroacetonitrile; F: t-Butyl dichloroacetate: G: Trifluoromethyl n-butyl sulfone.

The methylation of aromatic nitro compounds by dimethyloxosulfonium methylide [64TL867; 66JOC243] or the methylsulfinyl carbanion [66JOC248] also can be regarded as a vicarious nucleophilic aromatic substitution of hydrogen. Indeed, in the reaction of nitrobenzene with dimethyloxosulfonium methylide the formation of *ortho*- and *para*- nitrotoluene is facilitated by elimination of dimethyl sulfoxide (Scheme 35) [66JOC248].

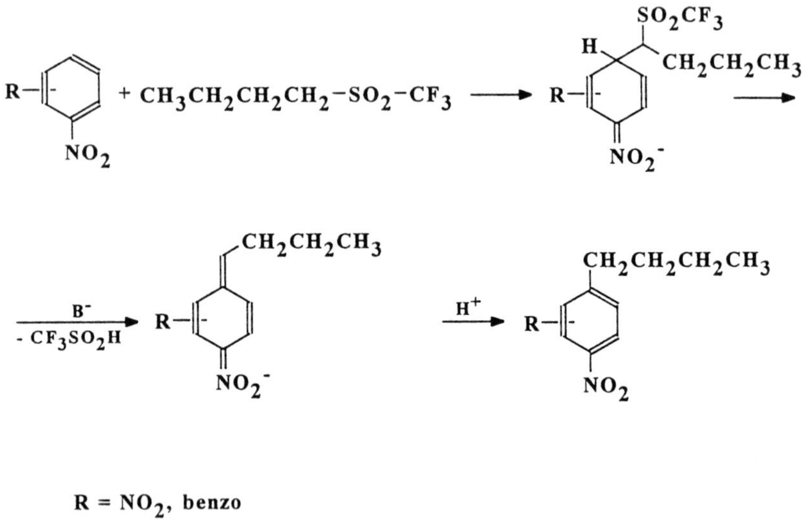

R = NO$_2$, benzo

Scheme 34

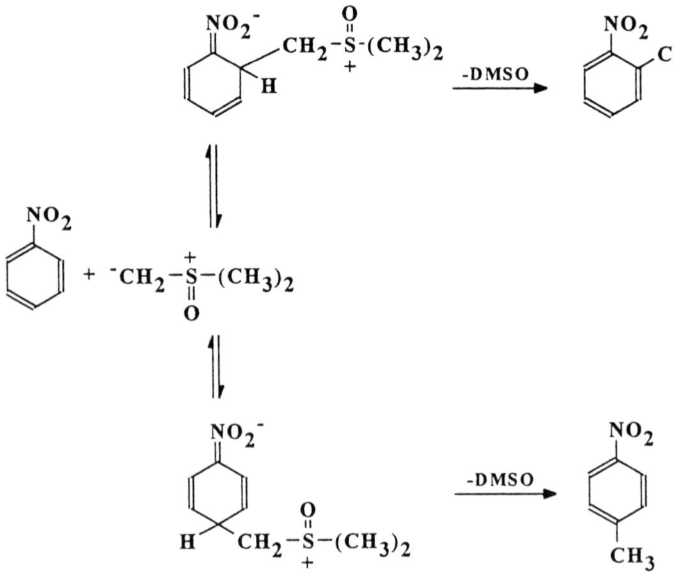

Scheme 35

The vicarious S_N^H methodology proved also to be effective for functionalization of *ortho*-nitrobenzophenone by carbanions. Especially in case of tertiary carbanions derived from α-substituted chloromethyl phenyl sulfones the replacement of the ring hydrogen takes place selectively in the *para*-position to the nitrogroup (Scheme 36; Table V) [89LA825]. Isomeric *meta*- and *para*-nitrobenzophenone behave differently toward these reagents: the addition of carbanions to the carbonyl group is often competing with vicarious nucleophilic substitution of hydrogen giving rise to a mixture of products [89LA825].

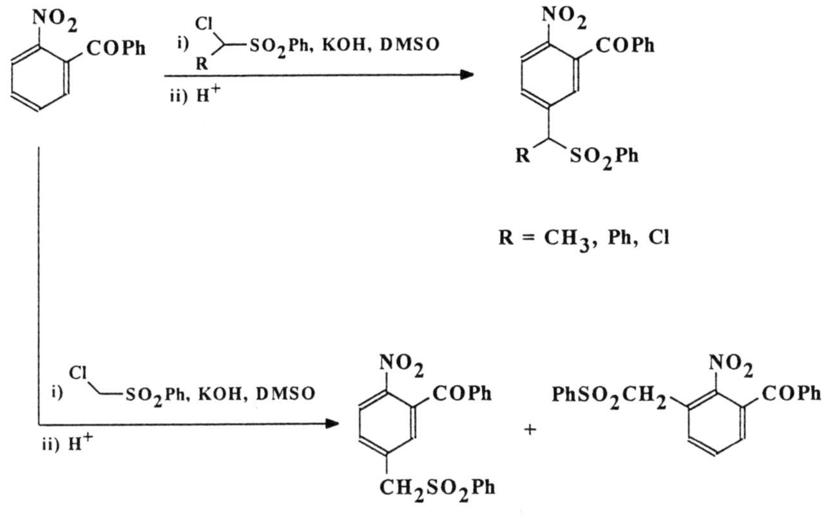

Scheme 36

The vicarious S_N^H methodology was also successful, when reacting α-chloroalkyl phenyl sulfones with *ortho*-, *meta*-, and *para*-nitrobenzoic acids and their esters. The results obtained show that the negatively charged carboxylate substituent does not disturb the reaction course, although in some cases the S_N^H products have been isolated in only moderate yields (Scheme 37; Table V) [86S50].

TABLE V
Vicarious Nucleophilic Substitution of Hydrogen in Aromatic
Ketones and Carboxylic Acid by Action of Carbanions

Nitroarene	Carbanion precursor[1]	S_N^H Product(s)	Yield (%)	References
o-Nitrobenzophenone	A	2-Nitro-5-[(phenylsulfonyl)methyl]benzophenone	22	89LA825
		2-Nitro-3-[(phenylsulfonyl)methyl]benzophenone	31	
o-Nitrobenzophenone	B	2-Nitro-5-[1-(phenylsulfonyl)ethyl]benzophenone	49	89LA825
o-Nitrobenzophenone	C	2-Nitro-5-[α-phenyl-(phenylsulfonyl)methyl]benzophenone	86	89LA825
o-Nitrobenzophenone	D	5-[Chloro(phenylsulfonyl)methyl]-2-nitrobenzophenone	50	89LA825
o-Nitrobenzoic acid	E	2-Nitro-5-[1-phenylsulfonyl)propyl]benzoic acid	18	86S50
m-Nitrobenzoic acid	A	5-Nitro-2-[(phenylsulfonyl)methyl]benzoic acid	23	86S50
		2-[(Phenylsulfonyl)methyl]-3-nitrobenzoic acid	34	
p-Nitrobenzoic acid	A	3-[(Phenylsulfonyl)methyl]-4-nitrobenzoic acid	65	86S50
t-Butyl 4-nitrobenzoate	A	t-Butyl 3-[(Phenylsulfonyl)methyl]-4-nitrobenzoate	62	86S50

[1]A: Chloromethyl phenyl sulfone: B: α-Chloroethyl phenyl sulfone;
C: α-Chloro benzyl phenyl sulfone; D: dichloromethyl phenyl sulfone;
E: α-Chloropropyl phenyl sulfone.

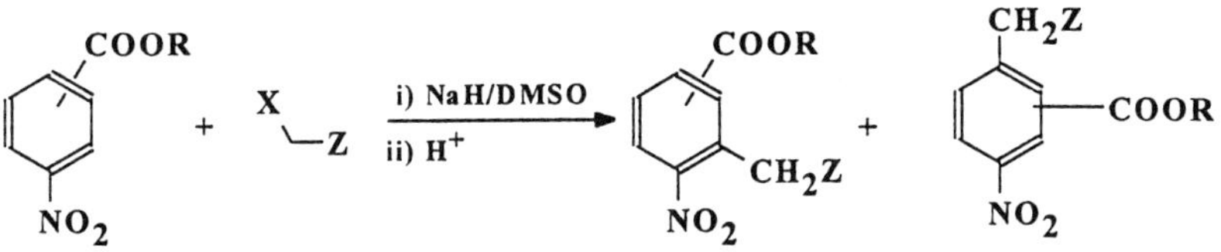

R = H, CH$_3$, t-Bu; X = Cl, OPh, SPh; Z = SO$_2$Ph, CN

Scheme 37

For an intramolecular version of vicarious nucleophilic aromatic substitution of hydrogen in nitroarenes see Chapter 2,VI,D.

6. Copper-Mediated Vicarious Nucleophilic Substitution. A novel mechanistic approach for nucleophilic substitution of hydrogen in arenes, the so-called copper-mediated vicarious nucleophilic substitution, has recently been advanced [90S942; 93MI3]. It has been inspired by the fact that iodoarenes cross-couple with 1,3-dinitro and 1,3,5-trinitrobenzenes in quinoline in the presence of copper or copper(I) oxide to produce biaryls in 10-20% yields [68ACSA2338]. Modification of the process by using copper *t*-butoxide (*t*-BuOCu) and dimethoxyethane as solvent has considerably improved the yields of the coupling products [79JCS(CC)296; 82JCS(P1)2299].

Based on the concept of the vicarious nucleophilic substitution of hydrogen in nitroaromatics, as well as taking into account the σ–adduct formation between 1,3,5-trinitrobenzene and phenols [82CRV77], the reaction of *meta*-dinitrobenzene with *para*-iodo-phenol has been rationalized as a copper(I)-promoted S_N^H process involving the formation of a zwitterionic "Meisenheimer-Wheland" intermediate, which then eliminates a proton and iodide anion to give the final S_N^H product (Scheme 38) [90S942]. The mechanism can be considered to be more acceptable than the one proceeding via 2,6-dinitrophenyl copper [68ACSA2338; 68ACSA2581]. The mechanism of such couplings still needs further clarification (Scheme 38) [68ACS2338; 68ACS2581; 93MI3]. The presence of copper *t*-butoxide was found to be crucial to cause the reaction, since it plays a vital role in stabilization of intermediate σ-adducts and facilitates elimination of the iodide anion. Copper is also responsible for regioselective displacement of hydrogen at the 2-position of 1,3-dinitrobenzene [90MI3; 90S942]. The fact that α-bromo- and α-iodocarbanions generated from typical vicarious reagents, such as iodomethyl phenyl sulfone, iodoacetonitrile, bromoacetate and bromomalonate esters, are able to cause the copper-mediated displacement of hydrogen in position 2 of 1,3-dinitrobenzene is a strong argument that the reaction indeed follows the pattern of copper-mediated vicarious aromatic nucleophilic substitution (Scheme 39; Table VI) [90MI3; 90S942; 93MI3].

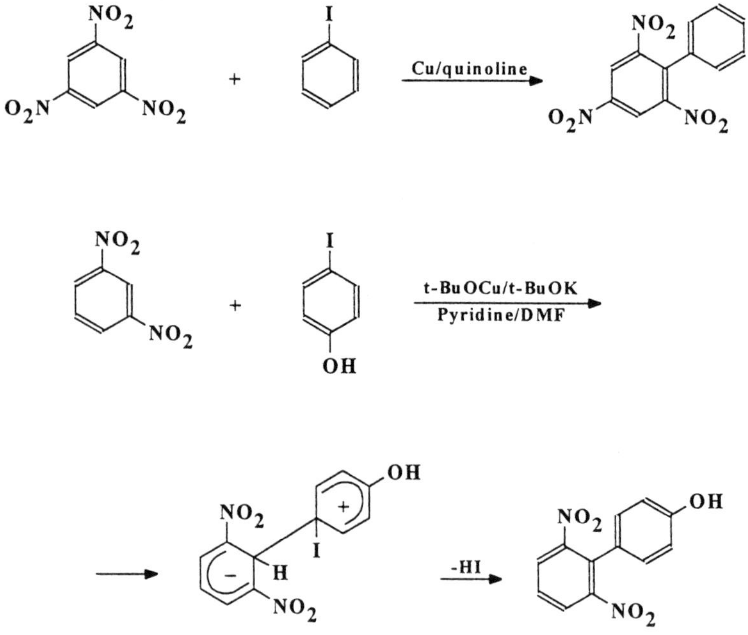

Scheme 38

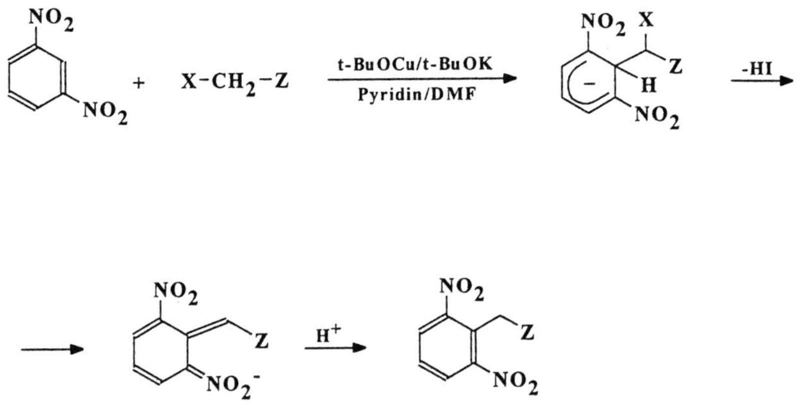

Chapter 2 S_N^H REACTIONS IN ARENES

Scheme 39

X = Br, I; Z = SO$_2$Ph, COOR, CN

TABLE VI
Copper-Mediated Vicarious Nucleophilic Substitution of Hydrogen in 1,3-Dinitrobenzenes

Reagent	S_N^H Product(s)	Yield (%)	References
p-Iodophenol	2',6'-Dinitrobiphenyl-4-ol	76	90S942
Iodomethyl phenyl sulfone	2,6-Dinitrobenzyl phenyl sulfone	76	90S942
Iodomethyl phenyl sulfoxide	2,6-Dinitrobenzyl phenyl sulfoxide	68	90S942
Bromoacetonitrile	2,6-Dinitrobenzyl cyanide	82	93MI3
Methyl α-bromoacetate	Methyl 2,6-dinitrophenylacetate	50	93MI3
Dimethyl bromomalonate	Dimethyl (2,6-dinitrophenyl)malonate	83	93MI3

B. Reactions with O-Nucleophiles

The smoothly proceeding conversion of 1,3,5-trinitrobenzene into 2,4,6-trinitrophenol by treatment with an alkaline solution of potassium ferricyanide seems to be the first reported example of a nucleophilic displacement of hydrogen in aromatics (Scheme 40) [1882LA344].

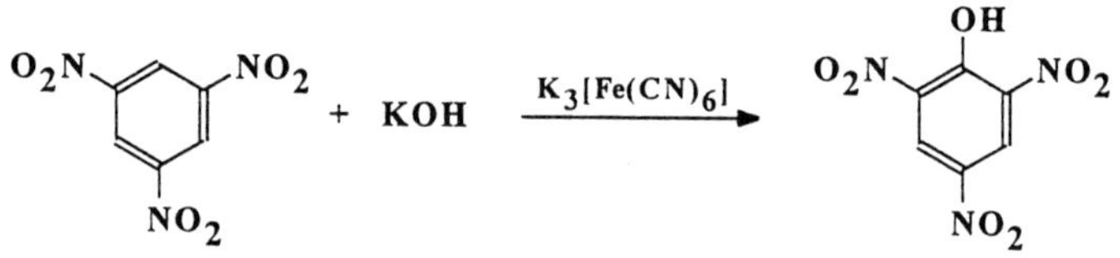

Scheme 40

Also, treatment of nitrobenzene by alkali has long been known to be a useful synthetic procedure for the preparation of *ortho-* and *para*-nitrophenol (see Chapter 1, Scheme 2) [1899CB3486; 52MI1]. The oxidant in the process is nitroarene itself, being reduced into azoxy and/or nitroso compounds [1899CB3486; 80CC394].

Reinvestigation of the hydroxylation reaction in liquid ammonia in argon atmosphere has shown that *ortho-* and *para*-nitrophenol are formed in the ratio of 3:1. The yields of the products were 70–80% relative to the converted nitrobenzene, however only a low conversion (about 10%) has been reached after 6-7 hours. Passing molecular oxygen through the reaction mixture resulted in an approximately threefold increase in conversion degree and in quantitative yields of *ortho-* and *para*-nitrophenol (ratio 4:1) [82ZOR1898]. Interestingly, 4-X-nitrobenzenes, containing nucleofugal groups, such as X = Cl, Br, and I, also undergo under oxidative conditions nucleophilic substitution of hydrogen, without nucleophilic replacement of X (Scheme 41) [85ZOR1150]. In case of

X = F or NO$_2$, however, conventional nucleophilic replacement of these groups takes place, leading to *para*-nitrophenol.

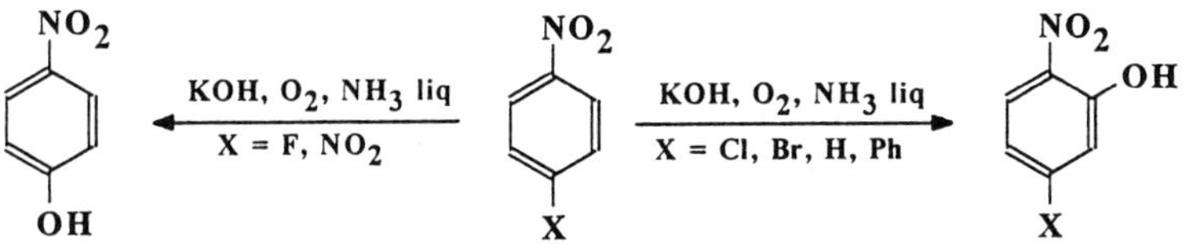

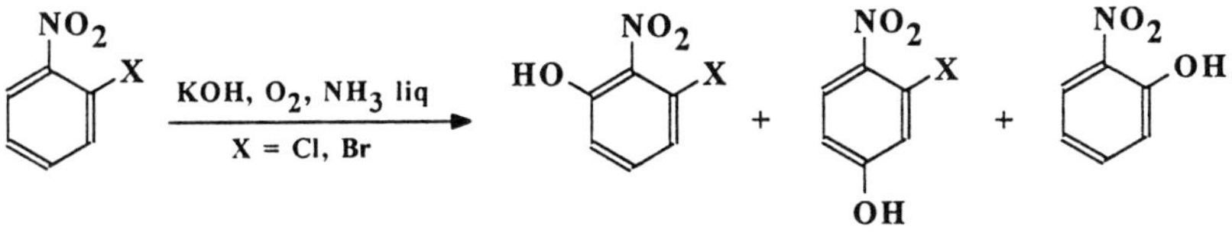

Substituent X	Ratio of products		
	31	32	33
X = Cl	4.8	1	3.3
X = Br	3.9	1	2.5

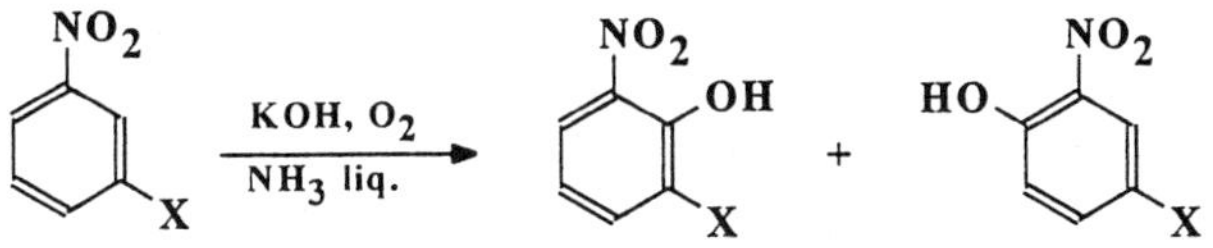

Scheme 41

Using the same oxidative conditions as mentioned above, hydroxylation of 2-X-nitrobenzenes (X= F, Cl, Br) gave for X= Cl, Br replacement of both halogen and hydrogen atoms in *ortho*- and *para*-positions relative to the nitro group (Scheme 41). For X=F hydroxy-defluorination is found; no hydroxy-dehydrogenation was observed.

Oxidative hydroxylation of 3-X nitrobenzenes by action of potassium hydroxide in liquid ammonia shows that each of both *ortho*-positions relative to the nitro group undergoes hydroxy-dehydrogenation, without substitution of group X (Scheme 41) [86ZOR806; 87ZOR1039].

1-Nitronaphthalene reacts with potassium hydroxide in liquid ammonia in the presence of oxygen to give a mixture of 1-nitro-2-naphthol and 1-nitro-4-naphthol; the ratio of these compounds is striking depending on the reaction temperature, quantity of potassium hydroxide, and the presence of water in ammonia (Scheme 42, Table VII) [87ZOR1039].

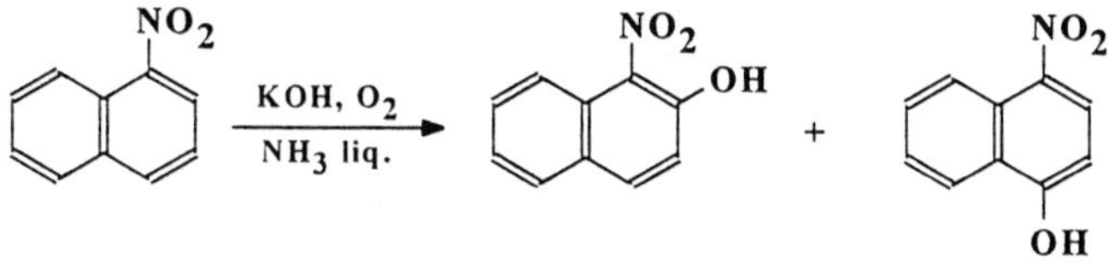

Scheme 42

Although nitrobenzene and nitronaphthalene seem to behave similarly, distinguished differences in the mechanism of their interaction with KOH in liquid ammonia in the presence of oxidants are observed. Experiments with oxygen-labeled potassium hydroxide K^{18}OH and unlabeled molecular oxygen have revealed that the content of the ^{18}O-label in the hydroxy group (as shown by mass spectroscopy) in the 1-nitronaphthols is much lower than in the K^{18}OH (Table VII) [85ZOR1150; 87ZOR1039]. In similar experiments with nitrobenzene 95% of the ^{18}O label has been found in the corresponding nitrophenols. This result leads to the conclusion that the hydroxy group in the 1-nitro-2- and 4-naphthols obtained not only originates from KOH, but also from the oxidative process with participation of molecular oxygen.

TABLE VII

Results of the Reaction between 1-Nitronaphthalene and
$K^{18}OH$ in Liquid Ammonia in the Presence of Oxygen $^{16}O_2$ [1]

Reaction temp., (°C)	Molar excess of $K^{18}OH$	Contents of the ^{18}O label in the reaction products relative to that in $K^{18}OH$		Ratio between 1-nitro-2-naphthol and -4-naphthol
−33	1.5	1-Nitro-2-naphthol	19	1:1
		1-Nitro-4-naphthol	44	
−33	2.5	1-Nitro-2-naphthol	47	3:1
		1-Nitro-4-naphthol	40	
−33	4.0	1-Nitro-2-naphthol	64	6:1
		1-Nitro-4-naphthol	48	
−55	2.5	1-Nitro-2-naphthol	82	9:1
		1-Nitro-4-naphthol	70	

[1] See ref. [87ZOR1039].

The mechanism presented in Scheme 43 for hydroxylation of nitrobenzene is in good agreement with the ^{18}O-labeling experiments, but seems only partly applicable to the oxidative hydroxylation of 1-nitronaphthalene.

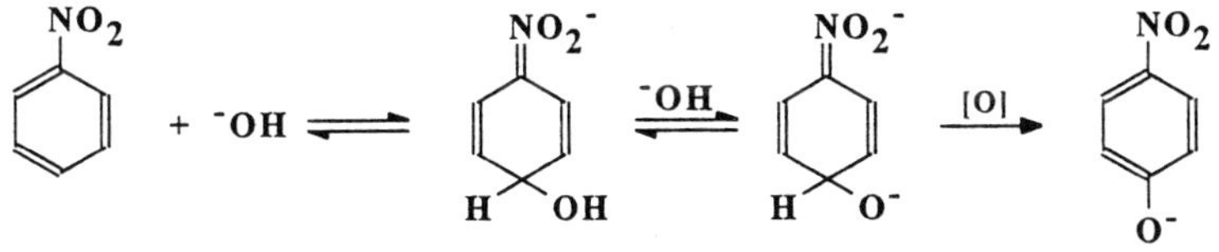

Scheme 43

Several plausible reaction mechanisms have been advanced to explain the participation of molecular oxygen in the formation of nitronaphthols. They involve the intermediacy of the radical anion of nitronaphthalene **34** which recombines with molecular oxygen or oxygen radical anion to give **35** or **36**, respectively. Interaction between nitroarene and the oxygen radical anion $O_2^{•-}$ or dianion O_2^{2-} can also result in the formation of species **35** and **36** (Scheme 44) [81ZOR2402].

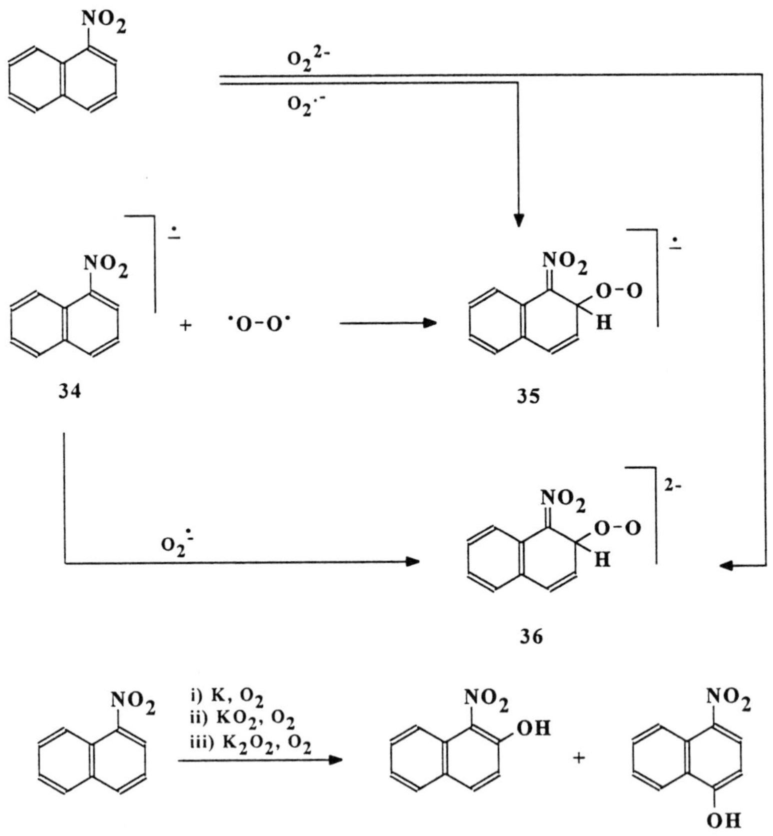

Scheme 44

The intermediary formation of such intermediates is substantiated by the experimental data. Treatment of 1-nitronaphthalene with one equivalent of potassium, converting it into the naphthalene radical anion species **34,** followed by its oxidation with oxygen indeed results in the formation of isomeric mixture of 1-nitro-2- and 1-nitro-4-naphthol in the ratio 1:1. The same result has been obtained when 1-nitronaphthalene reacted with potassium superoxide or potassium peroxide and oxygen (Scheme 44) [81ZOR2402].

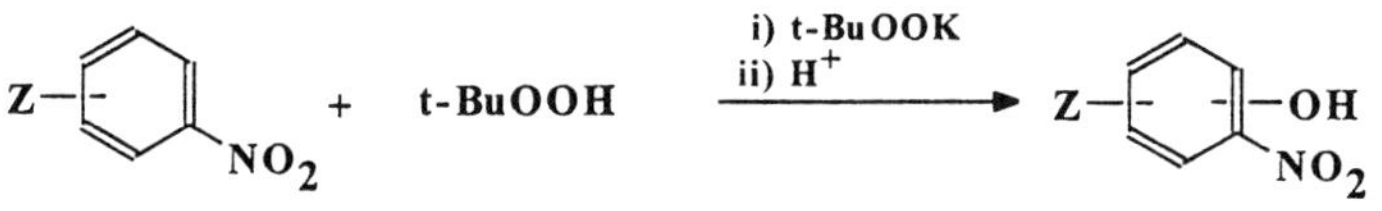

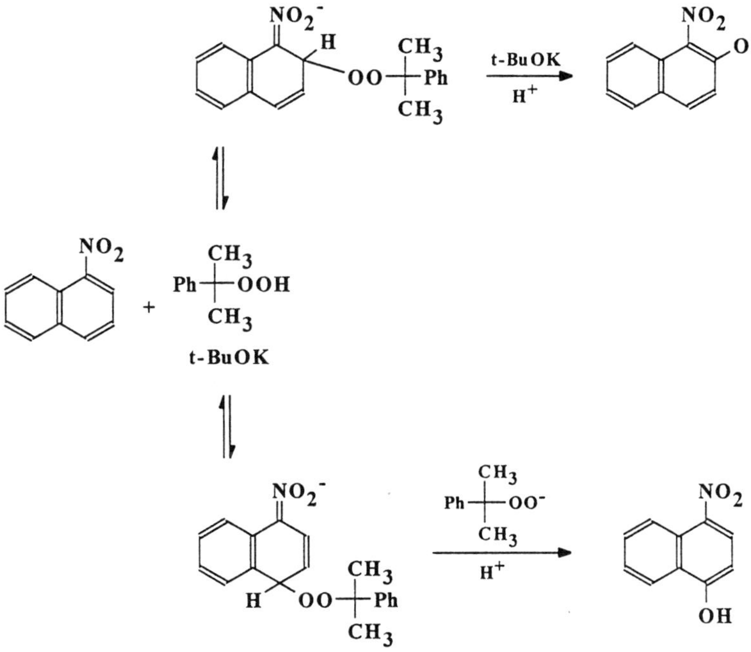

Scheme 45

A new synthetic approach to nitrophenols and nitronaphthols using the S_N^H methodology has recently been developed [92PJC3]. Readily available and relatively stable *t*-butyl hydroperoxides and cumyl hydroperoxides were found to react with nitroarenes in liquid ammonia containing potassium *t*-butoxide, affording *ortho*- or/and *para*-hydroxylation products. The ROO⁻ anions (R= *t*-butyl, cumyl) derived from these alkylhydroperoxides react with nitroarenes via the usual vicarious mechanistic pathway, that is, nucleophilic addition followed by base-induced elimination of ROH, yielding nitrophenols and nitronaphthols, respectively (Scheme 45; Table VIII) [90JOC4979; 90MI2]

TABLE VIII
Hydroxylation of Nitroarenes with *tert*-Butyl Hydroperoxide
in Liquid Ammonia in the Presence of *t*-BuOK[1]

Nitroarene	S_N^H Product(s)	Yield (%)
Nitrobenzene	4-Nitrophenol	45
3-Fluoronitrobenzene	2-Fluoro-4-nitrophenol	76
	2-Fluoro-6-nitrophenol	7
3-Chloronitrobenzene	2-Chloro-4-nitrophenol	79
	2-Chloro-6-nitrophenol	3
3-Cyanonitrobenzene	2-Cyano-4-nitrophenol	87
p-Chloronitrobenzene	5-Chloro-2-nitrophenol	27
	4-Nitrophenol	27
m-Dinitrobenzene	2,4-Dinitrophenol	96
2,4-Dinitrochlorobenzene	5-Chloro-2,4-dinitrophenol	93
1-Nitronaphthalene	1-Nitro-2-naphthol	87

[1]Ref. [90JOC4979].

C. Reactions with N-Nucleophiles

Based on the successful results of the Chichibabin amination of the π-deficient azaaromatics by potassium amide [78RCR1042], it was justified to expect that reactions of nitrobenzenes with ammonia or potassium amide will also lead to the formation of *ortho-* and/or *para*-amino nitro compounds. However, a very complicated mixture of products was obtained when reacting the parent compound nitrobenzene with potassium amide [32JCS1254; 75CIL520]. It has been considered for a long time that aromatic compounds are not able to undergo the Chichibabin amination reaction [92MI1]. A rare case was the direct amination of 4-nitrobenzophenone with potassium amide in liquid ammonia (Scheme 46), yielding 3-amino-4-nitrobenzophenone as by-product in a very low yield; the main reaction product is 4,4'-dibenzoylazobenzene [79JOC4705].

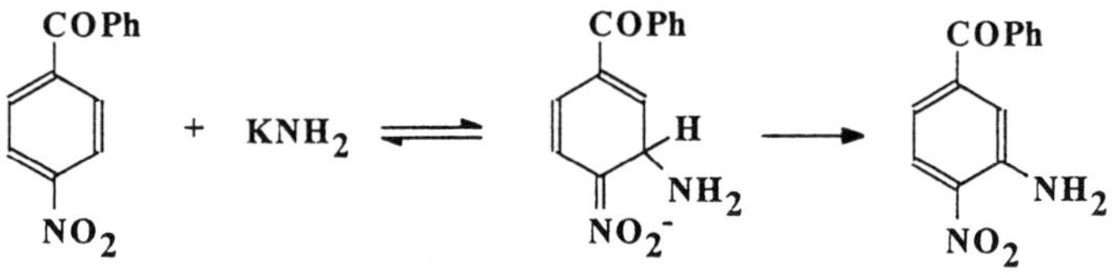

Scheme 46

However, "the vicarious nucleophilic substitution methodology" (see Chapter 2,III,A) proved to be an appropriate synthetic and facile tool for direct amination of nitroarenes, by using "vicarious" nucleophiles, such as hydroxylamine [06CB2533; 67ZOR1617; 73YZ1019], 4-amino- or 4-alkylamino-1,2,4-triazoles [86JOC5039; 88JOC3978], or sulfenamides [92JOC4784].

Amination of *meta*-dinitrobenzene by action of hydroxylamine is a long-known example of a reaction leading to the formation of 2,4-dinitro-1,3-phenylenediamine [06CB2533; 67ZOR1617; 73YZ1019]. The reaction is supposed to be initiated by the addition of hydroxylamine followed by the departure of the hydroxy anion and a proton. Thus, the hydroxy group acts as vicarious leaving group and promotes elimination of the hydrogen ion (Scheme 47) [91MI2]. Experiments with d_4-*meta*-dinitrobenzene showed that in the final product the deuterium is present not only in the ring, but also in the amino group, thus suggesting a migration of the sp^3-hydrogen in the intermediates to the amino group during the reaction; the exact mechanism of the rearomatization step remains still to be established (Scheme 47) [91MI2].

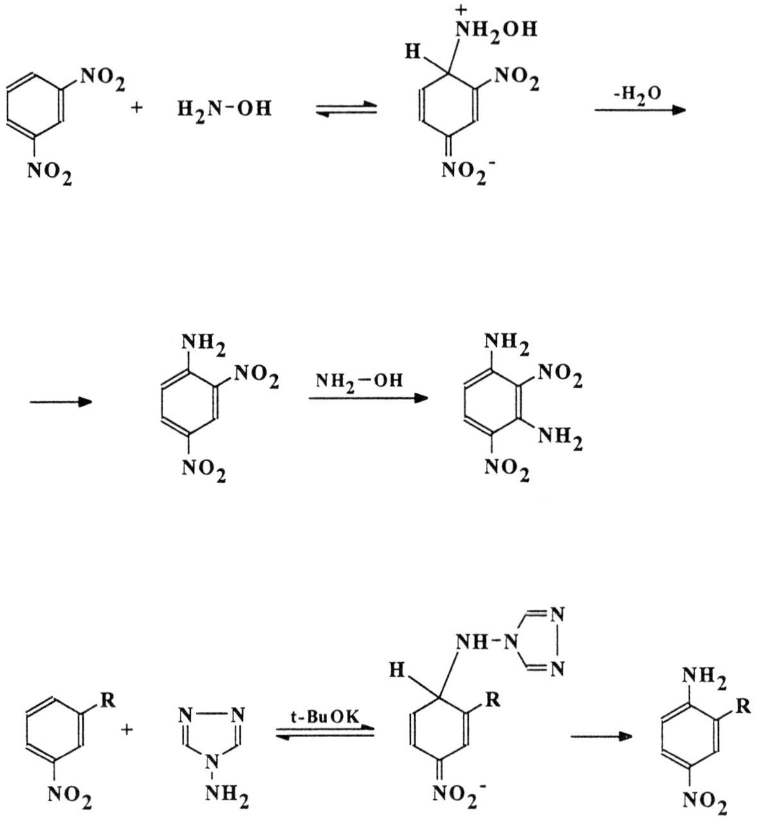

R = H, CH$_3$, F, Cl, OCH$_3$, NO$_2$

Scheme 47

Analogous amination reactions in the series of nitrobenzenes and nitronaphthalenes have been carried out with 4-amino- and 4-alkylamino-1,2,4-triazoles. In this case the aromatization is facilitated by elimination of the 1,2,4-triazole ring, acting as the vicarious leaving group (Scheme 47) [86JOC5039; 88JOC3978]. It is very interesting that the amination of nitrobenzenes with 4-amino-1,2,4-triazoles is regioselective, affording exclusively 4-nitroanilines (Scheme 47, Table IX). Amination of 1- and 2-nitronaphthalenes with 4-amino-1,2,4-triazole was found to occur preferentially at the *ortho-* position relative to the nitro group [86JOC5039; 88JOC3978; 91MI2].

Sulfenamides of general structure RSNH$_2$ also proved to be effective reagents to aminate aromatic nitro compounds [92JOC4784]. These reagents are easily available and stable enough in the form of anions, provided they contain electron-withdrawing substituents at the nitrogen atom, such as 2,4,6-trichlorobenzenesulfenyl, 2-benzothiazolsulfenyl, and N,N-tetramethylenethiocarbamoyl sulfenyl groups (Scheme 48).

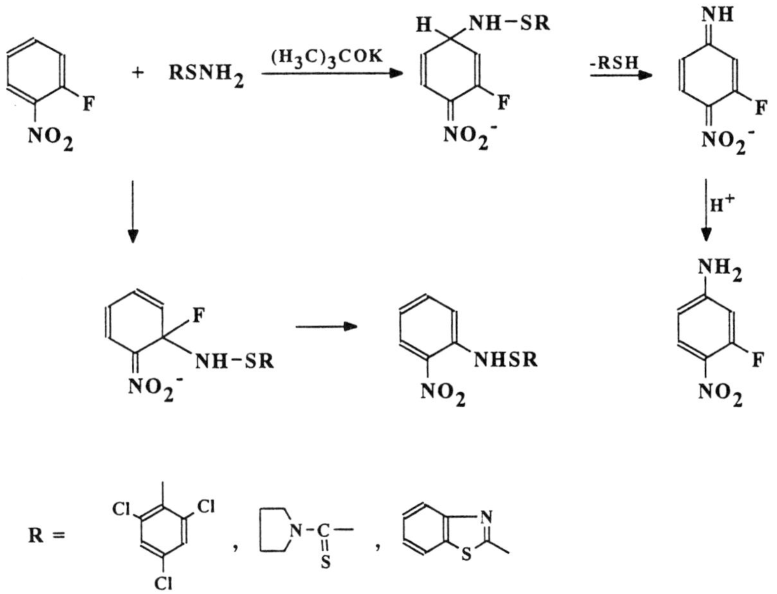

Scheme 48

TABLE IX
Amination of Nitroarenes via Vicarious Nucleophilic Substitution of Hydrogen in Arenes

Nitroarene	Aminating Reagent[1]	S_N^H Product(s)	Yield (%)	References
Nitrobenzene	A	2-Nitroaniline	14	92JOC4784
		4-Nitroaniline	71	
	B	2-Nitroaniline	34	92JOC4784
		4-Nitroaniline	35	
	D	4-Nitroaniline	58	86JOC5039
3-Nitrotoluene	D	2-Methyl-4-nitro-toluene	74	86JOC5039
1-Chloro-3-nitro-benzene	A	2-Chloro-4-nitro-aniline	86	92JOC4784
	D	2-Chloro-4-nitro-aniline	91	86JOC5039
3-Fluoronitro-benzene	D	2-Fluoro-4-nitro-aniline	47	86JOC5039
3-Methoxynitro-benzene	D	2-Methoxy-4-nitroaniline	36	86JOC5039
3-Nitrobenzoic acid	D	2-Amino-5-nitro-benzoic acid	60	86JOC5039
4-Chloronitro-benzene	B	5-Chloro-2-nitro-aniline	60	92JOC4784
1-Nitronaphthalene	C	2-Amino-1-nitro-naphthalene	71	92JOC4784
		4-Amino-1-nitro-naphthalene	6	
1,3-Dinitrobenzene	E	2,4-Dinitro-1,3-phenylene-diamine	52	67ZOR1617

[1]A: N,N-Tetramethylenethiocarbamoyl sulfenamide;
B: 2,4,6-Trichlorobenzenesulfenamide; C: 2-Benzothiazolesulfenamide;
D: 4-Amino-1,2,4-triazole; E: Hydroxylamine

The amination reaction of nitrobenzenes with sulfenamides is not as selective as with 4-amino-1,2,4-triazole (Scheme 48) [92JOC4784].

An interesting example of a S_N^H amination reaction has recently been reported to occur between nitrobenzene and aniline in the presence of tetramethylammonium hydroxide, TMA(OH) [92JA9237]. The characteristic of this reaction is that it proceeds under anaerobic conditions in the absence of an auxiliary leaving group or external oxidant and results in derivatives of 4-nitroso- (**38**) or 4-nitrodiphenylamine (**39**) (Scheme 49) [92JA9237].

A plausible mechanism for the formation of these S_N^H products involves the addition of the anilide ion, generated from aniline by the strong base TMA(OH), to nitrobenzene leading to the anionic adduct **37**. This adduct undergoes either an intramolecular oxidation by the nitro group giving rise to **38** and **39** or an intermolecular oxidative process with the starting material nitrobenzene to yield 4-nitrodiphenylamine **39** and nitrosobenzene. The latter is not directly observed in the reaction mixture since it is immediately condensed with aniline to produce azobenzene as a by-product (3–4%). The effect of varying the ratio between aniline and nitrobenzene on distribution of products **38** and **39** is in full agreement with the mechanism (Scheme 49) [92JA9237]. A somewhat similar process has been found to occur, when reacting nitrobenzene with acetanilide in DMSO in the presence of sodium hydroxide to give 4-nitrosodiphenylamine as the major product (60%), together with minor amounts of 4-nitrodiphenylamine (10%) and azobenzene (10%) [90TL3217].

The very early report on formation of phenazine and azoxybenzene from nitrobenzene and aniline is certainly another example of an amination which involves the *ortho-* nitrosodiphenylamine as intermediary compound which undergoes a subsequent oxidative reductive process (Scheme 50) [01CB2442].

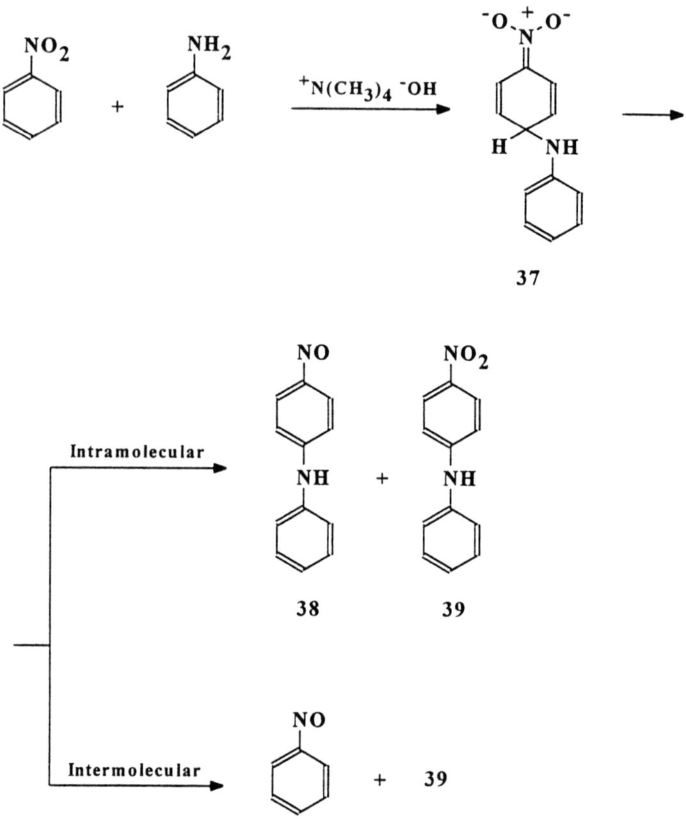

Scheme 49

Molar ratio aniline/nitrobenzene	Yield of product (%)	
	38	39
1.3	15	80
11.9	55	35
51.5	86	9

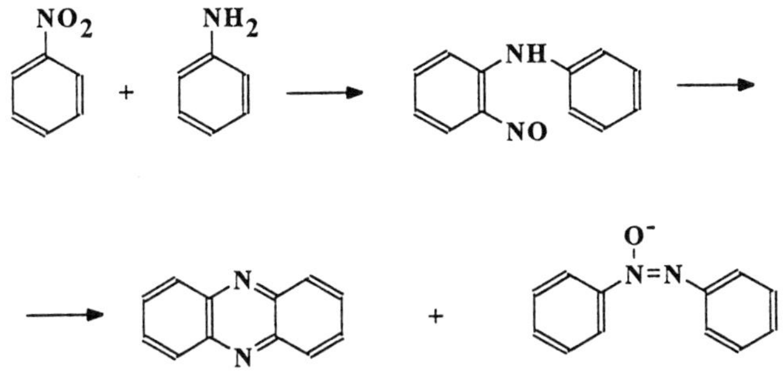

Scheme 50

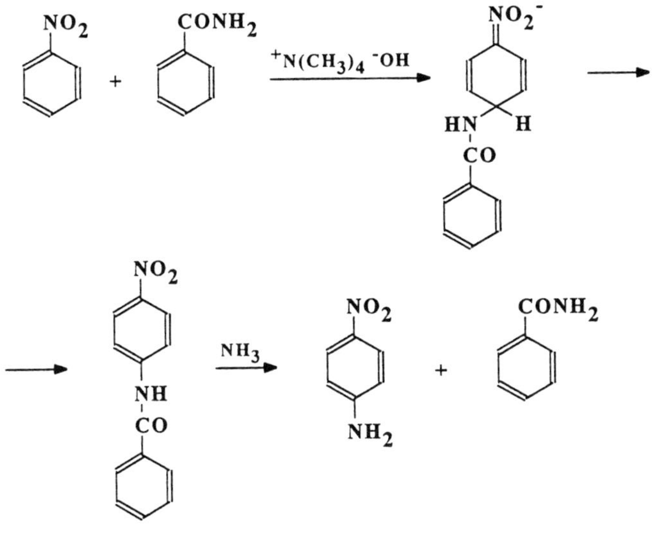

Scheme 51

A very elegant way for obtaining regiospecific amination of nitrobenzene has recently been found, when benzamide reacts with nitrobenzene and tetramethylammonium hydroxide. It results in the formation of N-(4-nitrophenyl)benzamide **40** in an excellent yield of 98% (Scheme 51) [93JOC6883]. This S_N^H process represents the first example of the direct formation of aromatic amide bonds. Aminolysis of the amide bond in **40** by treatment with methanolic ammonia affords 4-nitroaniline and the starting material benzamide. The overall S_N^H process can be regarded as an interesting and novel synthetic route to nitroanilines that competes with the conventional methodology of using halogenated nitroaromatics in amination reactions (Scheme 51) [93JOC6883].

D. *Reactions with P-Nucleophiles*

Substitution of hydrogen in 1,3,5-trinitrobenzene by trialkyl phosphite in DMSO leads to picrylphosphonates (Scheme 52). It is a rare example of an S_N^H reaction between nitroarenes and P-nucleophiles [77MI3; 77ZOB2480]. In a similar way picrylphosphonates are obtained by spontaneous oxidation in DMSO of the σ-adducts resulting from the reaction of 1,3,5-trinitrobenzene with dialkyl phosphites [78ZOB342; 79ZOB39] Whether 1,3,5-trinitrobenzene or DMSO is involved in the oxidation reaction has not been established.

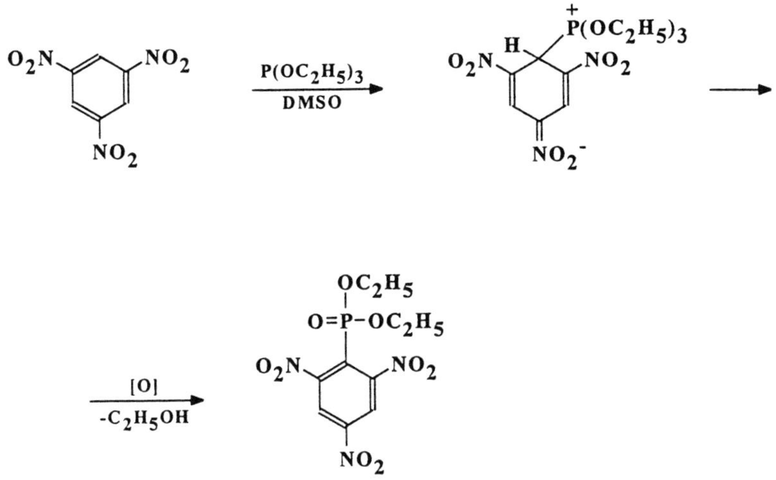

Chapter 2 S_N^H REACTIONS IN ARENES 67

Scheme 52

Experimental Procedures

Example 1. Preparation of *ortho*-nitrophenol from nitrobenzene

A. A mixture of nitrobenzene (24 g, 20 ml, 0.195 mol) with fine powdered and dry potassium hydroxide was heated up to 60-70°C and then it was kept at this temperature for 2 hours. After dissolving the reaction mixture in water unreacted nitrobenzene (13.2 g, 11 ml, 0.107 mol) was distilled off with a water steam. Acidification of the residue with hydrochloric acid followed by a water steam distillation gave *ortho*-nitrophenol (5 g, 0.036 mol, 42 %) [1899CB3486].

B. A suspension of potassium hydroxide in liquid ammonia (−33°C) was prepared by dissolving potassium (6.93 g, 0.178 mol) and water (3.50 g, 0.194 mol) in 300 ml of liquid ammonia. After addition of nitrobenzene (3.66 g, 0.030 mol), the oxygen was bubbled through the reaction mixture for 25 hours with volume velocity of approximately 15 liters per hour. The ammonia was evaporated and after heating the residue to room temperature it was treated with 100 ml of ether and 100 ml of water. The aqueous layer was separated and extracted with ether (2 x 100 ml) to get rid of unreacted starting material (1.15 g, 31%). The ether extracts were washed with water several times till pH ~ 7 and combined aqueous solutions were poured into 100 ml of 20% sulfuric acid previously cooled to –30°C. Aqueous solutions were extracted with ether (3 x 100 ml), the extracts were washed with water and dried over magnesium sulfate. The solvent was distilled off under reduced pressure and column chromatography of the residue on silica gel with benzene gave *ortho-* and *para-* nitrophenols with the ratio 3:1. The total yield 2.74 g (66%) [82ZOR1898].

Example 2. General procedure for alkylation of nitroaromatics

n-Butyl trifluoromethyl sulfone (1.425 g, 7.5 mmol) in dry DMF (1 ml) was slowly added to a stirred suspension of sodium hydride (360 mg, 7.5 mmol, 50% in oil) in DMF (2 ml) at room temperature under nitrogen. The mixture was stirred for 20–30 minutes until NaH disappeared and gas evolution ceased. At this point, a solution of nitroarene (5 mmol) in DMF (3 ml) was added in one portion. The resulting colored mixture was stirred for 1 hour (with cooling if necessary to maintain the ambient temperature), poured into ice-cold 5% HCl (220 ml) and extracted with methylene chloride (3 x 20 ml). The organic extracts were combined, washed with 10% aqueous NaCl (2 x 20 ml), and dried. The solvent was removed and the residue was chromatographed to give 4-*n*-butyl-1,3-dinitrobenzene (eluent: CCl_4), oil, yield 23%; 6-*n*-butyl-2,4-dichloronitrobenzene (eluent: hexane–ethyl acetate 30:1) oil, yield 31%; 2-*n*-butyl-1-nitronaphthalene (eluent: hexane) oil, yield 19%; 4-*n*-butyl-1-nitro- naphthalene (eluent: hexane) oil, yield 18% [90OPP575].

Example 3. Preparation of 2,4-dinitrophenylacetone

meta-Dinitrobenzene (168 mg, 1 mmol) and α-chloroacetone (93 mg, 1 mmol) were dissolved in DMF (3 ml) and cooled to –20°C (dry ice CCl$_4$ bath). At this point 1,8-diazabicyclo[5.4.0]undec-7-ene (DBU) (0.73 ml, 5 mmol) was added in a single portion and the mixture was allowed to reach room temperature during 1 hour. After acidic work-up and extraction with methylene chloride the product was separated by column chromatography on silica gel with hexane–ethyl acetate 1:1 as eluent. Recrystallization from ethanol-hexane 1:1 gave an analytical sample of 2,4-dinitrophenylacetone, melting point 66–67°C. Yield 153 mg (68%) [92PJC2005].

Example 4. Preparation of 5-chloro-2-nitrobenzyl cyanide

A solution of chloroacetonitrile (760 mg, 10 mmol) and *para*-chloronitrobenzene (1.57 g, 10 mmol) in 10 ml of ether was added dropwise to a vigorously stirred suspension of powdered NaOH (4g, 0.1 mol) in 20 ml of liquid ammonia. The reaction was carried out at –30 to –50°C for 1 hour. Ammonia was evaporated, the residue was dissolved in water and acidified with hydrochloric acid. The extracts were washed with water and dried and the solvent was evaporated. Column chromatography of the residue on silica gel with hexane-chloroform 1:1 gave a crude material which was then recrystallized from ethanol to give 5-chloro-2-nitrobenzyl cyanide, melting point 94–96°C [84JOC1 494; Copyright permission 1994 American Chemical Society].

Example 5. Preparation of methyl α-methyl-α–(4-nitrophenyl)propionate

To a mixture of 1.02 ml (10 mmol) of nitrobenzene and 2.14 ml (10.75 mmol) of 1-methoxy-1-(trimethylsiloxy)-2-methyl-1-propene in 15 ml of THF at –78°C was added 2.75 g of TASF dissolved in 5 ml of acetonitrile. The dropping funnel containing

TASF solution was washed down with 5 ml of THF. The cold bath was removed and the reaction warmed to $-10^{\circ}C$ and maintained at $-10^{\circ}C$ for 1 hour. After the solution was cooled to $-78^{\circ}C$, 0.512 ml (10 mmol) of bromine in 2 ml of cyclohexane was added dropwise. The cold bath was removed and the mixture was brought to room temperature. Stirring was continued for 1 hour. Twenty milliliters of saturated sodium bisulfite was added, and the product was extracted into ether. Isolation by flash chromatography on silica gel yielded 1.824 g (79%) of methyl α-methyl-α-(4-nitrophenyl)propionate [85JA5473; Copyright permission 1994 American Chemical Society].

Example 6. Preparation of 2',6'-dinitrobiphenyl-4-ol

Copper(I) chloride (500 mg, 5 mmol) was added to *t*-BuOK (651 mg, 5 mmol) in dimethoxyethane (DME) (5 ml) and the mixture is stirred for 3 hours at room temperature. Pyridine (12 ml) and 1,3-dinitrobenzene (698 mg, 5.15 mmol) were then added giving a reddish suspension. A solution of 4-iodophenol (990 mg, 4.5 mmol) and *t*-BuOK (505 mg, 4.5 mmol) in DME (5 ml) was added. The mixture was stirred at $85–90^{\circ}C$ for 2-3 hours, then quenched with 1.2 M aqueous HCl at room temperature, and extracted with CH_2Cl_2 (2 x 50 ml). The extract was washed with brine (50 ml), dried ($MgSO_4$), and evaporated. 2',6'-Dinitrobiphenyl-4-ol was isolated by column chromatography on silica gel using CH_2Cl_2 as eluent. Yield 819 mg (76%), melting point $188–189^{\circ}C$ [90S942].

Example 7. Preparation of 2,6-dinitrobenzyl phenyl sulfone

Copper(I) chloride (210 mg, 2.1 mmol) was added to a suspension of *t*-BuOK (494 mg, 4.4 mmol) in dimethoxyethane (DME) (20 ml) at $0^{\circ}C$. The mixture was stirred at room temperature for 30 minutes, then pyridine (0.7 ml) and 1,3-dinitrobenzene (504 mg, 3.0 mmol) were added. The reddish mixture was cooled to –40°C and a solution of iodomethyl phenyl sulfone (592 mg, 2.1 mmol) in DME (10 ml) was added slowly through a fine cannula over 10 minutes. The mixture was kept at –40°C for 1.5 hours, then allowed to warm to –20°C, quenched with 1.2 M aqueous. HCl (20 ml) and extracted with CH_2Cl_2 (2 x 50 ml). The extract was washed with brine (50 ml), dried ($MgSO_4$), and evaporated. 2,6-Dinitrobenzyl phenyl sulfone was isolated by column chromatography on silica gel using CH_2Cl_2 as eluent. Yield 435 mg (68%), melting point 133.5°C [90S942].

Example 8. Preparation of N-(4-nitrophenyl)benzamide and 4-nitroaniline

Benzamide (0.2 mol), and tetramethylammonium hydroxide (0.2 mol) was stirred in 100 ml of nitrobenzene at 65°C with air sweeping the surface of the reaction solution for 8 hours. Approximately 5 ml of water and 40 ml of nitrobenzene were collected in a Dean-Stark trap during this time. Another 100 ml of nitrobenzene was added and the solution was stirred overnight. A total of 70 ml of nitrobenzene was collected. The solution was cooled to ambient temperature and then 100 ml of water was added with good stirring. The solution was filtered and the precipitate was washed with more water and dried under suction to obtain a dark brown solid. The solid was washed with hot hexane to obtain 43.5 g (90%) [93JOC6883; Copyright permission 1994 American Chemical Society].

Example 9. Amination of nitrobenzene

A solution of nitrobenzene (615 mg, 5 mmol) and 4-amino-1,2,4-triazole (420 mg, 5 mmol) in DMSO (5 ml) was added to a solution of potassium t-butoxide in DMSO (15 ml) over 30 minutes at 24–27°C. Immediate formation of an intense yellow-brown was observed. After 4 hours at room temperature the reaction was quenched in saturated aqueous NH₄Cl (50 ml), and extracted with ether (3 x 50 ml). Evaporation of the solvent and purification of the residual yellow solid by flash chromatography (silica gel, 1:1 petroleum ether–ether) gave 4-nitroaniline in 58% yield, melting point 146–148°C [86JOC5039; Copyright permission 1994 American Chemical Society].

IV. ELECTROCHEMICAL S_N^H REACTIONS

As already discussed in previous chapters, S_N^H reactions require an oxidant in order to promote elimination of the hydride ion. In this respect, use of electrooxidative reactions for the replacement of an aromatic hydrogen seems to be rather attractive. Till now only a few successful examples of an S_N^H electrochemical functionalization of aromatic and heteroaromatic compounds have been reported [70ACSA2757; 75JCS(CC)262; 79JA4268; 83MI4; 84MI3; 92JCS(P1)333] (see also Chapter 3,V). For instance, oxidative cyanation of naphthalene and its mono- and dimethyl derivatives **41** with a platinum anode in methanol containing 0.4 M sodium cyanide has been found to occur preferentially at the free α-position to give the corresponding 1-naphthalene carbonitriles **44** (Scheme 53); 2-naphthalene carbonitriles are formed in some cases as minor products (Table X) [79JA4268]. The reaction has been shown to be initiated by electrochemically induced electron transfer from naphthalene substrates **41** to a platinum anode, followed by fast chemical reaction of the radical cation **42** with the cyanide anion, leading to the radical species **43**. The latter undergo further anodic oxidation and successive proton release to yield the aromatic cyanation products **44** (Scheme 53) [79JA4268].

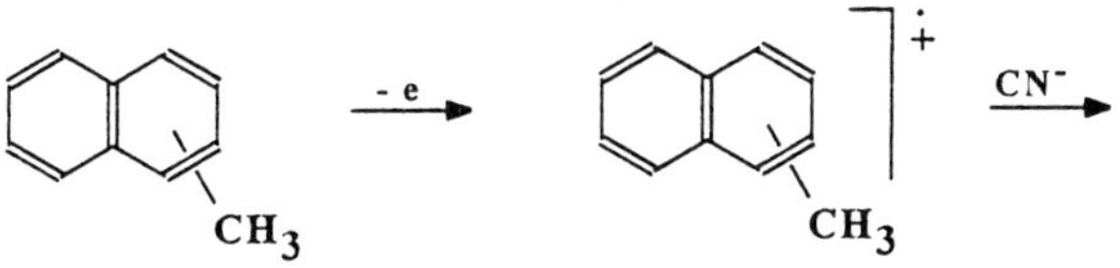

Scheme 53

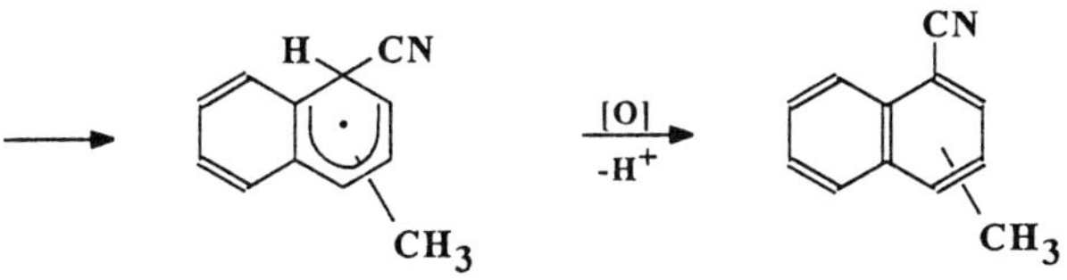

Scheme 54

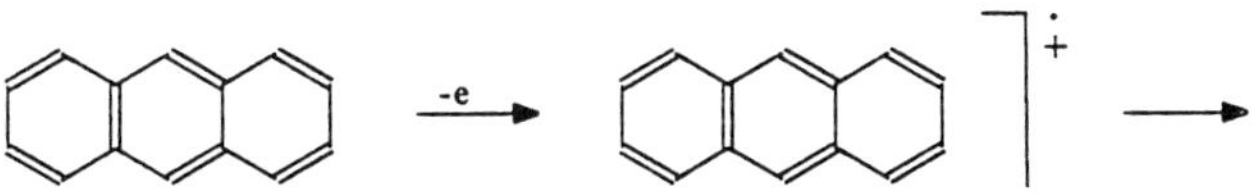

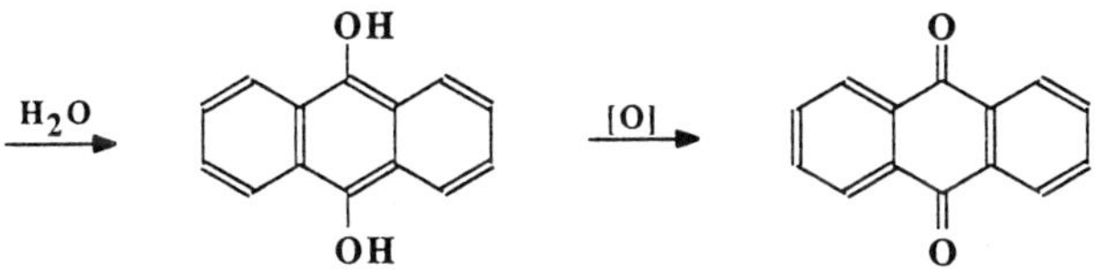

TABLE X

Yields and Products Obtained from Anodic Hydroxylation, Acetamidation, and Cyanation Reactions on Some Aromatics

Arene	E, V[1]	S_N^H Product(s)	CE,[4] (%)	References
Naphthalene	1.7[2]	Naphthalene-1-carbonitrile	74	79JA4268
		Naphthalene-2-carbonitrile	6	
1-Methyl-naphthalene	1.5[2]	4-Methylnaphthalene-1-carbonitrile	34	79JA4268
		5-Methylnaphthalene-1-carbonitrile	5	
		8-Methylnaphthalene-1-carbonitrile	23	
2-Methyl-naphthalene	1.5[2]	2-Methlnaphthalene-1-carbonitrile	60	79JA4268
		3-Methylnaphthalene-1-carbonitrile	4	
		6-Methylnaphthalene-1-carbonitrile	6	
		6-Methylnaphthalene-1-carbonitrile	7	
Acetophenone	2.4[3]	2-Hydroxyacetophenone	85	75JCS(CC)262
Ethyl benzoate	2.3[3]	Ethyl 2-hydroxybenzoate	43	75JCS(CC)262
		Ethyl 4-hydroxybenzoate	19	
Benzoic acid	2.3[3]	2-Hydroxybenzoic acid	55	75JCS(CC)262
Acetophenone	2.4[3]	2-Acetamidoacetophenone	70	75JCS(CC)262
		4-Acetamidoacetophenone	24	
Benzoic acid	2.3[3]	2-Acetamidobenzoic acid	54	75JCS(CC)262
		4-Acetamidobenzoic acid	11	
Ethyl benzoate	2.3[3]	Ethyl 2-acetamidobenzoate	18	75JCS(CC)262
		Ethyl 4-acetamidobenzoate	14	

[1]Controlled potential for preparative electrolysis at a platinum anode.
[2]Vs. SCE reference. [3]Vs. Ag/0.1 M AgNO$_3$. [4]Current efficiency.

Also, electrochemical oxidation of anthracene proceeds via intermediacy of the radical cation **45** which combines with water to give 9,10-dihydroxyanthracene **46**, and subsequently anthraquinone. This aromatic hydroxylation reaction can formally be regarded as nucleophilic displacement of hydrogen in anthracene by the hydroxide ion (Scheme 54) [70ACSA2757].

The electrochemical technique has also been successfully used for the direct introduction of the hydroxy and acetamido groups into *ortho*- and *para*-positions of aromatic carbonyl compounds. Hydrolysis of the acetamido group provides the corresponding *ortho* and *para* amino compounds (Table X, Scheme 55) [75JCS(CC)262].

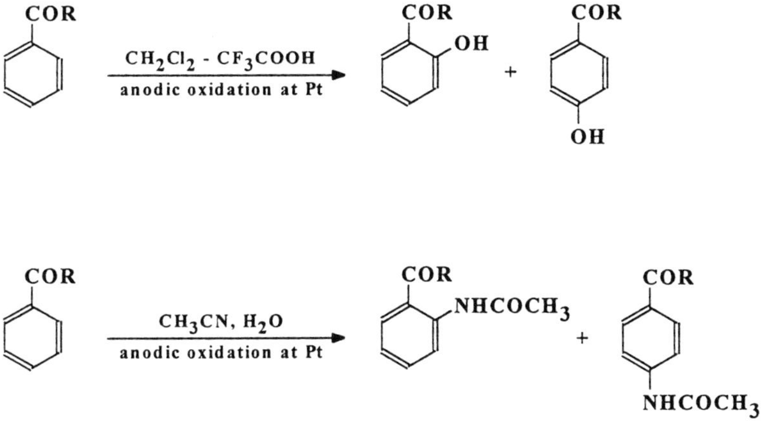

Scheme 55

Experimental Procedure

Example 1. Electrochemical cyanation of naphthalenes

A controlled potential electrolysis of a naphthalene was carried out in a three-compartment cell with SCE (saturated calomel electrode) as the reference electrode. The anolyte and catholyte were 0.4 M sodium cyanide in methanol. The amount of organic substrate was 5 mmol. A platinum (2 x 2 cm) sheet anode was used, and the anolyte (50 ml) was stirred magnetically. The anode compartment was kept under an atmosphere of nitrogen.

A background current was recorded before addition of naphthalene (negligible, up to 1.4 V). The potential was set in the region of the first voltametric wave (see Table X), and the reaction was performed until either 2 faradays/mol of naphthalene added was passed or the current dropped to some low value near the original background level.

To the electrolyzed mixture internal standards for gas-liquid chromatography (GLC) analysis were added; it was then treated with water and extracted with ether. The ethereal solution was concentrated and analyzed by GLC. The results are presented in Table X [79JA4268; Copyright permission 1994 American Chemical Society].

V. TROPYLIUM SALTS AND OTHER NONBENZENOID COMPOUNDS

Tropylium salts, being highly electrophilic aromatic systems, are certainly appropriate substrates for nucleophilic substitution of hydrogen. Indeed, the reaction of tropylium tetrafluoroborate **47** with vicarious carbanions **48**, bearing a chloro substituent at the carbanionic center, has been reported to proceed very smoothly. It yields the rather stable adducts **49**, which on treatment with a base are converted into the corresponding alkylidene cycloheptatrienes **50** (Scheme 56) [89LA95]. The reaction follows the pathway which is typical for vicarious nucleophilic aromatic substitution. This S_N^H reaction would be fully completed on protonation of the alkylidene derivatives **50** to produce substituted tropylium salts. A similar reaction of **47** with 9-phenylsulfonylfluorene results in a product, which, after elimination of phenylsulfonyl group and hydrogen, yields dibenzo[*i,k*]sesquifulvalene derivative **51**, which on protonation gives (9-fluorenyl)tropylium trifluoroacetate **52** (Scheme 56) [89LA95].

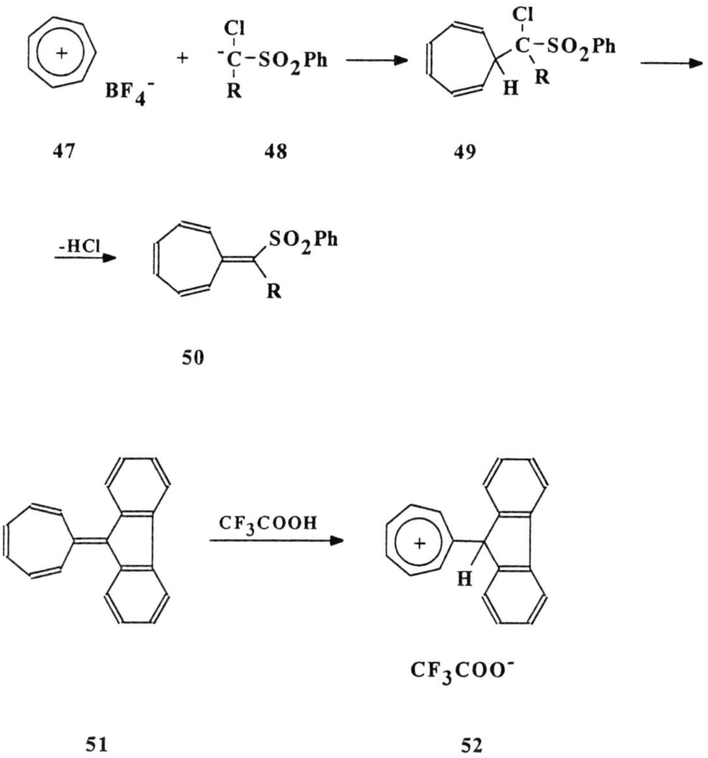

Scheme 56

Being activated by coordination with tricarbonyl complexes of molybdenum (chromium or tungsten), the coordinated tropylium cation **53** is able to add rather exotic nucleophiles such as, for instance, the anionic tricarbonylmanganese complex of cyclohexatriene **54**. It gives the hydrocarbon-bridged adduct **55**, the structure of which has unequivocally been established by X-ray diffraction analysis (Scheme 57) [92CB1369].

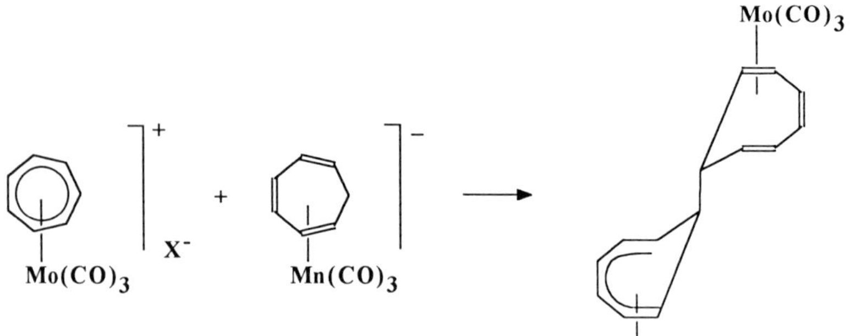

Scheme 57

It would be interesting to see if oxidation of compound **55** is possible in order to complete the nucleophilic displacement of hydrogen in the tropylium cation **53**; however, no attempts of performing this oxidation reaction were made [92CB1369].

Interesting examples of S_N^H reactions in nonbenzenoid systems are the displacement of hydrogen in 1,6-methano[10]annulenes [89CB493; 92HCA825]. When 2-nitro-1,6-methano[10]annulene reacts with vicarious reagents (chloroacetonitrile, methyl chloroacetate, chloromethyl *para*-tolyl sulfone), or with sodium methoxide in the presence of MnO_2, the displacement of hydrogen at C-3 takes place (Scheme 58) [89CB493]. Also, it is worth mentioning that dimethoxyethane (DME) which has first been used as solvent for these vicarious substitution reactions, instead of DMSO, appears to have a substantial advantage of easier handling on isolation of reaction products [89CB493].

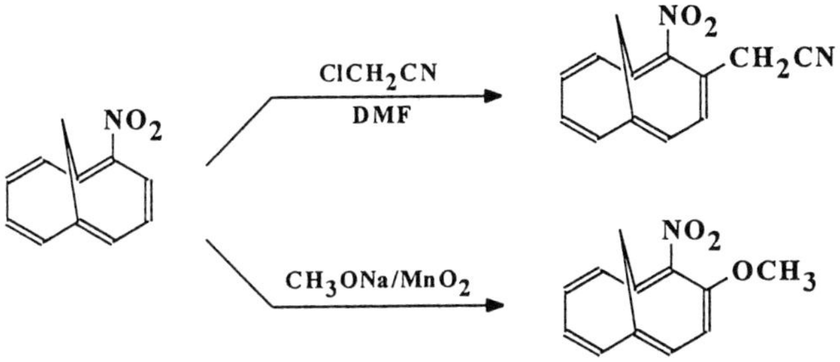

Scheme 58

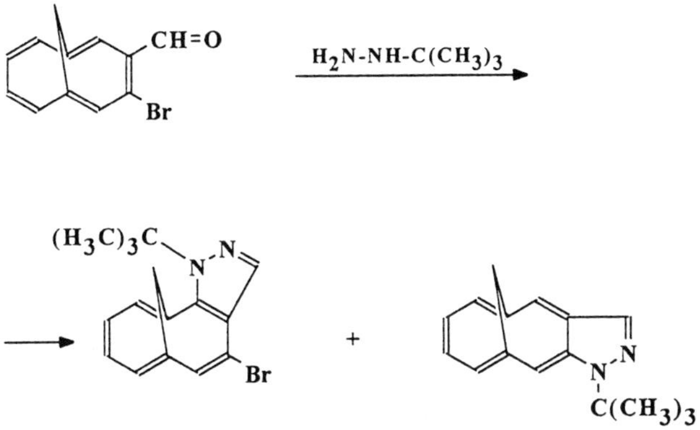

Scheme 59

An interesting observation was made in the reaction of 4-bromo-1,6-methano-[10]annulene-3-carbaldehyde with *t*-butylhydrazine in DMF at 25°C. It yielded the 2,3-annelated pyrazolo bromo compound **56** in 54% yield, showing that the Michael condensation followed by amino-dehydrogenation at C-2 is a more favored process than the intramolecular substitution of halogen at C-4. Both 2,3- and 3,4-pyrazolo-annelated

1,6-methano[10]annulenes are concurrently formed on reflux of the same reagents in DMF, but the yields of **56** and **57** proved to be rather low (5% and 4%, respectively) (Scheme 59) [92HCA825].

Experimental Procedures

Example 1. Preparation of (9-fluorenyl)tropylium trifluoroacetate

The Addition Step. To a stirred solution of potassium *t*-butoxide (146 mg, 1.3 mmol) in DMF (3 ml), cooled to –40°C under argon, a solution of 9-phenylsulfonylfluorene (397 mg, 1.3 mmol) in DMF (1 ml) was added with a syringe. After 5 minutes tropylium tetrafluoroborate (178 mg, 1 mmol) in DMF (2 ml) was added dropwise (within 5 minutes) while keeping the temperature below –40°C. The mixture was stirred for 15 minutes, poured into water (50 ml), and extracted with ethyl acetate (2 x 30 ml). The combined organic layers were washed with water (2 x 40 ml), dried with sodium sulfate, and evaporated. The residue was chromatographed with *n*-hexane–chloroform (4:1) to give 281 mg (71%) of 9-(7-cycloheptatrienyl)-9-(phenylsulfonyl)fluorene as crystals with melting point 173–175°C.

The Elimination Step. To a solution of 9-(7-cycloheptatrienyl)-9-(phenylsulfonyl)-fluorene (198 mg, 0.5 mmol) in THF (5 ml), potassium *t*-butoxide (280 mg, 2.5 mmol) was added in one portion. After two hours the mixture was poured into water (20 ml) and extracted with ethyl acetate (2 x 10 ml). The combined organic layers were dried with sodium sulfate and evaporated. A column chromatography on silica gel with *n*-hexane–chloroform (3:1 with a gradual increase of the content of chloroform till 1:1) followed by recrystallization from hexane gave pure dibenzo[*i,k*]sesquifulvalene as crystals with melting point 98°C. Yield 76 mg (61%). Treatment of a solution of dibenzo[*i,k*]-sesquifulvalene in chloroform with an excess of trifluoroacetic acid results in a full conversion of the former into (9-fluorenyl)tropylium trifluoroacetate (according to NMR analysis) [89LA95].

Example 2. Cyanomethylation of 2-nitro-1,6-methano[10]annulene

2-Nitro-1,6-methano[10]annulene (1.0 g, 5.35 mmol) was added to a solution of chloroacetonitrile (403 mg, 5.37 mmol) in 125 ml of absolute 1,2-dimethoxyethane and sodium methoxide (Merck), free of ethanol (738 mg, 13.66 mmol), was then added at 20°C under argon atmosphere with stirring. The resulting solution was kept dry at 20°C under stirring for 4.5 hours. After cooling with ice a buffer (pH= 4, 100 ml) was added followed by a dropwise addition of concentrated hydrochloric acid to reach a yellow-orange solution which was extracted with chloroform (3 x 100 ml). The combined chloroform extracts were washed with a buffer (pH= 4) and filtered through a layer of silica gel to get rid of polymeric impurities. Vacuum evaporation of the filtrate gave 3-cyanomethyl-2-nitro-1,6-methano[10]annulene, oil, yield 867 mg (72%) [89CB493].

VI. INTRAMOLECULAR S_N^H REACTIONS AND REARRANGEMENTS ACCOMPANIED BY S_N^H SUBSTITUTION

A. *The Smiles and Smiles-Truce Rearrangements*

The Smiles rearrangement proceeding through spirocyclic intermediates derived from the intramolecular nucleophilic attack at an *ipso*-position is sometimes accompanied by an intramolecular cyclization involving the addition of the side-chain nucleophilic moiety at a position *ortho* to that of the side-chain. For instance, carbanions of arylmesitylsulfones **58** have been found to undergo in basic medium both the Smiles rearrangement into diarylmethanes **59** and an intramolecular cyclization into tricyclic sulfones **61**, after oxidizing intermediate adducts **60** with bromine (Scheme 60) [71ZOR2388, 73ZOR156].

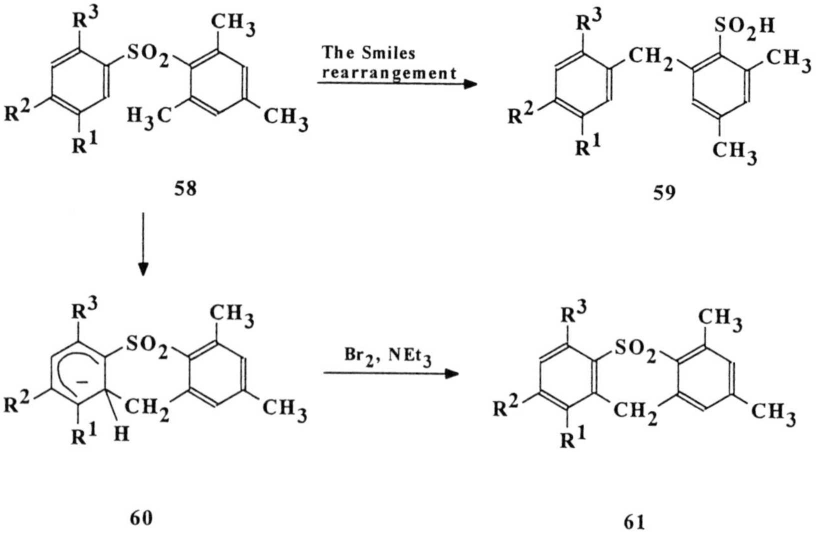

Scheme 60

B. *The Sommelet-Hauser Rearrangement and Related Reactions*

The Sommelet-Hauser rearrangement of benzylammonium salts, initiated by base-promoted formation of an ylide which then undergoes an intramolecular nucleophilic addition-dehydrogenation steps is an attractive method for functionalization of arenes [37CR56; 51JA4122; 58JOC358; 91CPB36; 91S996; 92JCS(P1)2851; 92JOC5034]. It can be illustrated by the regiospecific *ortho*-alkylation reaction performed on trimethylbenzylammonium salt **62**, leading to xylene derivative **65** (Scheme 61). The reaction is considered to proceed via the formation of the ylide **63** and the adduct **64**. The reaction does not require the addition of an oxidant since the quaternary ammonium moiety itself acts as an oxidant, being reduced in the course of this rearrangement. In a similar way salt **67**, obtained by quaternization of 4-acetyl-2-phenyl-1-trimethylsilylmethylpiperazine **66** with iodomethane, undergoes the fluoride-induced desilylation reaction followed by rearrangement of the ylide **68** into benzodiazonine **69** using the strong base 1,8-diaza-bicyclo[5.4.0]undec-7-ene (DBU) (Scheme 61) [92JCS(P1) 2851].

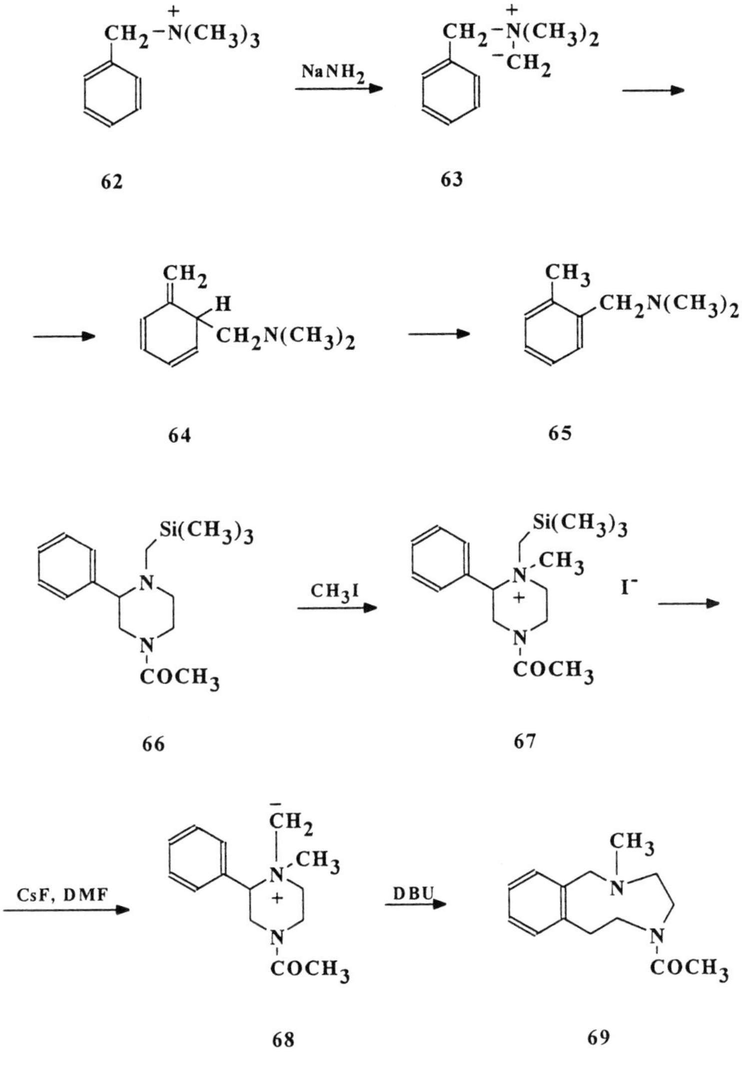

Scheme 61

Use of sulfonium ylides in the Sommelet-Hauser rearrangement makes it possible to obtain *ortho*-disubstituted arenes, such as the xylene **70** [72TL3587], the anilines **71** [57LA221; 72TL497; 72TL501; 78JA7600] and the phenols **72** [78JA7611] (Scheme 62). These intramolecular reactions do not require the presence of electron-withdrawing groups in the benzene ring and can even be applied to arenes bearing the electron-donating substituents CH_3, OCH_3 (Scheme 62).

C. *The von Richter Rearrangement*

The well-known von Richter rearrangement [1871CB21; 1871CB451; 1871CB553; 50JOC481], describing the formation of *meta*-halobenzoic acids **79** from *para*-halonitrobenzenes **73** by action of potassium cyanide, is clearly a reaction, in which the S_N^H process takes place at the *ortho*-position of the nitro group and being strongly favored above the $S_N(Ar)$ substitution of the halogen ion. In the negatively charged σ^H-adduct **74**, the sp^3-hydrogen has a somewhat acidic character, due to the presence of the cyano and the nitro group. After deprotonation the dianion **75** is formed, which can undergo an intramolecular cyclization into **76**. Ring opening leads to *ortho*-nitrosobenzamide **77**, which by an intramolecular cyclization yields indazolone **78**. Facile hydrolysis of **78** gives under expulsion of nitrogen compound **79** (Scheme 63). This mechanism has been substantiated by experiments performed in deuterated, ^{18}O-labeled water, and with ^{15}N-labeled compounds [60JA3797].

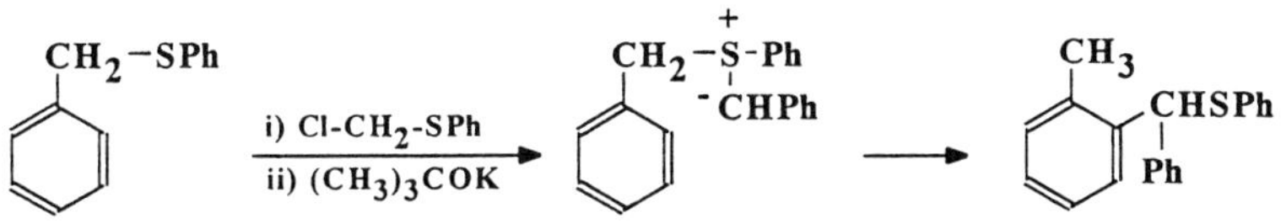

70

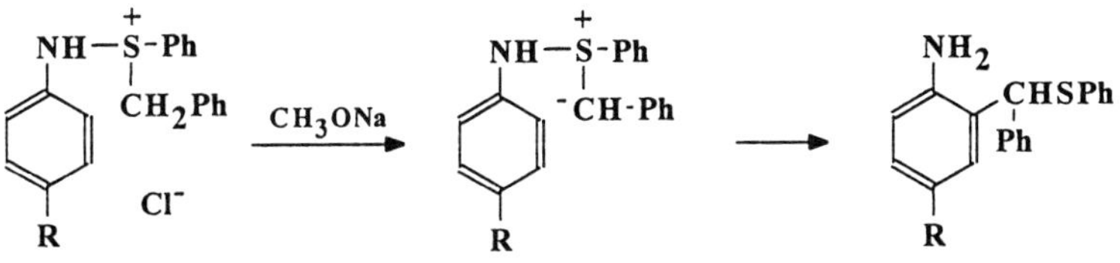

71

R = H, Cl, CH_3, $COOCH_3$, NO_2

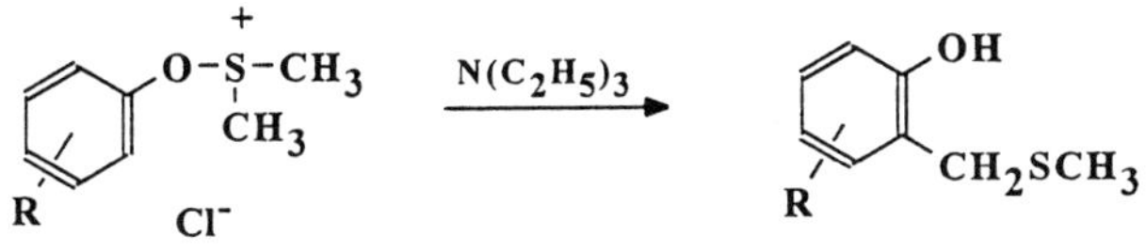

72

R = H, *p*-Cl, *p*-CH_3, *o*-CH_3, *o*-OCH_3, *o*-NO_2

Scheme 62

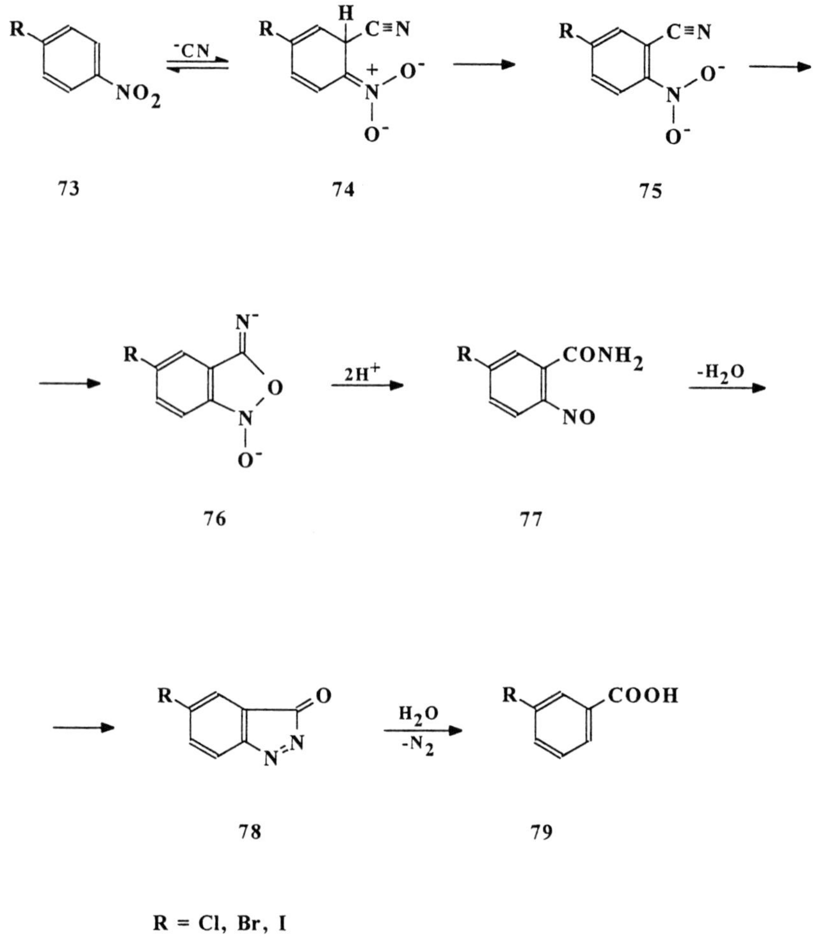

Scheme 63

R = Cl, Br, I

D. *Other Examples of Intramolecular* S_N^H *Reactions*

An intramolecular vicarious nucleophilic displacement of hydrogen in nitroaromatics can be illustrated by the cyclization of the N-methyl-N-(*meta*-nitrobenzyl) chloromethyl sulfonamide **80**, yielding two isomeric products **81** and **82** due to substitution of hydrogen at the position *ortho* and *para* to the nitro group (Scheme 64) [84TL4791].

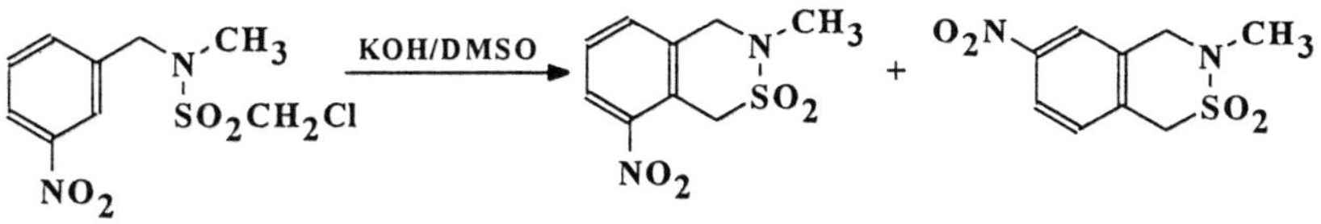

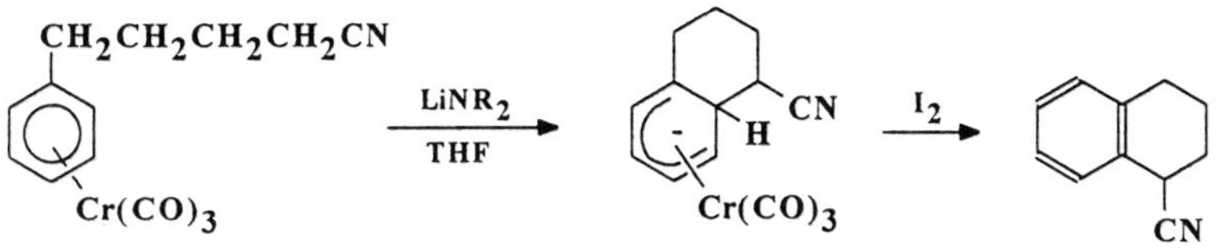

Scheme 64

η^6-Arenetricarbonylchromium(0) complexes are also found to be effective in intramolecular S_N^H reactions. For instance, the α-cyano stabilized anion derived from the carbonyl complex by treatment with lithium diisopropylamide, is successfully converted to the cyanotetrahydronaphthalene in 89% yield by iodine oxidation (Scheme 64) [77MI2].

Experimental Procedure

Example1. Preparation of 5-acetyl-2-methyl-2,3,4,5,6,7-hexahydro-1H-2,5-benzodiazonine (**69**)

4-Acetyl-1-methyl-2-phenyl-1-trimethylsilylmethylpiperazinium iodide **69** (1.3 g, 3 mmol) and 1,8-diazabicyclo[5.4.0]undec-7-ene (2.16 g, 15 mmol) was placed in a 30-ml flask equipped with a magnetic stirrer, a scepter, and a test tube connected with the flask by a short piece of rubber tube. CsF (2.0 g, 13 mmol) was placed in the test tube. The apparatus was dried under reduced pressure and was flushed with N_2. DMF (30 ml) was added via syringe, and then CsF was added from the test tube. The mixture was stirred for 48 hours at room temperature, poured into 1% aqueous $NaHCO_3$ (100 ml), and extracted with ethyl acetate (4 x 100 ml). The extract was washed with 1% aqueous $NaHCO_3$ (3 x 50 ml), dried ($MgSO_4$), and concentrated under reduced pressure. Preparative chromatography of the residual oil on an HPLC column (Merck Hibar LiChrosorb, 250 x 10 mm, ether) gave 5-acetyl-2-methyl-2,3,4,5,6,7-hexahydro-1H-2,5-benzodiazonine, boiling point 175°C/0.6 Torr. Yield 488 mg (70%) [92JCS(P1)2851].

Nucleophilic Substitution of Hydrogen in Heteroaromatics

I. AZINES

The Chichibabin amination of pyridine has been known for a long time as a typical and outstanding example of the amino-dehydrogenation reaction [14MI1; 71MI4; 78RCR1042]. During the last two decades considerable efforts have been made to modify and to improve the amination procedure [78RCR1042; 83AHC95; 85JCA23; 85T237; 86CCA33; 87KG1011; 88AHC2], as well as to develop new methods for substituting hydrogen in azaaromatic compounds by a variety of nucleophilic reagents [88T1].

Due to the electron-withdrawing effect of the aza group, being comparable with that of the nitro group, the *ortho-* and *para*-position relative to the ring nitrogen show an enhanced activity toward nucleophilic reagents. The addition of all kind of nucleophiles at the C-2, C-4, or C-6-position of the pyridine ring or at other azaaromatics is well documented in the literature [64AHC285; 65AHC145; 71MI4; 78ACR147; 78RCR1042; 82CRV77; 82CRV223; 83AHC95; 83AHC305; 84MI1; 85MI1; 85T237; 86CCA33; 86CCA89; 87KG1011; 88AHC199; 88H291; 88T1; 89AHC73; 91MI1; 93ACSA95]. From those studies it became evident that an important requirement for reaching a successful nucleophilic substitution of hydrogen in azaaromatic systems is to find an appropriate oxidant and suitable conditions for the smooth oxidation of the intermediary σ^H-adducts into the final products.

A. *Reactions with Uncharged Nucleophiles*

It was recently found that the liquid ammonia/potassium permanganate system can act as a very effective and useful reagent for the amination of highly π-deficient azaaromatics (for reviews see [86CCA33; 87KG1011; 93ACSA95]). Especially a great number of azines activated by nitro or aza-groups have been found to react easily with liquid ammonia (at −33 to −40°C) in the presence of potassium permanganate into amino compounds. These S_N^H reactions usually occur with polyaza six-membered heterocycles, that is, triazines [78RTC273; 85S884], tetrazines [81JOC3805], pteridines [75RTC45; 81JOC3805; 82JHC1527; 86JHC477; 86JHC843], and pyrimido[5,4-*e*]-1,2,4-triazines [87JHC1657] or with azines activated by one or several nitro groups, such as 3-nitropyridines [91LA875] and 3,5-dinitropyridines [93LA7], 5-nitropyrimidines [83JOC1354], a series of nitroquinolines [85JHC353; 87JOC5643; 90MI1; 91PJC323], and a series of nitronaphthyridines [83JHC9; 83RTC359; 83RTC511; 85JHC761; 86CCA33; 93LA471] (Tables XI and XIII). It is of interest to mention that unsubstituted pyridine, pyridazine, pyrimidine, and pyrazine are not reactive in liquid ammonia/potassium permanganate [82JHC1285], but are easily aminated when the system potassium amide/liquid ammonia/potassium permanganate is used as aminating agent (see further text).

3-Nitropyridine, when dissolved in liquid ammonia, gives after treatment with potassium permanganate a mixture of 2-, 4-, and 6-amino-3-nitropyridines as well as a small amount of 2,6-diamino-3-nitropyridine (Table XI) [91LA875]. 3-Nitropyridines, which contain nucleofugal groups Cl, OCH$_3$ at α- and γ-positions relative to the ring nitrogen and the nitro group, are aminated in a highly regioselective manner. Interestingly under the applied reaction conditions no substitution of the nucleofugal group takes place (Table XI) [93ACSA95]. Comparison of quantumchemical calculations on the regioselective amination of some 3-nitropyridines with the observed regioselectivity suggests that the amination reaction is controlled by Coulomb interaction (Table XII) [91LA875].

TABLE XI
Yields and Products, Obtained on Amination of Some Monocyclic
Azines with Liquid Ammonia (−33°C)/KMnO$_4$

Nitroazine	S$_N^H$ Product(s)	Yield (%)	Reference
3-Nitropyridine	2-Amino-3-nitropyridine	33	91LA875
	4-Amino-3-nitropyridine	24	
	6-Amino-3-nitropyridine	19	
	2,6-Diamino-3-nitro-pyridine	2	
2-Methoxy-3-nitropyridine	6-Amino-2-methoxy-3-nitropyridine	62	91LA875
6-Methoxy-3-nitropyridine	2-Amino-6-methoxy-3-nitropyridine	75	91LA875
6-Chloro-3-nitropyridine	2-Amino-6-chloro-3-nitropyridine	57	91LA875
	2,4-Diamino-6-chloro-3-nitropyridine	12	91LA875
4-Nitropyridazine	5-Amino-4-nitro-pyridazine	18	88JHC831
4-Nitro-6-phenyl-pyridazine	5-Amino-4-nitro-6-phenylpyridazine	93	88JHC831
5-Nitropyrimidine	2-Amino-5-nitro-pyrimidine	45	83JOC1354
2-Methyl-5-nitro-pyrimidine	4-Amino-2-methyl-5-nitropyrimidine	53	83JOC1354

TABLE XI (Continued)

Nitroazine	S_N^H Product(s)	Yield (%)	Reference
2-Methylthio-5-nitro-pyrimidine	4-Amino-2-methylthio-5-nitropyrimidine	72	83JOC1354
4-Methoxy-5-nitro-pyrimidine	2-Amino-4-methoxy-5-nitropyrimidine[1]	50	83JOC1354
	6-Amino-4-methoxy-5-nitropyrimidine[2]	65	83JOC1354
1,2,4-Triazine	5-Amino-1.2.4-triazine	95	85S884
3-Methyl-1,2,4-triazine	5-Amino-3-methyl-1,2,4-triazine	63	85S884
3-Phenyl-1,2,4-triazine	5-Amino-3-phenyl-1,2,4-triazine	89	85S884
3-Methoxy-1,2,4-triazine	5-Amino-3-methoxy-1,2,4-triazine	80	85S884
3-Methylthio-1,2,4-triazine	5-Amino-3-methylthio-1,2,4-triazines	87	85S884
3-Methyl-1,2,4,5-tetrazine	6-Amino-3-methyl-1,2,4-tetrazine	80	81JOC3805
3-Phenyl-1,2,4,5-tetrazine	6-Amino-3-phenyl-1,2,4,5-tetrazine	74	81JOC3805

[1] At the reaction temperature –60 °C. [2] When a reaction mixture was allowed to stand for 5 minutes at 20°C before coolling to –40 °C and treatment with $KMnO_4$.

TABLE XII

Charge Distrubutions q_r in Molecules of Some 3-Nitro Substituted Pyridines and Superdelocalizability Indices S_r^N Calculated for Their Reactions with Liquid Ammonia[1]

Pyridine	r	S_r^N	q_r	Preferable positions *(r)* for NH$_3$ attack under		
				orbital control	charge control	experimental conditions
3-Nitro-	2	0.107	0.136	2~4~6>5	2>6>4	2>4~6
	4	0.106	0.065			
	5	0.086	-0.140			
	6	0.106	0.106			
2-Methoxy-3-nitro-	4	0.106	0.099	6~4>5	6>4	6
	5	0.079	-0.193			
	6	0.107	0.154			
6-Methoxy-3-nitro-	2	0.109	0.185	4~2>5	2>4	2
	4	0.110	0.092			
	5	0.079	0.161			
6-Chloro-3-nitro-	2	0.078	0.146	4>2>5	2>4	2
	4	0.089	0.071			
	5	0.069	-0.148			

[1] Superdelocalizability indices determining regioselectivity in reactions which are under orbital control (see ref. [91LA875]).

Also 3,5-dinitropyridines have been found to undergo the amino-dehydrogenation reaction rather smoothly, although the yields of the S_N^H products being obtained are relatively low [85JOC484; 93LA7].

5-Nitropyrimidine is easily aminated into 2-amino-5-nitropyrimidine in a 45% yield [83JOC1354]. When position 2 of the pyrimidine ring is blocked by methyl, phenyl, or other substituents, the position of amination changes from C-2 to C-4 (Table XI). Worth mentioning is the fact that 2-methylthio-5-nitropyrimidine also undergoes amination at C-4, without nucleophilic replacement of the methylthio group, showing a unique feature of the liquid ammonia/potassium permanganate amination system (Scheme 65).

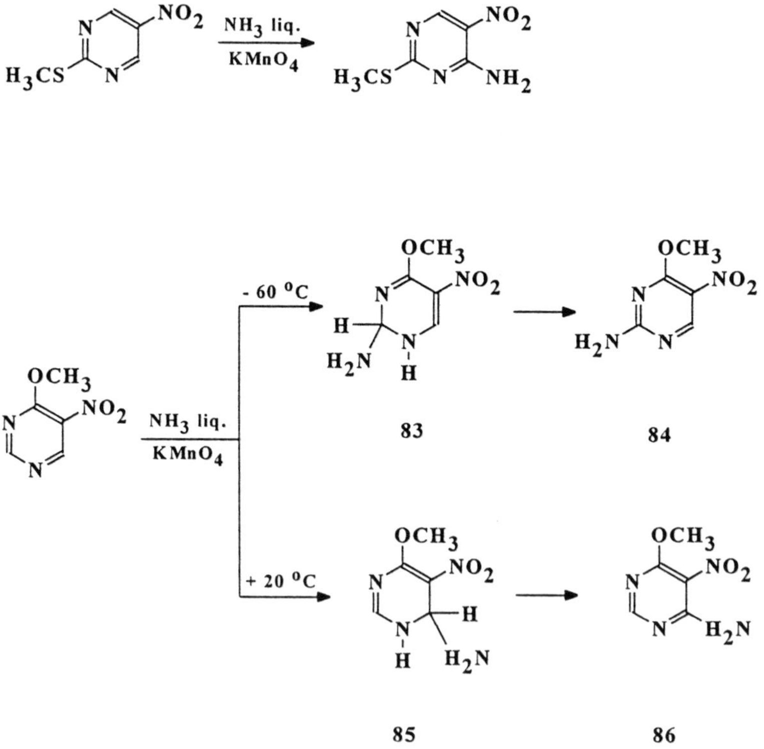

Scheme 65

Temperature dependency of the amination has been found in the reaction of 4-methoxy-5-nitropyrimidine with liquid ammonia/potassium permanganate. At −60°C to −70°C the reaction gave 2-amino-4-methoxy-5-nitropyrimidine **84** in 50% yield. However, 6-amino-4-methoxy-5-nitropyrimidine **86** was found to be formed in 65% yield, when the solution of 4-methoxy-5-nitropyrimidine in liquid ammonia was allowed to stand for 5 minutes at 10–20°C, cooled to −40°C and then treated with potassium permanganate (Scheme 65; Table XI). It is evident that at the very low temperature the kinetically favored C-2 adduct **83** is formed, but at room temperature the thermodynamically C-4 adduct **85** is obtained [83JOC1354].

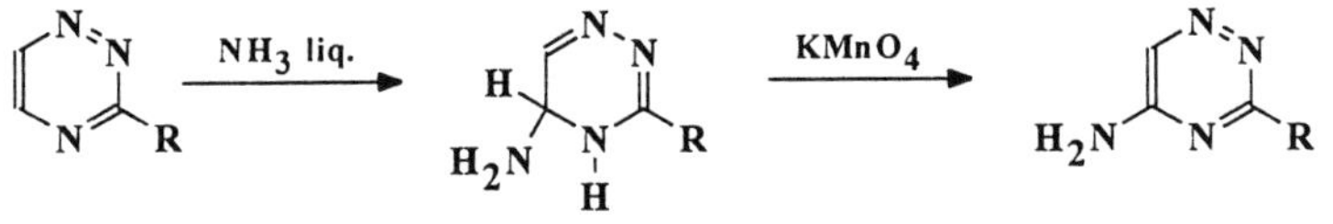

R = H, Alk, Ar, OCH_3, SCH_3

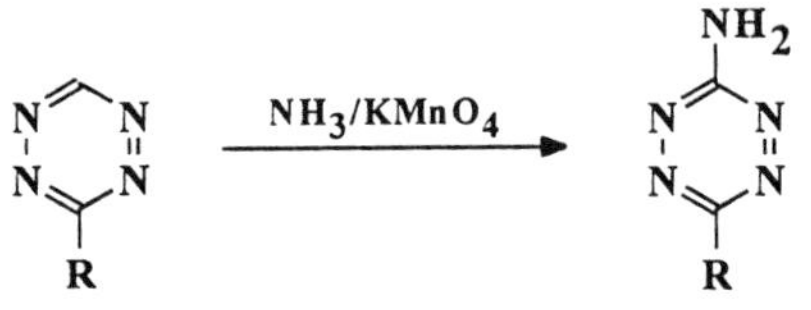

Scheme 66

In 1,2,4-triazines **87** and 1,2,4,5-tetrazines **90** the electron-deficient character of the azine ring is so enhanced due to the presence of several aza groups, that these compounds easily form detectable σ^H-adducts with liquid ammonia. It has been established by NMR spectroscopy that in solutions of **87** in liquid ammonia 5-amino-4,5-dihydro-1,2,4-triazines **88** are formed, which are easily oxidized by potassium permanganate into the corresponding 5-amino-1,2,4-triazines **89** in 63–95% yields (Scheme 66; Table XI) [85S884]. Also, oxidative amination of 3-alkyl or 3-aryl 1,2,4,5-tetrazines **90** in liquid ammonia in the presence of potassium permanganate proved to be an excellent route to obtain 6-amino-1,2,4,5-tetrazines **91** (Scheme 66, Table XI) [81JOC3805].

Nitroquinolines are also excellent substrates for the S_N^H amination [85JHC353; 87JOC5643; 91PJC323]. When dissolved in liquid ammonia, they give amino adducts easily detectable by NMR [85JHC353; 87JOC5643]. From the results presented in Table XIII, the favored position for attack by ammonia is always the one adjacent to the nitro group, even when the nitro group is present in the benzene ring. So, 8-chloro-5-nitroquinoline undergoes amination at C-6, while 6-chloro-8-nitroquinoline is aminated at C-7 (Scheme 67; Table XIII). In both cases the S_N^H reaction dominates over conventional substitution of halogen [91PJC323]. According to frontier molecular orbitals (FMO) calculations the regioselectivity of the amination reaction in the series of nitroquinolines can be explained nicely by the HOMO-LUMO interaction of the reaction partners [87JOC5643]. In this respect, it differs from the charge-controlled amination of 3-nitropyridines [91LA875].

The S_N^H amination of 5- and 8-nitroisoquinolines was found to follow the same pattern of substitution, as observed in the series of nitroquinolines [90LA653]. Like in the nitroquinolines, the substitution in 1-chloro-5-nitro- and 5-chloro-8-nitroisoquinolines indeed occurs at the *ortho*-position of the nitro group and not at the *ortho*- or *para*-position of the ring nitrogen (Scheme 67; Table XIII) [90LA653]. The data of FMO calculations indicates that the amination of nitroisoquinolines is controlled by the interaction between the highest occupied molecular orbital (HOMO) of ammonia and the lowest unoccupied molecular orbitals (LUMO) of nitroisoquinolines [90LA653]. Analysis of the charge distribution in these isoquinolines excludes a charge-controlled amination; otherwise the 1-amino compounds should be preferentially formed (Table XIV). The orbital-controlled aminations of the nitroquinolines and nitroisoquinolines are in contrast with the amination of nitropyridines where just the charge distribution in the pyridine ring is suggested to explain the observed regioselectivity [91LA875].

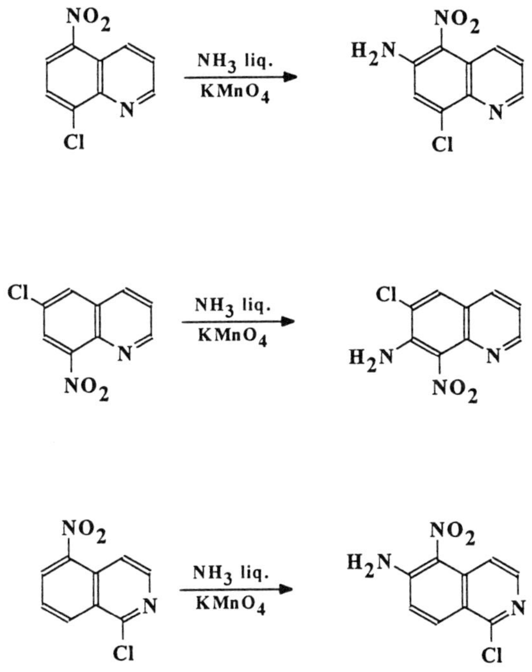

Scheme 67

TABLE XIII

Yields and Products, Obtained on Amination of Some Bicyclic Nitroazines with Liquid Ammonia (−33 °C)/KMnO$_4$

Nitroazine	S_N^H Product(s)	Yield, %	References
3-Nitroquinoline	4-Amino-3-nitroquinoline	65	85JHC353
4-Nitroquinoline	3-Amino-4-nitroquinoline	86	85JHC353
5-Nitroquinoline	6-Amino-5-nitroquinoline	33	87JOC5643
8-Chloro-5-nitroquinoline	6-Amino-8-chloro-5-nitroquinoline	88	91PJC323
6-Chloro-8-nitroquinoline	7-Amino-6-chloro-8-nitroquinoline	30	91PJC323
5-Nitroisoquinoline	6-Amino-5-nitro-isoquinoline	86	90LA653
1-Chloro-5-nitro-isoquinoline	6-Amino-1-chloro-5-nitro-isoquinoline	65	90LA653
5-Chloro-8-nitro-isoquinoline	7-Amino-5-chloro-8-nitroisoquinoline	78	90LA653
5,7-Dinitroquinoline	8-Amino-5,7-dinitro-quinoline	40	91PJC323
6,8-Dinitroquinoline	5-Amino-6,8-dinitro-quinoline	43	87JOC5643

TABLE XIII (Continued)

Nitroazine	S_N^H product(s)	Yield (%)	References
6-Nitroquinoxaline	2-Amino-6-nitro-quinoxaline	51	92LA899
	5-Amino-6-nitro-quinoxaline	32	92LA899
2-Chloro-7-nitro-quinoxaline	3-Amino-2-chloro-7-nitroquinoxaline	65	92LA899
3-Nitro-1,5-naphthyridine	4-Amino-3-nitro-1,5-naphthyridine	74	83RTC359, 83JHC9
2-Chloro-3-nitro-1,5-naphthyridine	4-Amino-2-chloro-3-nitro-1,5-naphthyridine	32	83RTC359, 83JHC9
3-Nitro-1,6-naphthyridine	4-Amino-3-nitro-1,6-naphthyridine	55	83RTC359, 83JHC9
2-Chloro-3-nitro-1,6-naphthyridine	4-Amino-2-chloro-3-nitro-1,6-naphthyridine	33	83RTC359, 83JHC9

Amination of 6- and 7-nitroquinoxalines with liquid ammonia/potassium permanganate gave substitution of the amino group in both rings, from which the ones, containing the amino group in the diazine ring are the main products (Scheme 68, Table XIII) 92LA899; 93ACSA95]. The fact that 2-chloro-7-nitroquinoxaline undergoes the amination-dehydrogenation reaction at C-3 to give 3-amino-2-chloro-7-nitroquinoxaline **92** in 65% yield provides a clear evidence for the preferential replacement of the hydrogen (not halogen!) atom (Table XIII; Scheme 68). It is worth noting that a similar reaction of 2-chloro-7-nitroquinoxaline with an excess of piperidine, yielding 7-nitro-

2,3-dipiperidinoquinoxaline **94** as the major product, in addition to obvious product **93**, was regarded as very unusual (Scheme 68) [91TL1311].

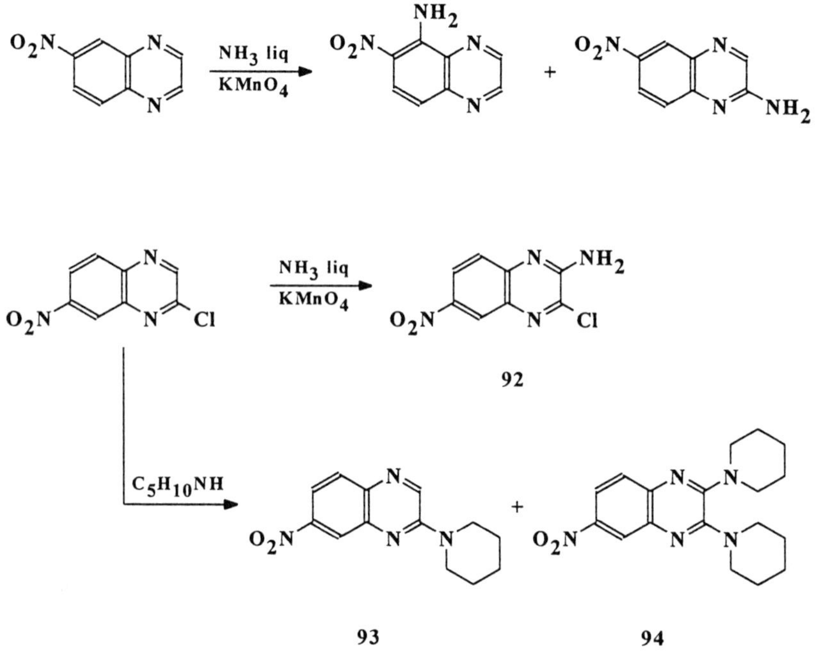

Scheme 68

The amination of 3-nitro-substituted 1,5-, 1,6-, and 1,8-naphthyridines has been extensively investigated. They are successfully aminated to the corresponding 4-amino-3-nitronaphthyridines in 32–74% yields (Table XIII) [83JHC9; 83RTC359; 83RTC511]. In case nucleofugal groups are present in the ring, substitution of the ring hydrogen is preferential to the substitution of this nucleofugal group, even when this group is present at an activated position.

TABLE XIV
Charge Distrubutions q_r in Molecules
of Some Nitroisoquinolines[1]

Isoquinoline	r	q_r	Preferable positions (r) for NH_3 attack under	
			charge control	experimental conditions
3-Nitroisoquinoline	C-1	0.0970	1>3>6>8	6
	C-3	0.0611		
	C-4	-0.1037		
	C-5	-0.0795		
	C-6	0.0557		
	C-7	-0.1006		
	C-8	0.0352		
1-Chloro-5-nitro-isoquinoline	C-1	0.1468	1>3>6>8	6
	C-3	0.0708		
	C-4	-0.0985		
	C-5	-0.0822		
	C-6	0.0608		
	C-7	-0.0972		
	C-8	0.0403		
5-Chloro-8-nitro-isoquinoline	C-1	0.0925	1>3>5>7	7
	C-3	0.0569		
	C-4	-0.1005		
	C-5	0.0480		
	C-6	-0.0473		
	C-7	0.0473		
	C-8	-0.0308		

[1]Ref. [90LA653].

Pyrido[2,3-*b*] and pyrimido[4,5-*b*]-annelated pyrazines have the profound tendency to add not only one but two molecules of nucleophilic reagents [75RTC45; 76OMR607; 79JHC301; 88MI1]. Thus, pteridine can react with ammonia and with primary and secondary alcohols to afford both 1:1 and 1:2 adducts (Scheme 69) [71JCS(B)2423; 75RTC45; 76OMR607]. The ratios between these mono and diadducts are strongly dependent on the temperature. At low temperature the C-4 addition is strongly favored, and use of potassium permanganate in liquid ammonia at about −40°C enabled the conversion of a number of pteridines into the corresponding 4-amino derivatives. For instance, 2-chloropteridine, when reacted with liquid ammonia in the presence of KMnO$_4$ gave 4-amino-2-chloropteridine (Scheme 69) [82JHC1527; 86JHC477]. Not any indication for an amino-dechlorination at C-2 was observed [82JHC1527].

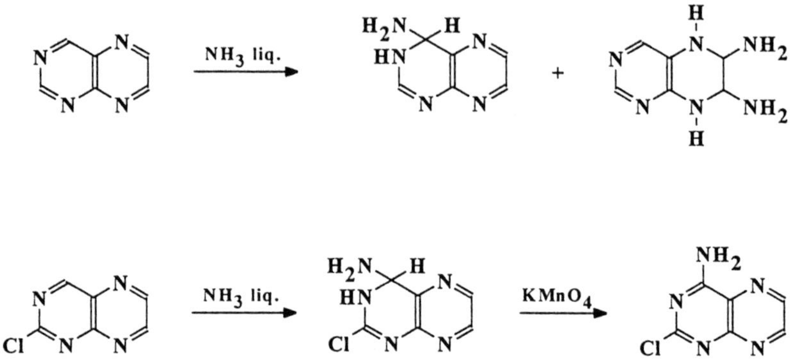

Scheme 69

Another interesting example of S_N^H reactions in the pyrimidopyrazine series is the displacement of hydrogen in 1,3-dimethyllumazines by action of alkylamines [90KG1575; 92KG1202]. Treatment of 1,3-dimethyllumazine with ammonia or alkylamines at 10–15°C in the presence of [Ag(C$_5$H$_5$N)$_2$]MnO$_4$ gave amination in the pyrazine ring at C-7 (yield 20–35%) [90KG1575]. In the reaction of 6-chloro-1,3-dimethyllumazine with 1,2-diaminoethane the S_N^H displacement at C-7 occurs concurrently with the conventional S$_N$(AE) substitution of the chloro atom at C-6 (Scheme 70) [92KG1202].

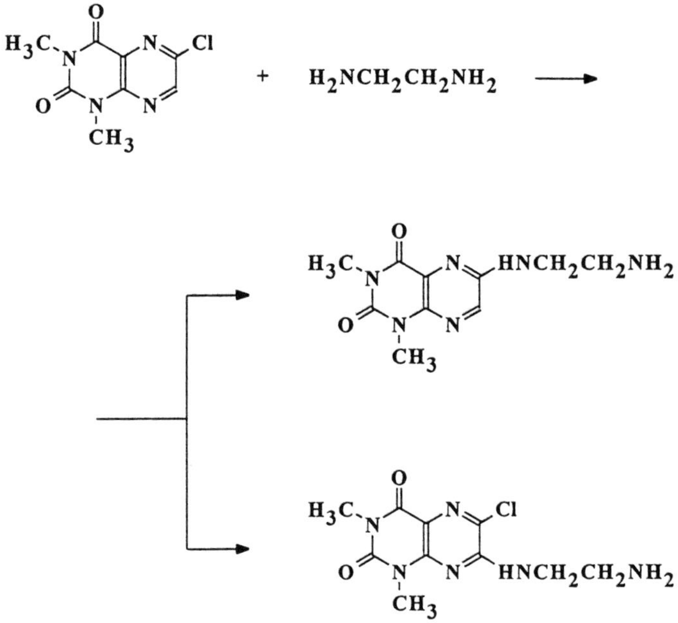

Scheme 70

Among other polyazaaromatic systems, capable of reacting with uncharged nucleophiles into S_N^H products, one should mention 5-azacinnoline [76KG976; 77KG1554] and deazaflavin derivatives **95** [92JHC763]. In the amino-dehydrogenation reaction of 5-azacinnoline with alkyl, arylamines, and hydrazines H-4 is selectively replaced under oxidative conditions (Scheme 71) [76KG976]. Also, direct coupling of 5-deazaflavins **95** with *n*-octyl and β-phenylethylamine succeeded by heating of **95** with neat amines at 90°C for 18 hours, affording the corresponding 5-amino-5-deazaflavins **96** in moderate yields (14–66%) (Scheme 71) [92JHC763]. The reaction has been suggested to proceed via the intermediacy of 5-amino-1,5-dihydro-5-deazaflavin which is then oxidized either by another molecule of 5-deazaflavin or air oxygen. The moderate yields of the S_N^H products **96** may be explained by a competitive oxidative reaction, which converts the amines used into the corresponding oxo compounds, thus partly consuming the starting deazaflavins **95** [92JHC763].

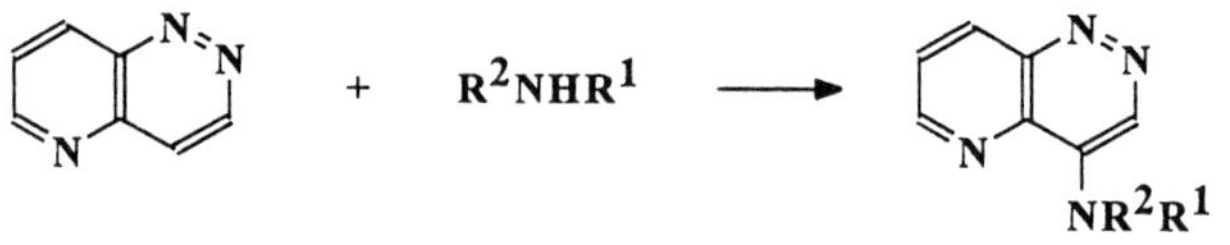

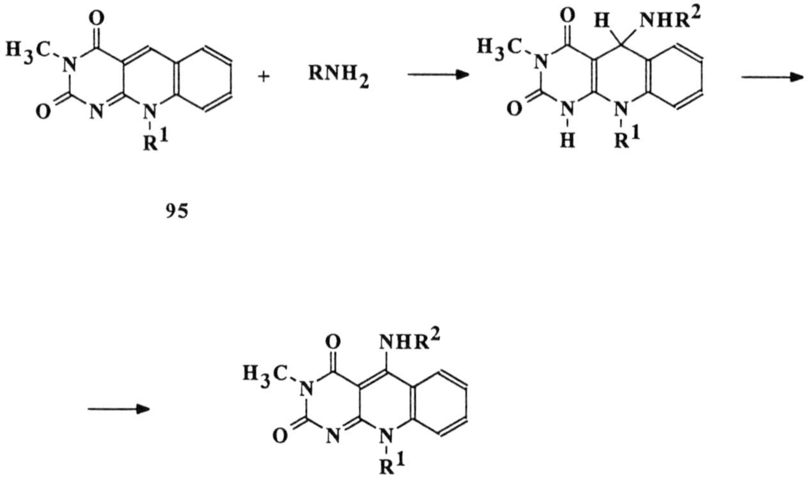

95

96

$R = C_2H_5$, n-C_4H_9, n-C_8H_{17} $R^2 = n$-C_8H_{17}, CH_2CH_2Ph

Scheme 71

A number of publications on the S_N^H reactions concern the arylation and hetero-arylation reaction in the series of diazanaphthalones [73MI2; 74KG993; 77KG684; 77ZOR204]. In particular, quinoxalin-2-one **97** was found to react successfully with arylamines on reflux in acetic acid in the presence of ammonium nitrate into 3-(4-aminoaryl)quinoxalin-2-ones **99** (Scheme 72; Table XV).

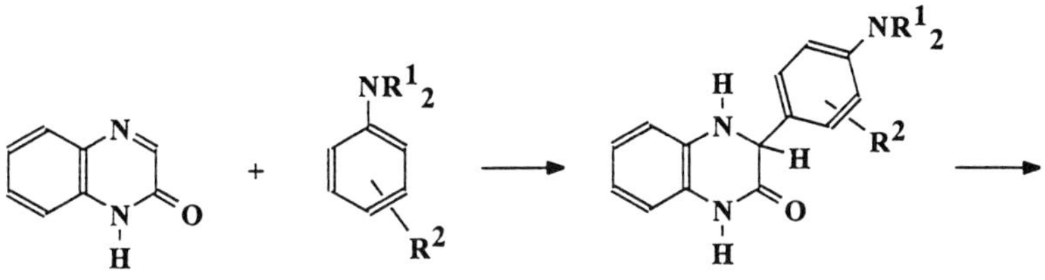

97 **98**

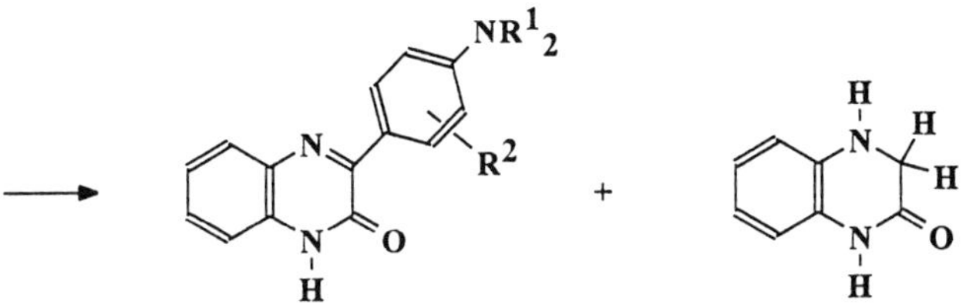

99 **100**

$R^1 = CH_3, C_2H_5, CH_2C_6H_5; R^2 = H, CH_3$

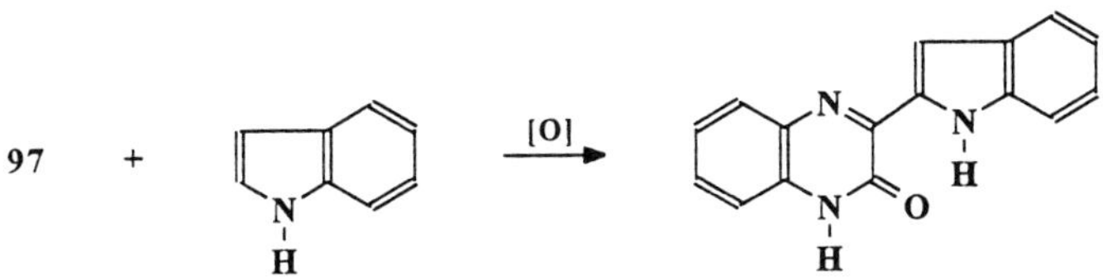

Scheme 72

According to NMR analysis of the reaction mixtures and kinetic measurements an amount of 3,4-dihydroquinoxalin-2-one **100** which is formed in argon atmosphere, being equivalent to the amount of **99**, indicates that intermediate adducts **98** are oxidized by quinoxalin-2-one **97** (Scheme 72) [74KG993; 77KG684]. Analogously quinoxalin-2-one undergoes heteroarylation with pyrroles and indoles (Scheme 72) [75MI3].

Also the arylation of cinnolin-3-one [75MI3] and quinazolin-2-one [73MI2] with arylamines was found to proceed smoothly in a melt with sulfur at 165–170°C or in DMF solution (Scheme 73; Table XV). The isomeric diazanaphthalones, such as quinazolin-4-one, phthalazin-4-one, and cinnolin-4-one, proved to be unreactive under the same conditions [77ZOR204] (for discussion on the reactivity of diazanaphthalones see Chapter 4).

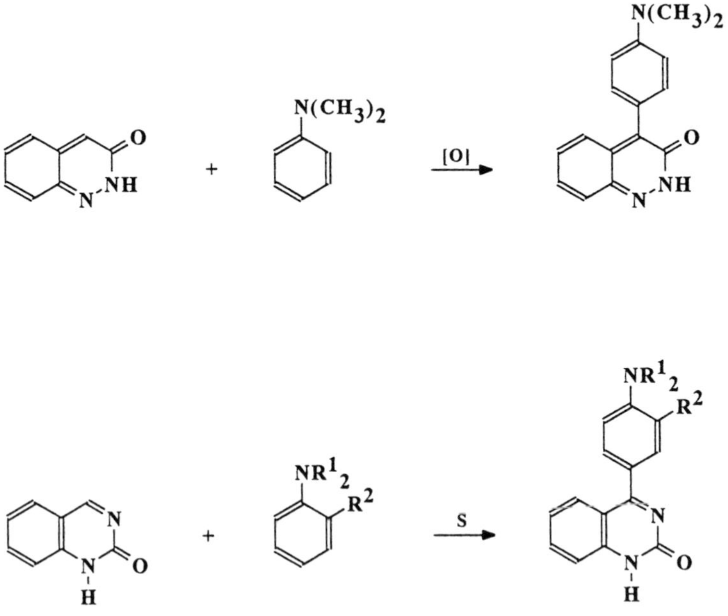

R^1 = H, CH_3, C_2H_5, C_3H_7; R^2 = H, CH_3, OC_2H_5, Cl

Scheme 73

TABLE XV
Nucleophilic Displacement of Hydrogen in Diazanaphthalones
by C-Nucleophiles

Diazanaphthalone	Reagent	S_N^H Product	Yield (%)	References
Quinoxalin-2-one	Aniline[1]	3-(4-Aminophenyl)-quinoxalin-2-one	50	74CGS933
	o-Toluidine[1]	3-(4-Amino-3-methyl-phenyl)quinoxalin-2-one	55	74CGS933
	N,N-Dimethyl-aniline[1]	3-(4-N,N-Dimethylamino-phenyl)quinoxalin-2-one	70	74CGS933
	Indole[2]	3-(3-Indolyl)quinoxalin-2-one	52	75MI3
Cinnoline-3-one	N,N-Dimethyl-aniline[2]	4-(4-N,N-Dimethylamino-phenyl)cinnolin-2-one	42	77ZOR204
Quinazolin-2-one	Aniline[3]	4-(4-Aminophenyl)-quinazolin-2-one	54	73MI2
	o-Toluidine[3]	4-(4-Amino-3-methyl-phenyl)quinazolin-2-one	56	73MI2
	N,N-Dimethyl-aniline[3]	4-(4-N,N-Dimethylamino-phenyl)quinazolin-2-one	75	73MI2
	Indole[4]	4-(3-Indolyl)-quinazolin-2-one	86	73MI2

[1]Reflux in acetic acid for 2 hours in the presence of 1 equiv. of ammonium nitrate. [2]Reflux in acetic acid for 3 hours. [3]On heating in a melt with 2 equiv. of sulphur at 165–175°C for 2 hours. [4]In a melt with 2 equiv. of sulphur at 135–140°C for 2 hours.

B. Reactions with Anionic Nucleophiles

Among the reactions involving the interaction between an uncharged azine and an anionic nucleophile leading to the displacement of hydrogen, the Chichibabin amination reaction seems to be the best studied. This reaction is of great industrial importance and has been discussed in detail in a number of reviews [78RCR1042; 83AHC95; 85JCA23; 85T237; 86CCA33; 87KG1011; 88AHC2]. The Chichibabin amination has long been associated with heterogeneous conditions (sodium amide in inert aprotic solvents), elevated temperatures (130–160°C), and moderate yields of the amino compounds due to the formation of by-products. It is illustrated by the heterogeneous amination of acridine with sodium amide on reflux in xylene, resulting in the formation of 9,9-diacridanyl in addition to the expected 9-aminoacridine (Scheme 74) [72KG1673].

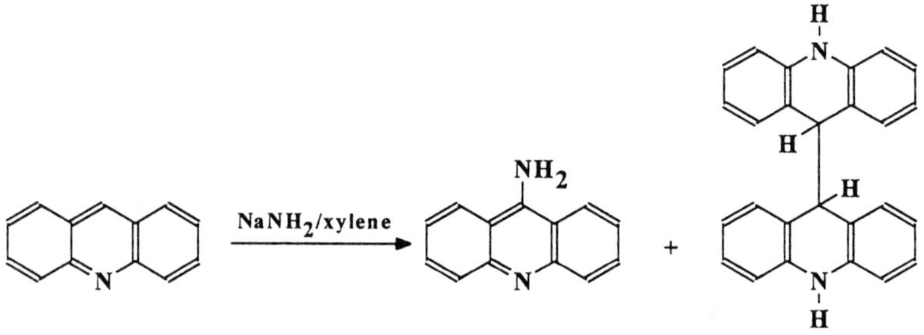

Scheme 74

Rapid advancements in the study of low-temperature reactions of azines in liquid ammonia resulted in a very effective procedure for homogeneous amination [83AHC95; 85T237; 86CCA33; 87KG1011; 88AHC2; 88T1; 90AHC117]. The low-temperature amination reaction is usually carried out in liquid ammonia containing potassium amide at −40 to −60°C followed by treatment of the reaction mixture with potassium permanganate. The reaction has been shown to proceed via an addition-elimination pathway, $S_N^H(AE)$, with the intermediary formation of a Meisenheimer σ-adduct [87KG1011].

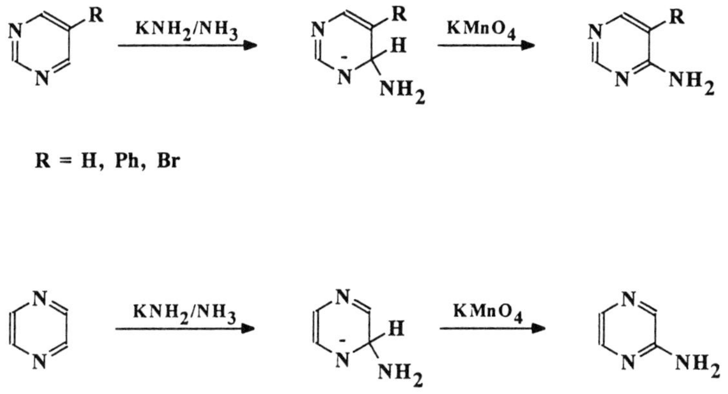

R = H, Ph, Br

Scheme 75

By extensive research work there has been collected unequivocal ^{1}H and ^{13}C NMR data, evidencing the existence of anionic σ-adducts, being formed by addition of the amide ion to diazines, triazines, and tetrazines [72JA682; 73RTC1232; 76RTC113; 78RTC288; 79JHC301; 83AHC95; 83AHC305; 83RTC367; 84MI1; 85JHC353; 85T237]. These σ^H-amino adducts are easily oxidized by potassium permanganate affording the amination products in good yields. For instance, pyrimidine is aminated in 72% yield when pyrimidine is dissolved in liquid ammonia containing potassium amide and solid potassium permanganate is added [82JHC1285]. The reaction has been successfully applied to aminate some other pyrimidines, pyrazine, pyridazine, and quinoline (Scheme 75; Table XVI) [82JHC1285; 85JHC353]. The amination of quinoline was found to be a temperature-dependent reaction, and its results can be interpreted with the concept of kinetic or thermodynamic control. Indeed, when the amination was carried out at $-65^{\circ}C$, 2-aminoquinoline was formed exclusively, while amination at room temperature gave only the 4-amino compound (Table XVI) [85JHC353].

TABLE XVI
Yields and Products, Obtained on Amination of Some Azines with Potassium (Sodium) Amide in Liquid Ammonia ($-33^{\circ}C$) in the Presence of Potassium Permanganate

Azine	S_N^H Product(s)	Yield (%)	References
Quinoline	2-Aminoquinoline[1]	52	85JHC353
	4-Aminoquinoline[2]	64	85JHC353
Pyrimidine	4-Aminopyrimidine	72	82JHC1285
5-Phenylpyrimidine	4-Amino-5-phenylpyrimidine	75	86MI2
	2-Amino-5-phenylpyrimidine	2	
5-Bromopyrimidine	4-Amino-5-bromopyrimidine	28	86MI2
Pyrazine	Aminopyrazine	65	82JHC1285
Pyridazine	4-Aminopyridazine	91	82JHC1285
4-Cyano-6-phenyl-pyridazine	5-Amino-4-cyano-6-phenyl-pyridazine	24	88JHC831

[1]At $-65^{\circ}C$. [2]At room temperature.

The hydroxylation of pyridine [1883CB2152], quinoline and isoquinoline [30MI1], acridine, and phenanthridine [72KG1673] occurs in a somewhat similar way as the Chichibabin amination reaction, although it requires much higher temperatures (about $300^{\circ}C$) to be carried out (Scheme 76).

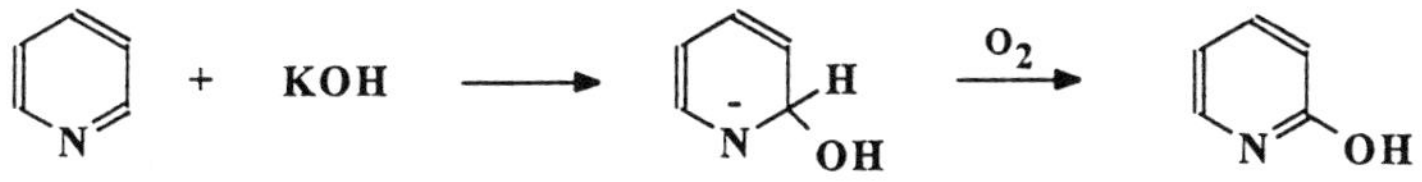

Scheme 76

The ambident phenolate anions [83MI3] containing electron-donating substituents in the benzene ring, when reacting with acridine in DMF at 130–140°C in the presence of air bubbling through the reaction mixture, behave themselves as C-nucleophiles, affording hydroxyaryl acridines. Sodium salts of nitrophenols or of hydroxybenzoic acids do not yield S_N^H products (Scheme 77; Table XVII) [76KG266].

TABLE XVII
Nucleophilic Substitution of Hydrogen in Acridine
by Action of Phenolate Anions[1]

Phenol	S_N^H Product	Yield (%)
Phenol	9-(4-Hydroxyphenyl)acridine	33
o-Cresol	9-(4-Hydroxy-3-methylphenyl)acridine	35
m-Cresol	9-(4-Hydroxy-2-methylphenyl)acridine	51
p-Cresol	9-(2-Hydroxy-5-methylphenyl)acridine	27
o-Bromophenol	9-(4-Hydroxy-3-bromophenyl)acridine	91
1-Naphthol	9-(1-Hydroxy-2-naphthyl)acridine	74
2-Naphthol	9-(2-Hydroxy-1-naphthyl)acridine	72
8-Hydroxyquinoline	9-(8-Hydroxy-5-quinolyl)acridine	81

[1]See ref. [76KG266].

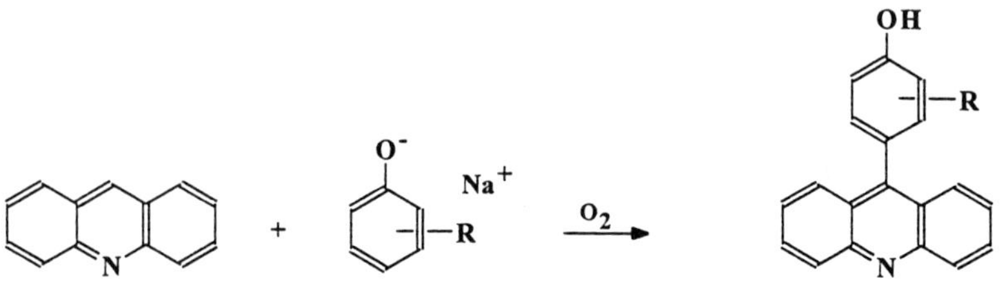

R = H, o-CH$_3$, o-Br, m-OH, m-OCH$_3$, 2,3-benzo, 3,4-benzo

Scheme 77

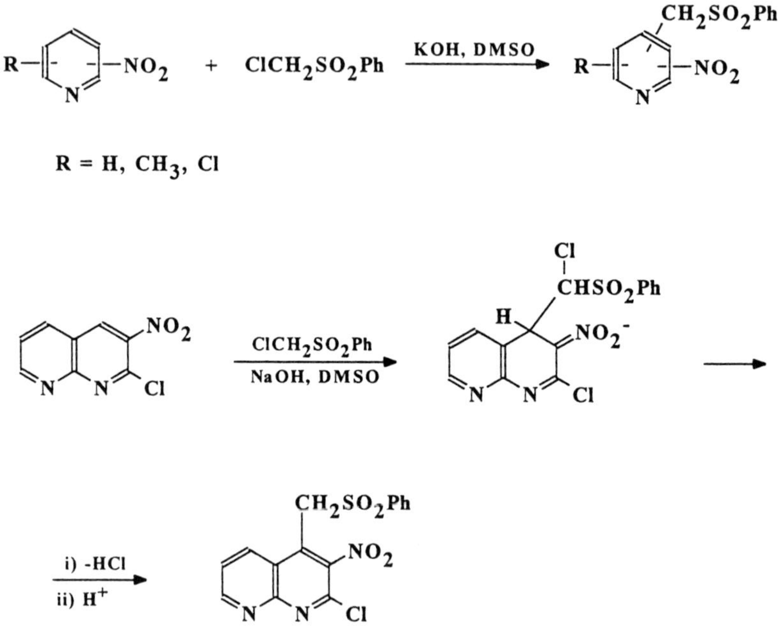

R = H, CH$_3$, Cl

Scheme 78

TABLE XVIII

Vicarious Nucleophilic Substitution of Hydrogen

in Some Azines

Azine	Reagent[1]	S_N^H Product(s)	Yield (%)	References
3-Nitropyridine	A	3-Nitro-4-(phenylsulfonylmethyl)pyridine	66	84LA8
		3-Nitro-6-(phenylsulfonylmethyl)pyridine	22	
2-Methyl-4-nitropyridine	A	2-Methyl-4-nitro-3-(phenylsulfonylmethyl)pyridine	60	84LA8
2-Chloro-3-nitropyridine	A	2-Chloro-3-nitro-4-(phenylsulfonylmethyl)pyridine	48	84LA8
		2-Chloro-3-nitro-6-(phenylsulfonylmethyl)pyridine	7	
2-Methoxy-5-nitropyridine	C	6-(*n*-Butyl)-2-methoxy-5-nitropyridine	47	90OPP575
3,6-Dichloropyridazine	B	3,6-Dichloro-4-(*p*-tolylsulfonylmethyl)pyridazine	98	92TL4787
1,2,4-Triazine	A	5-(Phenylsulfonylmethyl)-1,2,4-triazine	72	88LA627
3-Methyl-1,2,4-triazine	A	3-Methyl-5-(phenylsulfonylmethyl)-1,2,4-triazine	78	88LA627

TABLE XVIII (Continued)

Azine	Reagent[1]	S_N^H Product(s)	Yield (%)	References
3-Methylthio 1,2,4-triazine	A	3-Methylthio-5-(phenylsulfonylmethyl)-1,2,4-triazine	76	88LA627
Pyrido[3,4-*b*]-pyrazine	B	2-(*p*-Tolylsulfonylmethyl)-pyrido[3,4-*b*]-pyrazine	44	88T1721
7-Methylpteridine	B	4-(*p*-Tolylsulfonyl-methyl)-7-methylpteridine	78	88JPR789

[1]A: Chloromethyl phenyl sulfone; B: Chloromethyl *p*-tolyl sulfone; C: Trifluoromethyl *n*-butyl sulfone.

The vicarious S_N^H substitution by carbanionic species is also usefully applied for functionalization of an azine ring. As is already explained in Chapter 2,III,A, the characteristic feature of the vicarious reagent is the presence of a good leaving group X on the carbon in the carbanionic reagent; X facilitates in the σ^H-adducts removal of the hydrogen on the sp^3-carbon atom of the ring as HX. Numerous examples of vicarious nucleophilic substitution have been discovered in the reactions of nitro-activated azines with halogenmethylsulfones, nitroalkanes, and other CH-active compounds (Table XVIII) [83MI2]. For instance, a number of nitropyridines [84LA8] and 3-nitro-1,8-naphthyridines [91JHC1075] easily undergo the vicarious nucleophilic substitution of hydrogen mainly at positions *ortho* to the nitro group by action of chloromethyl phenyl sulfone (Scheme 78; Table XVIII). Also 2-chloro-3-nitro-1,8-naphthyridine is successfully substituted at C-4 by treatment with chloromethyl phenyl sulfone in the presence of a base (Scheme 78) [91JHC1075].

3-Mono- or 3,6-disubstituted 1,2,4-triazines are found to undergo S_N^H substitution at C-5, when reacting with vicarious carbanions, derived from chloromethyl aryl sulfones

[91CB577], as well as with "conventional" carbanions, generated on treatment of ethyl cyanoacetate, benzyl cyanide, or acetophenone with sodium hydride (Scheme 79) [83TL3277; 87CPB1378]. An interesting application of the vicarious S_N^H reaction in the series of triazines is the very convenient and elegant method for the direct preparation of 5-acyl-1,2,4-triazines (Scheme 80) [84TL4795].

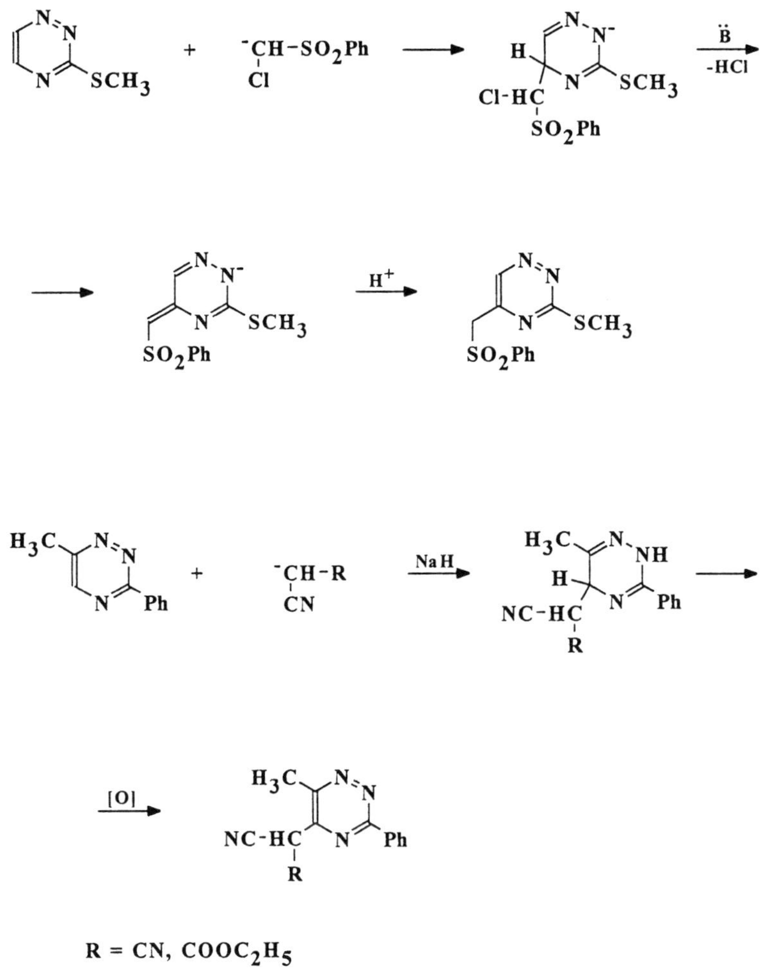

R = CN, $COOC_2H_5$

Scheme 79

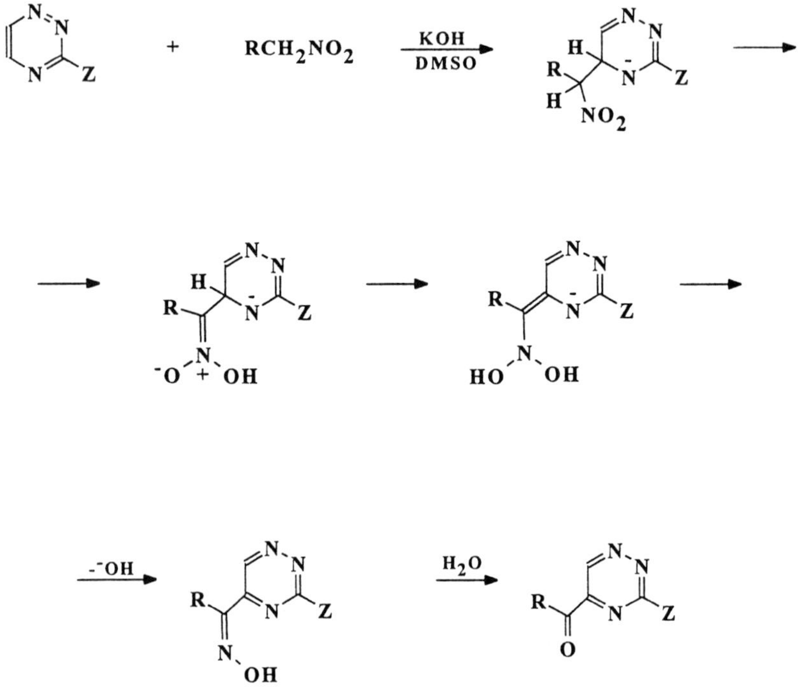

R = H, CH$_3$, C$_2$H$_5$; Z = H, CH$_3$, SCH$_3$, Ph

Scheme 80

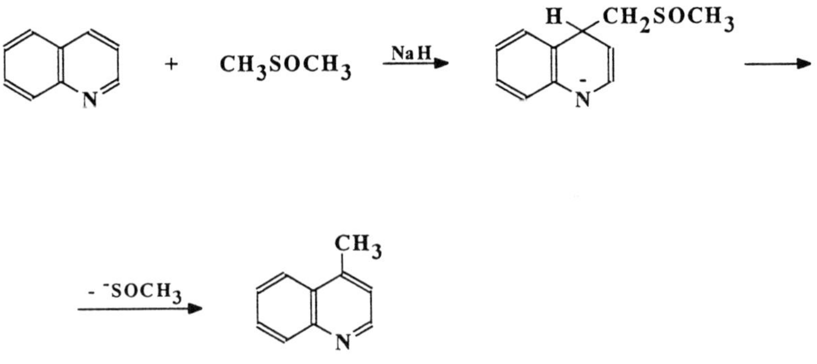

Scheme 81

The carbanions derived from acetylacetone and ethyl cyanoacetate [57CB2215], as well as from such CH-active methylheterocycles as α- and β-picolines, quinaldine, lepidine, and 2-methylbenzthiazole [70KG1384], were found to react with acridine in a melt of sulfur into the 9-substituted products (Scheme 82).

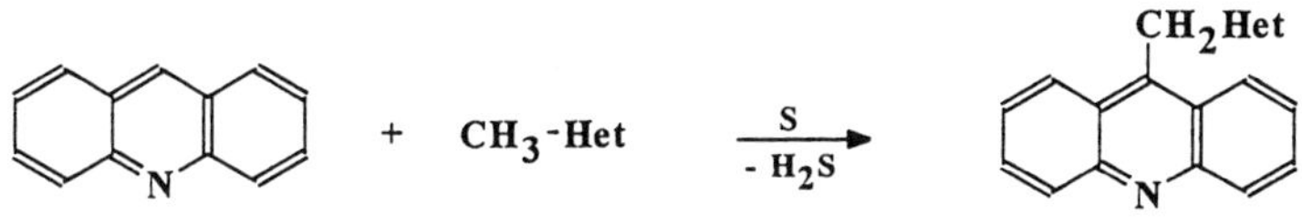

Het = 2-pyridyl, 3-pyridyl, 2-quinolyl, 3-quinolyl, 2-benzothiazolyl

Scheme 82

An interesting example of a S_N^H displacement reaction in an azine ring is the recently reported reaction of 3-(α)-chlorobenzylpyridazine **101** with alkoxides, resulting in the formation of 6-alkoxy- and 4-alkoxy-3-benzylpyridazines **102** and **103**, respectively, rather than the expected ethers **104** (Scheme 83) [92LA19]. For example, reaction of **101** with sodium methoxide in methanol gives 6-methoxy-3-benzylpyridazine (**102**, R= CH$_3$), as the main product (yield 55%) together with the 4-methoxy isomer **103** (R= CH$_3$) (yield 12%) and the ether **104** (R= CH$_3$) in 19% yield [92LA19]. The interesting feature of this reaction is that the displacement of hydrogen in the pyridazine ring is facilitated by elimination of halogen from the side chain substituent. This *tele*-substitution reaction can be considered as a vicarious substitution of hydrogen, in which not the reagent but the side-chain substituent in the pyridazine substrate bears the vicarious leaving group. A plausible mechanism for this reaction is shown in Scheme 83, and is similar to that suggested for the displacement of hydrogen in nitrobenzylidene chlorides by action of alkoxides [30CB749; 64JCS2806] and carbanions [79AJC1949] (see Chapter 2, Sections III,A and III,B).

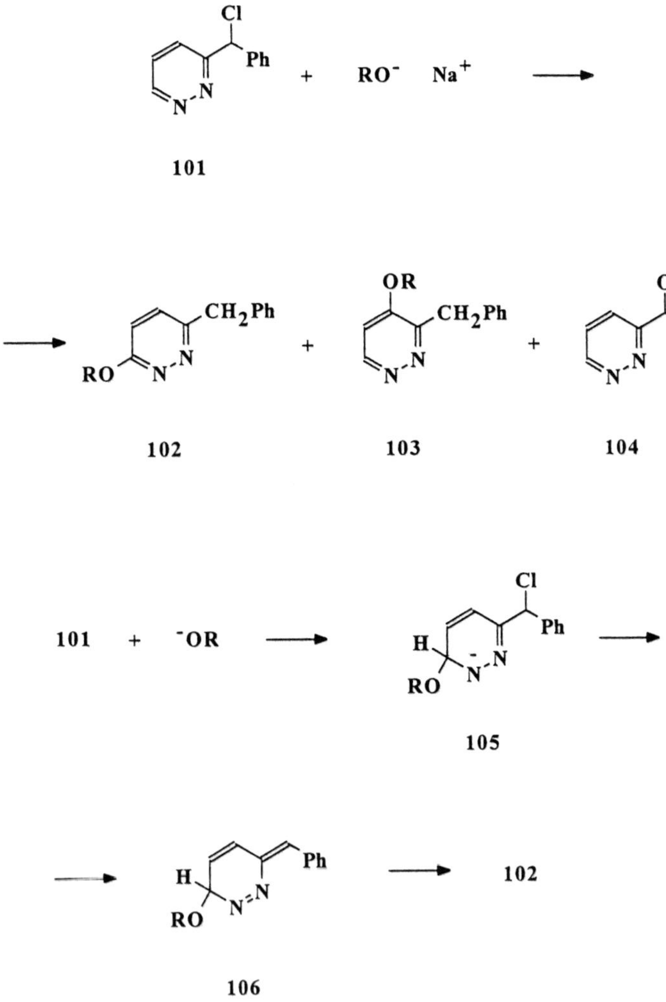

Scheme 83

C. Reactions with Organometallic Compounds

One of the simple methods to reach substitution of hydrogen in an azine ring is use of organolithium compounds. In the series of pyridines, the addition of alkyl or aryllithiums to the pyridine ring followed by aromatization of the dihydro adducts formed have been successfully applied for structural modification of acridines [30LA80; 31LA174; 50JA2181], pyridine and 3-substituted pyridines, such as nicotinamides [83TL4735], and 3-pyridyloxazolines [82JOC2633; 84H1091; 84JCS(P1)2227; 84JOC56; 86H125]. For example, reaction of pyridine with n-butyllithium gives 2-n-butylpyridine in good yield (Scheme 84). The reaction proceeds via the intermediate C-2-adduct **107** identified by ^{1}H and ^{13}C NMR spectroscopy [93BCJ1028]. Also the 1,4-dihydro-adducts **108**, formed by reacting N-phenyl nicotinamide **107** with alkyllithiums, are smoothly oxidized with potassium permanganate into 4-substituted nicotinamides **109** in 60–90% yields (Scheme 84) [83TL4735]. Pyridyloxazoline **110** reacts with alkyllithium to give after oxidation with 2,3-dichloro-5,6-dicyano-1,4-benzoquinone (DDQ) compound **111**, being substituted in the pyridine ring in position 6 (Scheme 84) [82H13].

Grignard reagents have been found to react with 3,6-disubstituted 1,2,4-triazines into 5-substituted 2,5-dihydro-1,2,4-triazines which on oxidation with potassium permanganate give the final S_N^H products (Scheme 85) [85H2807; 87CPB1378; 89AHC73; 92H1].

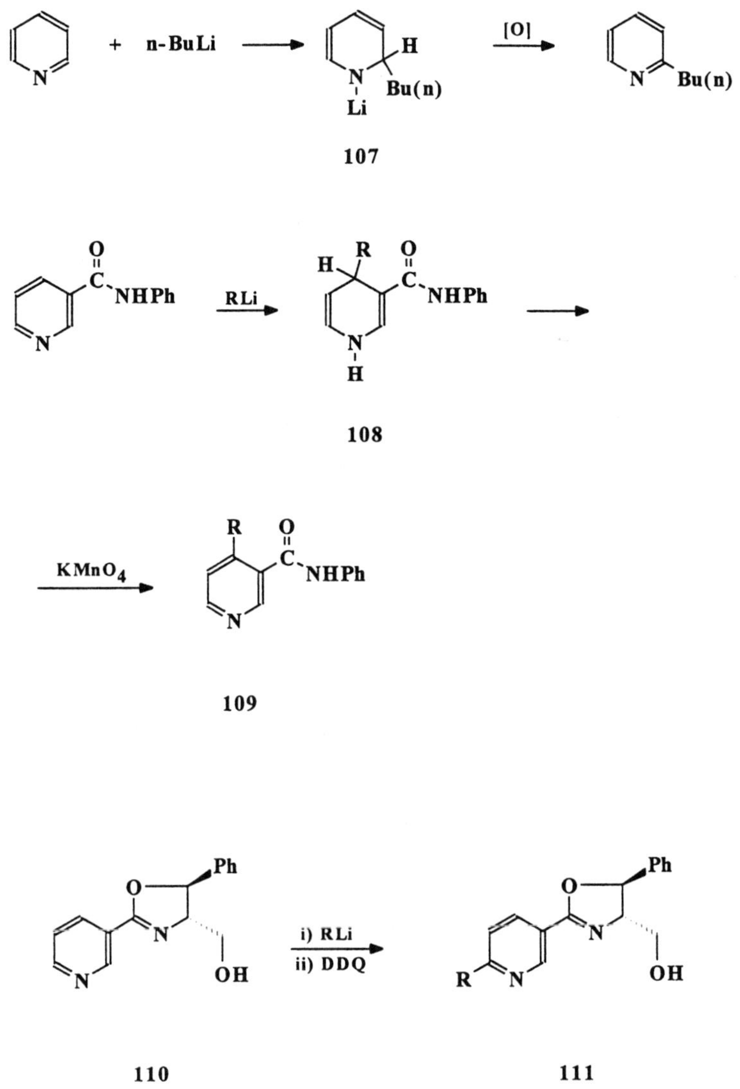

Scheme 84

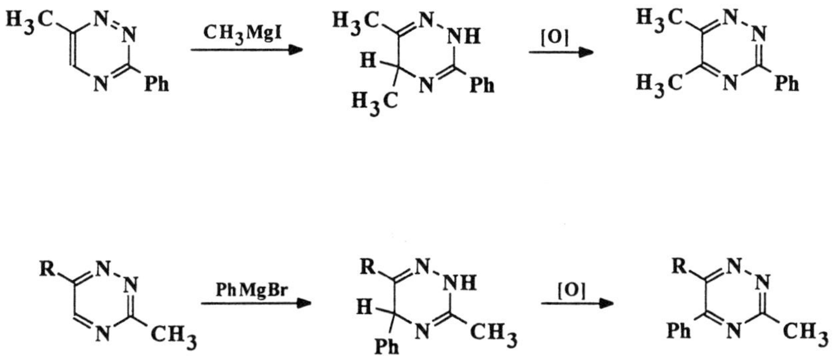

Scheme 85

Experimental Procedures

Example 1. Amination of 2-methoxy-5-nitropyridine

2-Methoxy-5-nitropyridine (308 mg, 2 mmol) was dissolved in 30–40 ml of liquid ammonia. Potassium permanganate (620 mg) was then added, and the mixture was stirred for 5 hours. After evaporation of ammonia, 50 ml of water was added to the residue, and the mixture was continuously extracted with chloroform for 30 hours. The residue obtained after evaporation of the solvent from the combined extracts was crystallized from heptane to give 2-amino-6-methoxy-3-nitropyridine, melting point 167–169°C. Yield 255 mg (75%) [91LA875].

Example 2. Amination of pyrimidine

Pyrimidine (320 mg, 4 mmol) was dissolved in 30 ml of liquid ammonia, containing 16 mmol of potassium amide [prepared by dissolving potassium (624 mg, 16 mmol) in the liquid ammonia and adding a trace of ferric nitrate]. After 10 minutes solid potassium permanganate (1 g, about 6 mmol) was added portion by portion and the solution was stirred for another 10–20 minutes. After addition of ammonium sulfate to decompose the excess of potassium amide, methanol was added, and ammonia was evaporated. Further work-up (column chromatography) gave 4-aminopyrimidine, melting point 154–156°C. Yield 346 mg (72%) [82JHC1285].

Example 3. Amination of 4-methoxy-5-nitropyrimidine

4-Methoxy-5-nitropyrimidine (310 mg, 2 mmol) was added in small portions to a solution of potassium permanganate (220 mg, 1.05 redox equiv.) in 20 ml of ammonia at −60 to −70°C with stirring. The reaction mixture was kept for 15–20 minutes at −60°C, and then methanol (20 ml) was added through a dry ice/acetone condenser. 2-Amino-4-methoxy-5-nitropyrimidine was isolated by column chromatography on silica gel with ethanol as eluent. Yield 170 mg (50%), melting point 225–227°C [83JOC1354; Copyright permission 1994 American Chemical Society].

Example 4. Amination of 4-nitro-6-phenylpyridazine

4-Nitro-6-phenylpyridazine (211 mg, 1 mmol) was added to a solution of potassium permanganate (420 mg, 2 redox equiv.) in 25 ml of liquid ammonia at −45°C. After 0.5 hour, 20 ml of cold chloroform was added to the brown mixture. The ammonia was evaporated and the residue was extracted with warm chloroform and ethyl acetate. The extracts were filtered, concentrated and purified by column chromatography on silica gel with chloroform–methanol 6:1 as eluent to give 5-amino-4-nitro-6-phenylpyridazine as yellow crystals, melting point 215–218°C. Yield 210 mg (93%) [88JHC831].

Example 5. Amination of 1,2,4-triazine

To a solution of 1,2,4-triazine (80 mg, 1 mmol) in dry liquid ammonia (20 ml), an excess of potassium permanganate (0.2 g) is added in one portion and the mixture is stirred for an additional 30 minutes. Ammonia is evaporated and the residue is extracted with hot isopropanol. The crude product remaining after evaporation of the solvent is crystallized from ethanol to give 5-amino-1,2,4-triazine, melting point 230–231°C. Yield 91 mg (95%) [85S884].

Example 6. Amination of 3-nitroquinoline

3-Nitroquinoline (261 mg, 1.5 mmol) was dissolved in liquid ammonia (20 ml) and an excess of potassium permanganate (2–3 equiv.) was added in small portions under stirring. After additional stirring for two hours at −33°C the ammonia was evaporated off, to the remaining residue 20 ml of water was added, and the mixture was continuously extracted with ether. The crude residue left after evaporation of the ether was crystallized from ethanol, yielding 180 mg (65%) of 4-amino-3-nitroquinoline, melting point 274–275°C [85JHC353].

Example 7. Amination of quinoline

A. Preparation of 2-aminoquinoline. Quinoline (260 mg, 2 mmol) was added to a solution of liquid ammonia (20 ml) containing an excess of potassium amide (2.5

equiv.) at −65°C. After additional reaction time of 15 minutes, the reaction mixture was quenched with ammonium sulfate, the ammonia was evaporated off and the residue was extracted with benzene. Evaporation of benzene gave 2-aminoquinoline as light yellow crystals, melting point 132°C. Yield 152 mg (52%) [85JHC353].

B. Preparation of 4-aminoquinoline. Quinoline (260 mg, 2 mmol) was added to the solution of liquid ammonia (20 ml) containing potassium amide (2.5 equiv.) at −45°C and this mixture was sealed in a glass tube. After keeping this mixture for about one hour at room temperature, the tube was cooled to about −45°C, then opened and potassium permanganate (3.5 equiv.) was added in small portions under stirring. After 15 minutes the reaction mixture was quenched with ammonium sulfate. The ammonia was evaporated and the whole mixture was extracted with ether. Gas chromatography of the ethereal solution showed two peaks in the ratio 1:14 belonging to 2- and 4-aminoquinolines, respectively. The residue, obtained after evaporating off the ether solvent, was recrystallized from benzene, yielding 185 mg (64%) of 4-aminoquinoline, melting point 149–150°C [85JHC353].

Example 8. Amination of 5-nitroisoquinoline

To a solution of 5-nitroisoquinoline (350 mg, 2 mmol) in 35–40 ml of liquid ammonia potassium permanganate (1 g) was added, and the mixture was stirred for 6 hours. After evaporation of liquid ammonia 50 ml of water was added to the residue, and the mixture was continuously extracted with chloroform for 20 hours. The residue obtained after evaporation of the solvent was recrystallized from methanol to give 6-amino-5-nitroisoquinoline as yellow needles, melting point 266–267°C. Yield 327 mg (86%) [90LA653].

Example 9. Amination of 3,6-dinitro-1,8-naphthyridine

3,6-Dinitro-1,8-naphthyridine (150 mg, 0.68 mmol) was dissolved in 25–30 ml of liquid ammonia and 300 mg of potassium permanganate was added. The mixture obtained was stirred at −33ºC for 1 hour. After evaporation of ammonia, 50 ml of water was added to the residue and the mixture was continuously extracted with chloroform for 30 hours. The residue obtained after evaporation of the solvent from the combined extracts was dissolved in 120 ml of boiling methanol, the solution was filtered, concentrated to 30 ml and cooled to give 4-amino-3,6-dinitro-1,8-naphthyridine as orange needles, melting point 315–317 ºC. Yield 80 mg (50%) [93LA471].

Example 10. Preparation of 2-chloro-3-nitro-4-(phenylsulfonylmethyl)-1,8-naphthyridine

To a stirred suspension of powdered sodium hydroxide (400 mg, 10 mmol) in dimethyl sulfoxide (3 ml) a solution of 2-chloro-3-nitro-1,8-naphthyridine (207 mg, 1 mmol) and chloromethyl phenyl sulfone (190 mg, 1 mmol) was added dropwise at room temperature. The reaction mixture was stirred for 2 hours at room temperature. Then the mixture was poured into diluted aqueous ammonia. A precipitate obtained was filtered off, washed with water, dried, and recrystallized from methanol to yield 2-chloro-3-nitro-4-(phenylsulfonylmethyl)-1,8-naphthyridine (338 mg, 90%), as white needles, melting point 263–264ºC [91JHC1075].

Example 11. Methylation of 6-methyl-3-phenyl-1,2,4-triazine

The Addition Step. A Grignard solution [prepared from 240 mg (10 mg atom) of Mg and 1.39 g (10 mmol) of methyl iodide in dry ether (35 ml)] was added dropwise to a solution of 1.28 g (7.5 mmol) of 6-methyl-3-phenyl-1,2,4-triazine in dry ether (35 ml) at room temperature under nitrogen and the mixture was stirred for 10 minutes. Saturated aqueous solution of ammonium chloride was added to the reaction mixture and the organic layer was separated, then dried over sodium sulfate. After removal of the solvent, the residue was recrystallized from hexane–ethyl acetate to give 1.2 g (86%) of 5,6-dimethyl-3-phenyl-2,5-dihydro-1,2,4-triazine, melting point 109–111°C, as colorless needles.

The Oxidation Step. Potassium permanganate (2.37 g, 15 mmol) was added to a solution of 5,6-dimethyl-3-phenyl-2,5-dihydro-1,2,4-triazine (940 mg, 5 mmol) in acetone (100 ml) all at once and the mixture was stirred for 12 hours. The reaction mixture was filtered, and the filtrate was concentrated under reduced pressure. The residue was recrystallized from hexane to give 720 mg (77%) of 5,6-dimethyl-3-phenyl-1,2,4-triazine, melting point 78–80°C, as pale yellow needles [87CPB1378].

Example 12. Methylation of isoquinoline

To a solution of 2.64 g (110 mmol) of sodium hydride in 100 ml of DMSO at 70°C was added 2.2 ml of isoquinoline (19 mmol) in 100 ml of DMSO. The reaction mixture was stirred at 70°C for 4 hours under a nitrogen atmosphere, 100 ml of water was added, and the reaction mixture was poured into 1.5 liter of water. The aqueous solution was extracted with benzene. Removal of the benzene yielded an oil which was analyzed by GLPC. The analysis indicated a quantitative yield of 1-methylisoquinoline. The oil was distilled at 62–66°C/4 mm and gave a picrate of 1-methylisoquinoline, melting point 227°C [66JOC243; Copyright permission 1994 American Chemical Society].

II. AZAAROMATIC SUBSTRATES CONTAINING QUATERNARY NITROGEN

As already discussed in previous chapters, nucleophilic displacement of hydrogen in heteroaromatic compounds usually requires the presence of strong electron acceptors in the aromatic ring, such as the aza or nitro groups, which facilitate the formation of σ-adducts [88T1]. In the series of azines, effective activation of the azaaromatic ring can also be reached by quaternization of the ring nitrogen. Enhanced reactivity of azinium cations toward nucleophilic reagents makes these charged molecules very attractive and appropriate substrates for S_N^H reactions. Among reactions commonly used for preparation or generation of these azinium cations, protonation, alkylation, and acylation are widely applied routes. A new class of azinium cations, such as N-fluoroazinium salts, has recently been shown to possess a unique ability to undergo a variety of S_N^H reactions (see Chapter 3,II,E). In the following sections subsequently the S_N^H substitutions in NH-azinium, N-alkylazinium, N-acylazinium, N-alkoxy, N-acyloxy, and N-fluoroazinium salts will be discussed.

A. NH-Azinium Salts

The NH-azinium salts can be used in S_N^H reactions only when the reacting nucleophiles are not too basic to cause deprotonation of the NH-azinium substrates. Therefore, weak basic nucleophiles, such as arylamines, phenylhydrazones, phenols, pyrroles, and indoles, but not the more basic alkylamines, enamines or organometallic compounds, are appropriate reagents for S_N^H substitutions in NH-azinium salts. It is extensively illustrated by the arylation, aminoarylation, or hydroxyarylation reactions of the protonated species of quinoline [65MI1], of quinazoline [76TL3537; 79CB1348], of quinoxaline [76ZOR2464] and acridine [71KG112; 72KG216; 75KG387; 76RCR454] with arylamines (Schemes 86–88; Tables XIX and XX).

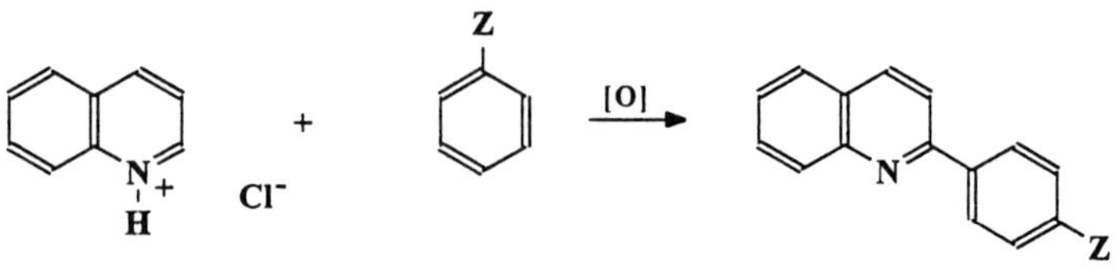

Z = NR₂, OH

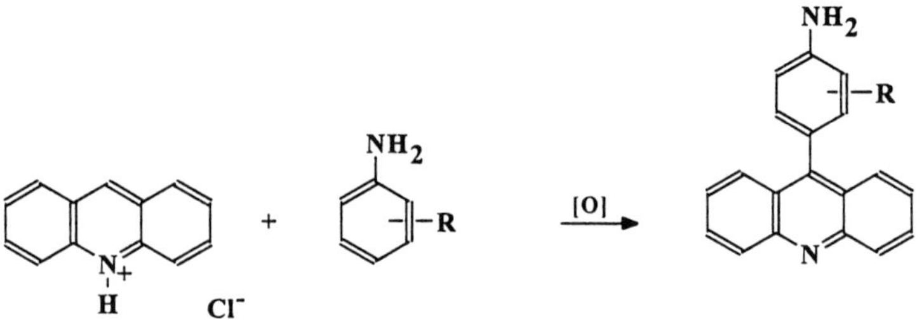

R = alkyl, hal, OR, NR₂, benzo

Scheme 86

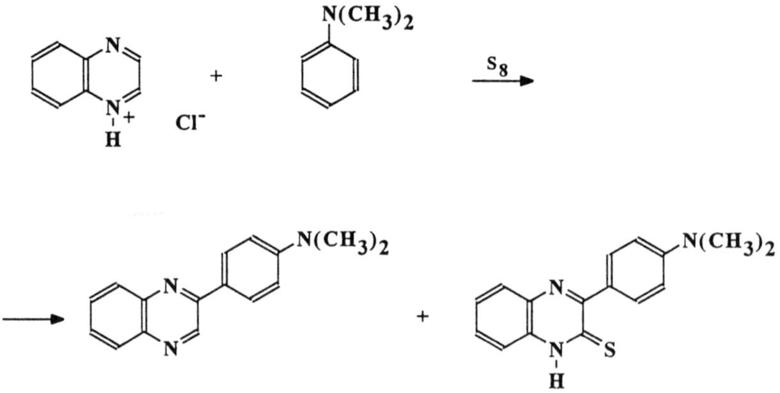

Scheme 87

TABLE XIX
Nucleophilic Substituton of Hydrogen in Acridine Hydrochloride by Action of Arylamines[1]

Arylamine	Temp. (°C)	Time (hours)	S_N^H Product	Yield[2] (%)	References
Aniline	130	2.0	9-(4-Aminophenyl)-acridine	95 (4)	71KG112
N-Methyl-aniline	120	1.5	9-(4-N-Methylamino-phenyl)acridine	92 (5)	71KG112
N,N-Dimethyl-aniline	130	2.0	9-(4-N,N-Dimethyl-aminophenyl)acridine	94 (4)	71KG112
o-Toluidine	125	2.0	9-(4-Amino-3-methyl-phenyl)acridine	90 (8)	71KG112
m-Toluidine	120	2.0	9-(4-Amino-2-methyl-phenyl)acridine	70 (20)	72KG216
p-Toluidine	130	7.0	9-(2-Amino-5-methyl-phenyl)acridine	25 (43)	72KG216
o-Phenylen-diamine	135	1.0	9-(3,4-Diamino-phenyl)acridine	80 (0)	72KG216
m-Phenylen-diamine	130	1.0	9-(2,4-Diamino-phenyl)acridine	91 (1)	72KG216
o-Chloroaniline	140	4.0	9-(4-Amino-2-chloro-phenyl)acridine	25 (20)	77MI1
p-Chloroaniline	140	4.0	9-(2-Amino-5-chloro-phenyl)acridine	20 (25)	77MI1

[1]In a melt with sulfur as oxidant (molar ratio of reagents: acridine hydrochloride–arylamine–sulfur 1:2:3. [2]Yield of 9-thioacridone which is formed as a by-product is given in parenthesis.

TABLE XX
Nucleophilic Addition and Substitution Reactions
on Quinazolinium Trifluoroacetate[1]

Nucleophile	Reaction time (hours)	Adduct or S_N^H products	Yield (%)
Phenol	2	4-(4-Hydroxyphenyl)-3,4-dihydroquinazoline	89
Anisole	3	4-(4-Methoxyphenyl)quinazoline[1]	80
Resorcin[3]	0.5	4-(2,4-Dihydroxyphenyl)-3,4-dihydroquinazoline	92
Mesitylene	36	4-(2,4,6-Trimethylphenyl)-3,4-dihydroquinazoline	80
Naphthalene[4]	24	4-(1-Naphthyl)-3,4-dihydroquinazoline	12
Anthracene[4]	24	4-(9-anthracenyl)quinazoline[2]	37
Pyrene[3]	12	4-(1-Pyrenyl)-3,4-dihydroquinazoline	90
Pyrrole[5]	10	4-(2-Pyrrolyl)-3,4-dihydroquinazoline	72
Indole[3]	16	4-(3-Indolyl)-3,4-dihydroquinazoline	49

[1]In trifluoroacetic acid at room temperature All products are isolated as hydrochloride salts. [79CB1348]. [2]With $K_3Fe(CN)_6$ as oxidant. [3]In a mixture of benzene–trifluoroacetic acid (2:1) at room temperature. [4]Reflux in a mixture of benzene–trifluoroacetic acid (2:1). [5]With quinazoline hydrochloride in methanol.

All these reactions take place under oxidizing conditions; suitable oxidants are sulfur (melt) or atmospheric oxygen (Tables XIX and XX) [88T1]. Therefore, it is understandable that besides the S_N^H substitutions, the formation of by-products containing oxygen or sulfur may concurrently occur. For instance, acridine hydrochloride reacts with a great variety of arylamines in a melt with sulfur at 120–140°C, affording 9-aminoarylacridines in good yields (Scheme 86; Table XIX). In many cases the formation of minor amounts of 9-thioacridone is observed. This thione becomes the major product in the reactions with *meta*- or *para*-substituted arylamines (*para*-toluidine, *para*-phenylenediamine, *para*-chloroaniline, *meta*-nitroaniline), when electronic or steric effects of the substituents hinder or even prevent addition to a carbon atom in the NH-acridinium cation; N-coordination of these amines with the acridinium salt facilitates the formation of thioacridone (Table XIX) [77MI1]. Indeed, heating of the NH-acridinium salt with *para*-phenylenediamine or *meta*-nitroaniline at 140°C for 4 hours gave no S_N^H products; instead 9-thioacridone was obtained in 25% yield [77MI1]. Analogously, when quinoxaline hydrochloride, N,N-dimethylaniline, and sulfur are heated up to 130°C in the molar ratio 1:2:3, aminoarylation of the pyrazine ring is accompanied by subsequent thionation at C-2 (Scheme 87) [76ZOR2464].

Quinazolinium trifluoroacetate is so active toward nucleophiles that it is able to undergo C-arylation not only by phenols, pyrrole, and indoles, but also by the aromatic hydrocarbons such as naphthalene, anthracene, and mesitylene. The interesting feature of this reaction is that the intermediary 3,4-dihydroquinazolines are stable and can easily be isolated (Table XX) [76TL3537; 79CB1348]. Although this reaction is regarded as an electrophilic attack of the quinazolinium cation on the arene [76TL3537; 79CB1348], it can also be argued that this reaction involves a two-step process in which addition is the first step and in which the second step involves oxidation; potassium ferricyanide proved to be the suitable oxidant for this aromatization process (Scheme 88; Table XX) [79CB1348].

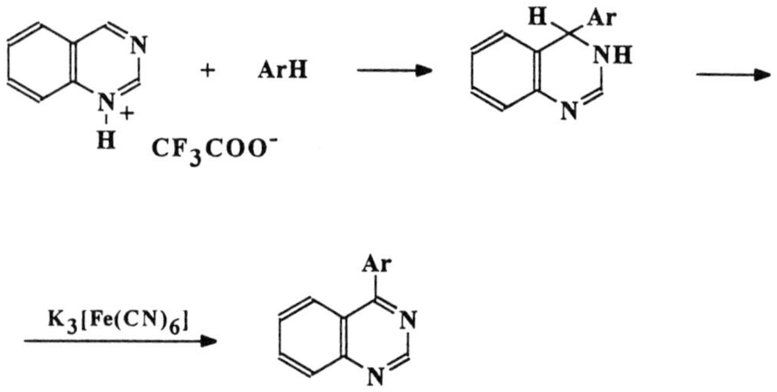

ArH = pyrrole, indole, phenols, mesitylene, naphthalene, anthracene

Scheme 88

An interesting series of papers has been published by F. Minisci and his co-workers concerning substitution of a ring hydrogen in NH-azinium cations by nucleophilic free radicals. This reaction and its modifications seem to be a general approach for functionalization of heteroaromatic bases [73S1; 84JA7146; 85T1199; 86MI1; 86MI3; 86T5973; 86TL3187; 87H731; 88S119; 89ACSA995; 89H489; 89JOC5224; 89MI1; 89TL4569; 90JHC79; 90KG579; 90T2525; 91JOC2866; 92BSB579; 92CL1295; 92TL687; 93JOC959; 93JOC4207]. Important substitution reactions, such as alkylation, hydroxymethylation, formylation, carbamoylation, or acylation of parent azaaromatics and their derivatives have been reported. The procedure involves treatment of solutions of the substrates in strong acidic medium with peroxide-induced free radicals from iodoalkanes, alkenes, methanol, formaldehyde, aldehydes, formamide, acids, dioxane, and THF (Scheme 89; Table XXI) [73S1; 84TL3897; 85T617; 85T4157; 86JOC4411; 86MI1; 86TL3187; 87JOC730; 87MI2; 88OPPI105; 89H489; 89MI2; 90JHC79; 93JOC959; 93JOC4207].

TABLE XXI
Substitution of Hydrogen in Protonated Azines
by Action of Nucleophilic Carbon-Centered Radicals

Aromatic substrate	Precursor and initiator of radical species	S_N^H Product(s)	Yield (%)	References
Pyridine	Cyclohexene, (Ph-COO)$_2$	2-(Cyclohexenyl-1)-pyridine	67	86TL3187
Pyridine	Dioxane, H$_2$O$_2$, Fe^{2+}	2-(1,4-Dioxanyl-2)-pyridine	65	87MI2
Pyridine	THF, H$_2$O$_2$	(Tetrahydrofuranyl-2)-pyridine	67	88OPPI105
Pyridine	1,3-Dioxalane, Fe^{2+}, PhC(CH$_3$)$_2$OOH	(1,3-Dioxalanyl-2)-pyridine	65	85ZOR193
Pyridine	1,3-Dioxalane, H$_2$O$_2$, Fe^{2+}	2-(1,3-Dioxalanyl-2)-pyridine	11	85ZOR193
		4-(1,3-Dioxalanyl-2)-pyridine	38	85ZOR193
4-Cyano-pyridine	Methanol, NH$_3$OSO$_3$, Fe^{2+}	2-Hydroxymethyl-4-cyanopyridine	78	85T617
4-Cyano-pyridine	•COOC$_2$H$_5$[1]	4-Cyano-2-ethoxy-carbonylpyridine	85	85AGE692
4-Methyl-pyridine	•COOC$_2$H$_5$[1]	2-Ethoxycarbonyl-4-methylpyridine	53	86T5973
4-Ethoxy-carbonyl-pyridine	•COOC$_2$H$_5$[1]	4-Ethoxycarbonyl-2-methoxy-carbonylpyridine	81	87H731

TABLE XXI (Continued)

Aromatic substrate	Precursor and initiator of radical species	S_N^H Product(s)	Yield %	References
Quinoline	Ethane-1,2-diol, $S_2O_8^{2-}$, Ag^{2+}	2-Hydroxymethyl-quinoline	36	71T3655
Quinoline	Iodocyclohexane, $Ph-N_2^+ X^-$	2-Cyclohexylquinoline 4-Cyclohexylquinoline	52	86JOC4411
2-Methyl-quinoline	Methanol, NH_3OSO_3	4-Hydroxymethyl-2-methylquinoline	90	85T4157
Isoquino-line	Iodoethane, $Ph-N_2^+ X^-$	1-Ethylisoquinoline	85	86JOC4411
Acridine	Iodocyclohexane, $(Ph-COO)_2$	9-Cyclohexylacridine	94	86JOC4411
Acridine	Benzaldehyde, t-BuOOH	9-Benzoylacridine	70	87MI2
4-Ethoxy-carbonyl-pyridazine	•$COOC_2H_5$[1]	4,5-Diethoxycarbonyl-pyridazine	65	85T1199
Pyrimidine	•$COOC_2H_5$[1]	4-Ethoxycarbonyl-pyrimidine	54	89MI1
Pyrazine	•$COOC_2H_5$[1]	2-Ethoxycarbonyl-pyrazine	89	86T5973

[1] According to the two-phase (methylene chloride/water) procedure with generation of the radical species from ethyl (methyl) pyruvate by action of hydrogen peroxide in the presence of $FeSO_4 \cdot 7H_2O$ [89MI1].

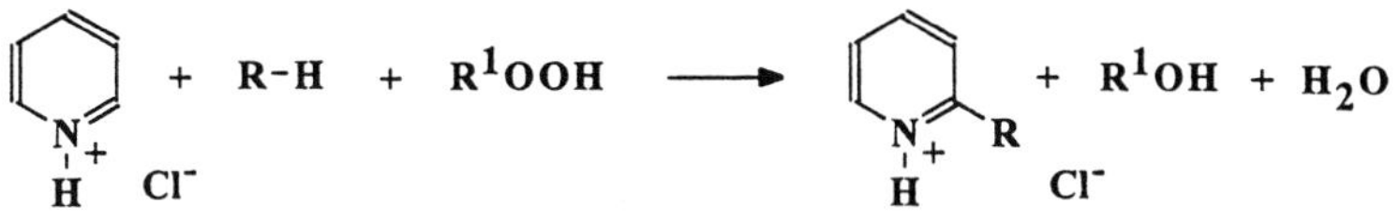

Scheme 89

R = CH_2OH, CHO, $CONH_2$, COR^2

Scheme 89

An example of a selective alkylation by olefins in the presence of benzoyl peroxide is the C-2 alkenylation of NH-quinolinium salts (Scheme 90) [86TL3187].

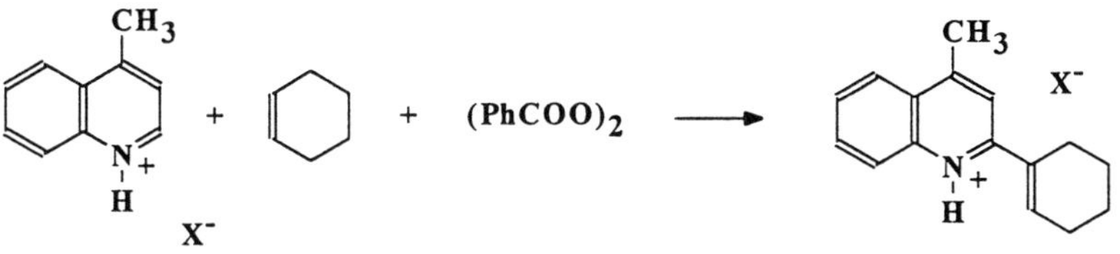

Scheme 90

A plausible mechanism for this Minisci-type nucleophilic substitution of ring hydrogen in protonated azaaromatics is characterized by three steps (Scheme 91) [89H489]:

1. Generation of carbon-centered radicals by action of benzoyl peroxides, persulfate, or hydroxylamine-O-sulfonic acid in the presence of iron (II) sulfate (3%) or other initiators of radical species;

2. Addition of nucleophilic radicals to α- or γ-positions of the protonated pyridine ring and subsequent deprotonation of the radical cation formed;

3. Aromatization of the dihydroazinyl radicals in an oxidation step (Scheme 91).

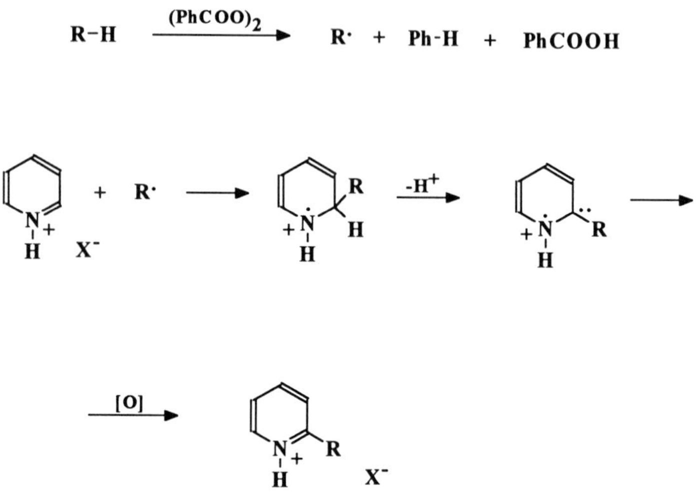

Scheme 91

The reaction has been extended to protonated pyrimidines [78RTC159; 89MI1], pyrazines [86T5973], and pyridazines [85T1199; 87H731; 89MI1] (Table XXI). A disadvantage of the procedure described above is that in many cases the regioselectivity is not so high as desirable, and that further substitution reactions cannot be avoided, since the product formed is again protonated in the strong acidic medium and therefore becomes an appropriate substrate for a further substitution by the highly reactive carbon-centered nucleophilic free radicals. In order to reach better regioselectivity and to suppress subsequent substitution, a very convenient two-phase-system procedure has been developed [87H731; 89JHC1751; 89MI1]. A two-phase system using dichloromethane–water has proved to be very suitable to prevent further substitution since the organic solvent used as the second layer removes the initially formed monosubstituted product from the acidic reaction medium (Scheme 92; Table XXI) [89MI1; 91T8573].

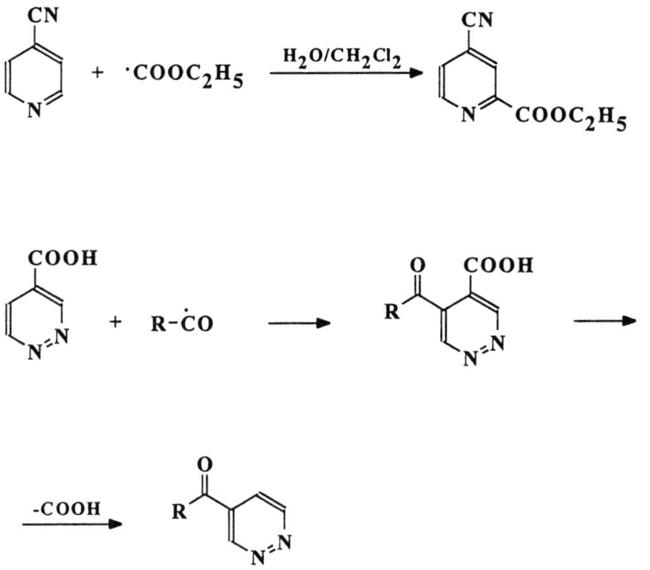

Scheme 92

It is sometimes very convenient to carry out the reaction with carboxydiazines. These compounds are highly reactive and after an S_N^H substitution they can easily be decarboxylated. Based on this methodology an elegant two-step synthesis of aroylated pyrazines and pyridazines has been developed (Scheme 92) [88S119; 89MI1; 92BSB579].

A rare and interesting example of the nucleophilic halogenation reaction on an aromatic system is the displacement of hydrogen H-10 in cinnolino[3,4-*b*]quinoxaline **112** on treatment with hydrochloric or hydrobromic acid. The reaction seems to involve protonation of **112** at N-12 followed by the addition of chloride (bromide) at C-10 of the NH-azinium salt **113**. Air oxidation of the adduct **114** gives the final S_N^H product **115** (Scheme 93) [87JCS(P1)507].

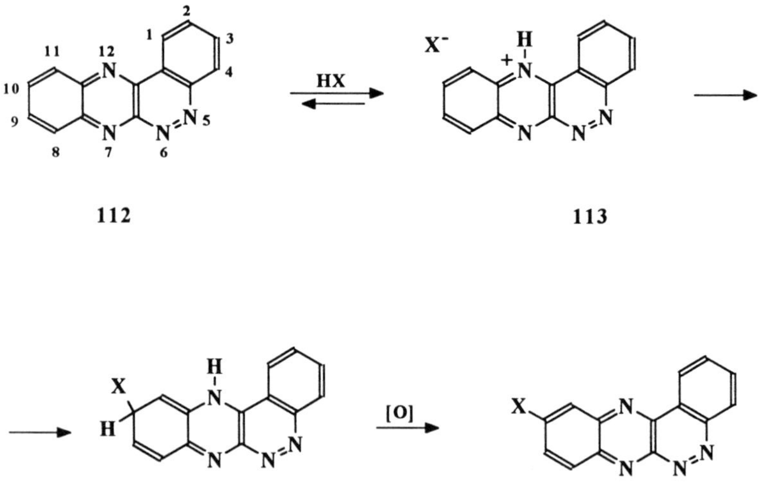

Scheme 93

112 ⇌ (HX) **113** → **114** → [O] → **115**

X = Cl, Br

Experimental Procedures

Example 1. A typical experimental procedure for alkylation of heteroaromatic bases by nucleophilic free radicals. Preparation of 2-ethyl-4,6-dimethylpyrimidine from 4,6-dimethylpyrimidine

4,6-Dimethylpyrimidine (216 mg, 2 mmol), propionic acid (740 mg, 10 mmol) and silver nitrate (31 mg, 0.2 mmol) were dissolved in a solution of 2 mmol of sulfuric acid and 20 ml of water. After being heated to 70–80°C, a solution of ammonium persulfate

(6 mmol) in water (10 mmol of ammonium peroxydisulfate was dissolved in 5 ml of water) was added during 1 hour with stirring at the 70–80°C. After the evolution of carbon dioxide had finished, heating was continued for an additional half hour. Then the solution was poured onto ice, and the resultant mixture was made alkaline with dilute aqueous sodium hydroxide and then extracted with chloroform. The extracts were dried ($MgSO_4$), the solvent was evaporated, and the residue was distilled *in vacuo* to give 2-ethyl-4,6-dimethylpyrimidine as an oil, boiling point 73–75°C/13 mm. Yield 196 mg (72%) [78RTC159].

Example 2. A typical procedure for α-hydroxylation of heteroaromatic bases by alcohols and hydroxylamine-O-sulfonic acid. Preparation of 2-hydroxymethyl-4-methyl-quinoline from lepidine

In a 25 ml flask equipped with magnetic stirrer reagents were introduced in the following order: lepidine (570 mg, 4 mmol), methanol (20 ml), water (15 ml), concentrated sulfuric acid (0.20 ml, 4 mmol), and eventually $FeSO_4 \cdot 7H_2O$ (1.16 g, 4 mmol). The solution was flushed with nitrogen for 5 minutes and poured in a thermostatic bath (20°C) for 10 minutes. Then hydroxylamino-O-sulfonic acid (450 mg, 4 mmol) was added. Gas evolution was observed during the reaction which was run for 4 hours. The reaction mixture was transferred in a separatory funnel and washed with water (10 ml). Sodium citrate (300 mg) was added and the solution basified at pH = 10 with 30% NH_3 solution and then extracted with methylene chloride (4 x 10 ml). The combined extracts were washed with water (10 ml) and dried. Separation from unconverting lepidine (280 mg, 49%) by means of thin layer chromatography resulted in 2-hydroxymethyl-4-methylquinoline, melting point 85°C. Yield 280 mg (81%) [85T617].

Example 3. Preparation of ethyl 2-pyrazinecarboxylate from pyrazine

A 30% solution of hydrogen peroxide (3.4 g, 30 mmol) was added with stirring to ethyl pyruvate at $-10^{o}C$ to $0^{o}C$. This solution was then added with stirring and cooling ($-5^{o}C$ to $0^{o}C$) to a mixture of pyrazine (800 mg, 10 mmol), concentrated sulfuric acid (3 g), water (8 ml), $FeSO_4 \cdot 7H_2O$ (8.3 g, 30 mmol), and 150 ml of methylene chloride. After 15 minutes of further stirring, the resulting mixture was poured into water at $2-3^{o}C$ and the aqueous phase was exhaustively extracted with methylene chloride. After drying the combined organic layer over anhydrous sodium sulfate, the solvent and excess of ethyl pyruvate were removed *in vacuo* to give spontaneously crystallizing pale yellow needles of ethyl 2-pyrazinecarboxylate, melting point $49-51^{o}C$. Yield 1.35 g (89%) [86T5973].

Example 4. Preparation of 2-cyclohexenyl-4-methylquinoline from lepidine

A solution of lepidine (300 mg, 2.1 mmol), trifluoroacetic acid (250 mg, 2.1 mmol) and benzoyl peroxide (830 mg, 6 mmol) in 20 ml of cyclohexene was refluxed for 40 hours. After addition of 50 ml of 5% aqueous sulfuric acid the reaction mixture was extracted with ethyl acetate. The aqueous solution was made basified with 10% sodium hydroxide and extracted with ethyl acetate. Evaporation of the solvent followed by column chromatography on silica gel with hexane–ethyl acetate (8:2) gave recovered lepidine (132 mg) and 2-cyclohexenyl-4-methylquinoline, oil, 193 mg (42%) [86TL3187].

Example 5. Preparation of 9-(4-aminophenyl)acridine from acridine hydrochloride

A mixture of 2.16 g (10 mmol) of acridine hydrochloride, 1.96 g (20 mmol) of aniline and 960 mg (30 mmol) of sulfur was stirred at 130°C, until the reaction mass no longer thickened and hydrogen sulfide evolution decreased appreciably (approximately 2 hours). The mixture was then cooled, and the resinous melt was washed with ether (100 ml) and treated three times with 10% hydrochloric acid (3 x 40 ml). The acid extracts were neutralized with ammonia, and the resulting precipitate was filtered, dried, and recrystallized from propanol to give 9-(4-aminophenyl)acridine, melting point 269–270°C. Yield 2.58 g (95%) [71KG112].

Example 6. Preparation of 4-(9-anthracenyl)quinazoline from quinazoline

Step 1. Quinazoline (1.04 g, 8 mmol), anthracene (1.42 g, 8 mmol), and trifluoro-acetic acid (910 mg, 8 mmol) were kept under reflux for 43 hours in 80 ml of benzene–tetrahydrofuran mixture (2:1). The solvents were evaporated *in vacuo* nearly to dryness and the residue was treated with absolute ethanol. Evaporation of ethanol at 60°C gave a solid product which, after treatment with acetone, was filtered off, giving 4-(9-anthracenyl)-3,4-dihydroquinazoline trifluoroacetate, melting point 230°C. Yield 2.37 g (73 %).

Step 2. Suspension of 4-(9-anthracenyl)-3,4-dihydroquinazoline trifluoroacetate (800 mg, 1.9 mmol) in 12 ml of water and 6 ml of aqueous 33% potassium hydroxide solution was treated with a solution of potassium ferricyanide (1.6 g, 5 mmol) in 10 ml of water. The reaction mixture was allowed to stay at room temperature for 5 days to yield a precipitate which was filtered off, washed with water, dissolved in methanol, and passed through Sephadex LH 20 column with methanol as eluent. Evaporation of methanol till the volume of 5 ml gave 4-(9-anthracenyl)quinazoline, melting point 160ºC. Yield 237 mg (47%) [79CB1348].

B. *N-Alkyl and N-Arylazinium Salts*

NH-Azinium salts have the disadvantage that in a S_N^H substitution process they can easily undergo NH proton transfer deactivating the azine system. The N-alkyl and N-arylazinium salts, on the contrary, are rather stable and, unlike the NH, N-acyl, N-cyano, and N-fluoro analogues, usually do not lose the N-alkyl(aryl) substituent in reactions with a variety of nucleophilic reagents [77MI1; 80H2015; 85MI1; 86CCA33; 87MI1; 88MI1; 88OPPI591; 88T1; 92H931]. It is a common point of view that N-alkylation of azaaromatic compounds is an irreversible process [78AHC71; 85H1513].

Nucleophilic alkylation, arylation, amination, aminoarylation, hydroxylation, nitromethylenation, and other types of S_N^H reactions have been successfully applied to a great number of N-alkyl or N-arylazinium salts.

1. *Reactions with O-Nucleophiles.* The formation of N-alkylazinone when quaternary N-alkylazinium salts and the hydroxide ion react in the presence of an oxidant is well documented in the literature [48JA2172; 60CB1579; 68JHC561; 71JCS(B)131; 73JCS(P1)2588; 74JCS(CC)308; 76CPB1813; 77CB3561; 78JOC3662; 79AHC1; 79MI2; 80ZOR2390; 87CPB3628; 89SC3523]. Many oxidizing agents (potassium ferricyanide, hydrogen peroxide, or potassium permanganate) can be used in this reaction, however potassium ferricyanide appears to be the superior one for the oxidative hydroxylation of N-alkylpyridinium [71JCS(B)131; 86JOC2068; 87CPB3628], N-alkylquinolinium [89SC3523], N-alkylisoquinolinium [73JCS(P1)2588; 78JOC3662; 86JOC2060], and N-alkylnaphthyridinium salts [68JHC561] (Scheme 94; Table XXII).

TABLE XXII
Oxidative Hydroxylation of N-Alkylazinium Salts with Aqueous Sodium Hydroxide/$K_3Fe(CN)_6$

Azinium salt	S_N^H Product	Yield (%)	References
1-Methylpyridinium iodide	1-Methyl-2-pyridone	49 20[1]	79MI2
1,3-Dimethylpyridinium iodide	1,3-Dimethyl-2-pyridone	70	79MI2
1-Ethyl-3-methyl-pyridinium iodide	1-Ethyl-3-methyl-2-pyridone	45	80ZOR2390
3-Methyl-1-*n*-propyl-pyridinium iodide	3-Methyl-1-*n*-propyl-2-pyridone	52	77CB3561
2-(2-Hydroxyethyl)-isoquinolinium iodide	2-(2-Hydroxyethyl)-1-isoquinolone	64	60CB1579

[1]With potassium permanganate as oxidant.

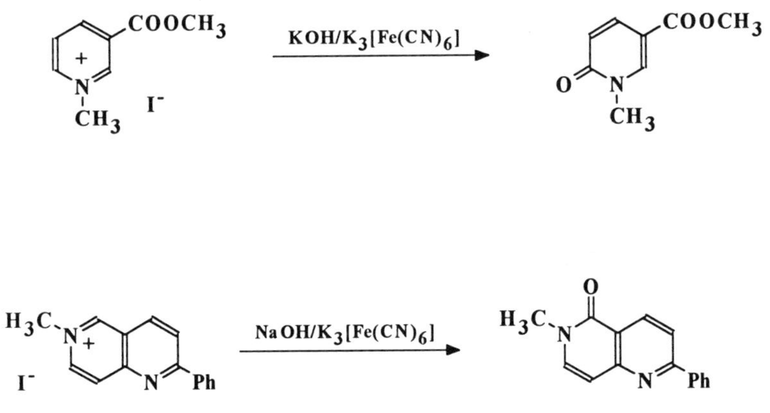

Scheme 94

There is also reported the successful use of an oxidizing enzyme isolated from rabbit liver to convert the 3-aminocarbonyl-1-methylpyridinium ion into 3-aminocarbonyl-1-methyl-6-pyridone [48JA2172].

2. *Reactions with N-Nucleophiles.* The replacement of the ring hydrogen in N-alkylazinium salts by the imino group, which can be performed in liquid ammonia in the presence of potassium permanganate, ferric nitrate, or other oxidants, is a relatively new method [88T1; 86CCA33]. This version of the Chichibabin amination has been successfully applied to 1-alkylpyridinium [84TL3763; 86JHC1015], 1-methylquinolinium [85JHC765; 87KG1011], 1-methylpyrimidinium [87JHC1377], 1-methyl-1,5-naphthyridinium [85JHC765], 1-methyl-1,8-naphthyridinium [85JHC765], 7-methyl-1,7-naphthyridinium salts [85JOC3435], and other N-alkylazinium salts (Scheme 95; Table XXIII, for a review see [87KG1011]).

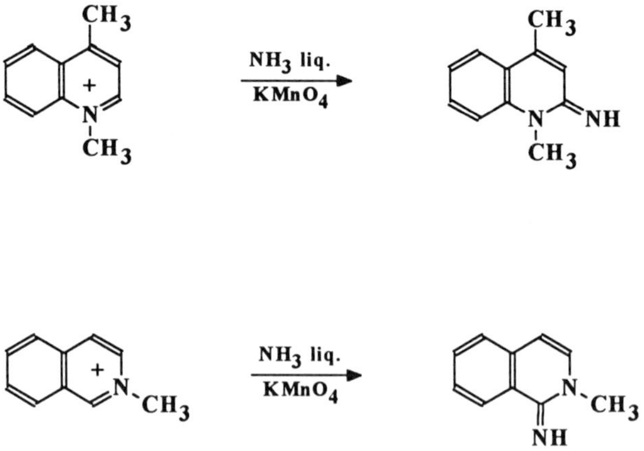

Scheme 95

TABLE XXIII
Imino-Dehydrogenation of N-Methylazinium Salts
in Liquid Ammonia with Potassium Permanganate

Azinium salt	S_N^H Product	Yield (%)	References
3-Aminocarbonyl-1-methylpyridinium iodide	3-Aminocarbonyl-1,6-dihydro-6-imino-1-methylpyridine	75	84TL3763
3-Phenyl-1-methyl-pyridinium iodide	1,6-Dihydro-3-phenyl-6-imino-1-methylpyridine	80	86JHC1015
1-Methylquinolinium iodide	1,2-Dihydro-2-imino-1-methylquinoline	90	85JHC765
1,4-Dimethyl-quinolinium iodide	1,2-Dihydro-2-imino-1,4-dimethylquinoline	60	85JHC765
7-Methyl-1,7-naphthyridinium iodide	7,8-Dihydro-8-imino-7-methyl-1,7-naphthyridine	15	85JOC3435
1-Methyl-1,8-naphthyridinium iodide	1,2-Dihydro-2-imino-methyl-1,8-naphthyridine	17	85JHC765

Extensive nuclear magnetic resonance (NMR) investigations on the amino adducts derived from N-alkylpyridinium, isoquinolinium, pyrimidinium, or pyrazinium salts and liquid ammonia have shown that the addition occurs quantitatively at the position adjacent to the quaternary nitrogen, yielding 1-alkyl-2-amino-1,2-dihydroazines [73JOC1949; 74RTC114; 75JA5531; 76JOC1303; 85JHC765; 85JOC3435; 86JHC1015; 87JHC1377; 87KG1011]. These aminodihydroazines are easily oxidized by potassium permanganate to the corresponding imino derivatives. This procedure of amination and oxidation is very easy and reproducible, and gives reasonable-to-good yields; it therefore forms a new synthetic and easily applicable tool for the preparation of N-alkyl-substituted iminoazines (Scheme 96; Table XXIII) [87KG1011; 88T1].

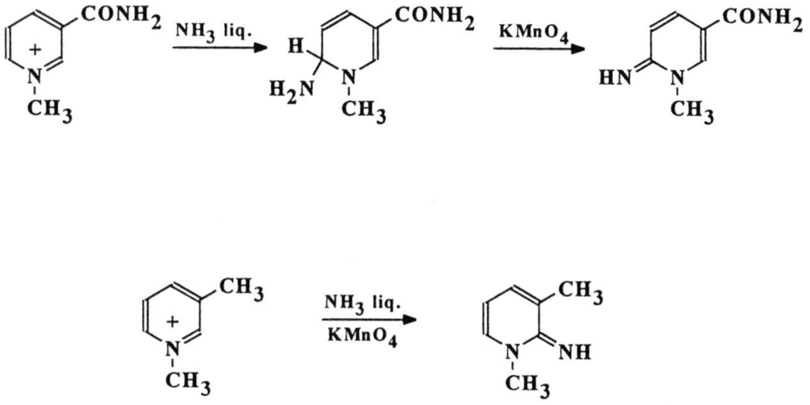

Scheme 96

Regioselectivity of the imino-dehydrogenation reactions depends on the nature of substituents present in the ring. In the reaction of 3-aminocarbonyl-1-methylpyridinium iodide with the liquid ammonia–potassium permanganate system the addition of ammonia takes place exclusively at C-6 of the pyridine ring due to the stabilizing effect of the electron-withdrawing group at C-3, while amination of 1,3-dimethylpyridinium iodide under the same oxidative conditions results in the corresponding 1,3-dimethyl-2-imino-1,2-dihydropyridine (Scheme 96) [86JHC1015; 87KG1011].

The S_N^H substitution in 10-methylacridinium salts by action of tertiary alkylamines is accompanied by the dealkylation of one alkyl group, as exemplified by the reaction of the N-methylacridinium iodide with N,N-diethylbenzylamine, yielding 9-(N-ethyl-N-benzylamino)-10-methylacridinium iodide **121** (29%) together with 10-methyl-9,10-dihydroacridine (54%) (Scheme 97) [92BCJ55]. The reaction is supposed to be initiated by the formation of σ^H-adduct **116** which dissociates into the radical cation **117** and the acridinyl radical **118**. The hydrogen atom transfer from **116** to **118** would lead to the acridan **119** and the radical cation **120**. Subsequent release of one of the alkyl substituents as free radical furnishes the aminoacridinium salt **121** (Scheme 97) [92BCJ55].

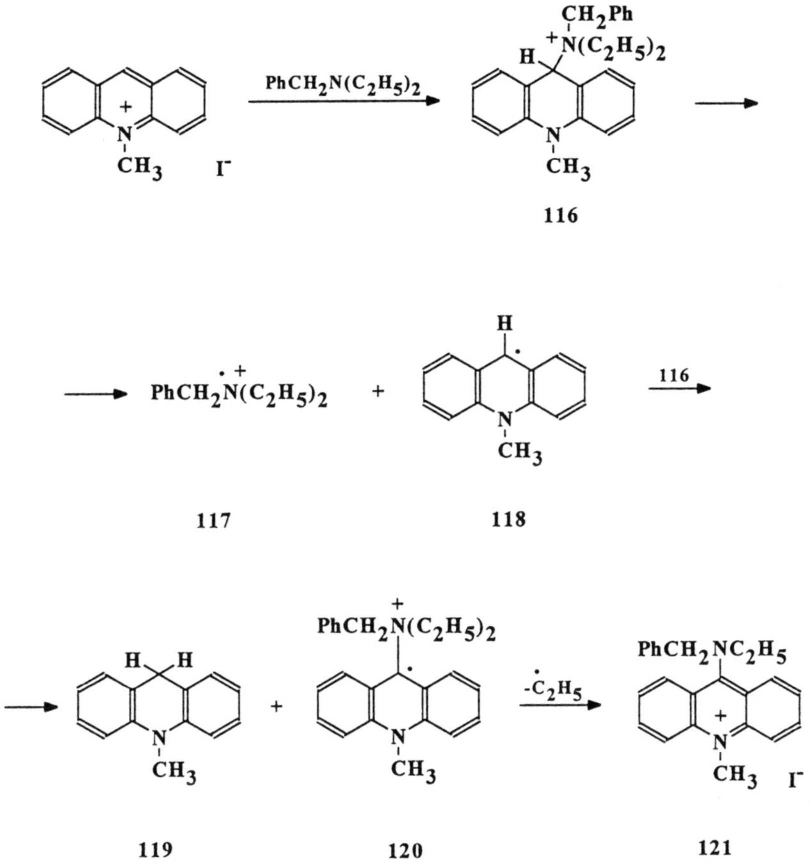

Scheme 97

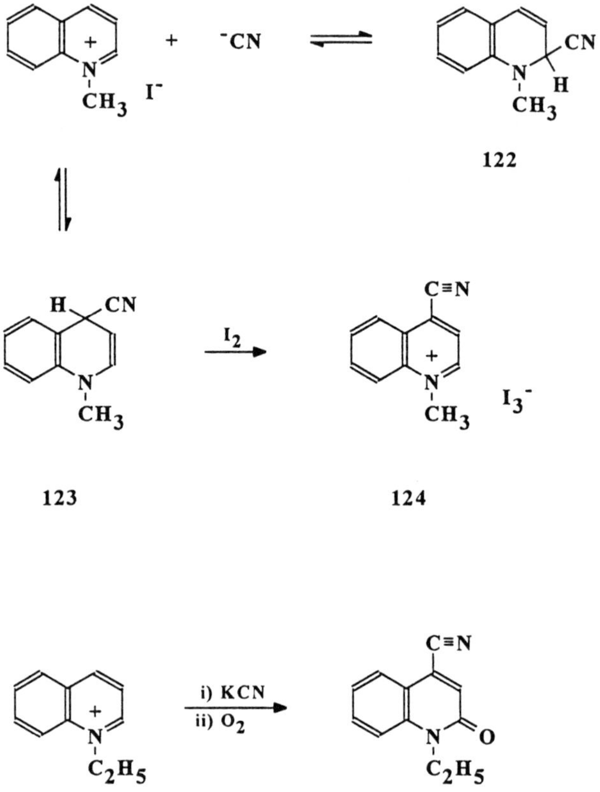

122

123

124

125

Scheme 98

3. *Reactions with C-Nucleophiles*. The cyanide ion easily adds to N-alkylpyridinium, quinolinium, and isoquinolinium ions to give 2- or 4-cyano adducts [09CB3776; 65TL4615; 72JCS(P1)2918; 81ZOR418; 84H2375]. The site selectivity was shown to be dependent mainly on the reaction temperature [81ZOR418]. A classical example is the reaction of N-methylquinolinium iodide with potassium cyanide, affording 2-cyano-1-methyl-1,2-dihydroquinoline **122** under kinetically controlled conditions (at –70 to –30°C), while 4-cyano-1-methyl-1,4-dihydroquinoline **123** is formed at temperatures

above −20°C (Scheme 98). The latter can be isolated and oxidized by iodine into 4-cyano-1-methylquinolinium triiodide **124** [81ZOR418]. It is of interest to note that air oxidation of the adduct derived from N-ethylquinolinium iodide and potassium cyanide gave 4-cyano-1-ethyl-2-quinolone **125** (Scheme 98) [09CB3776].

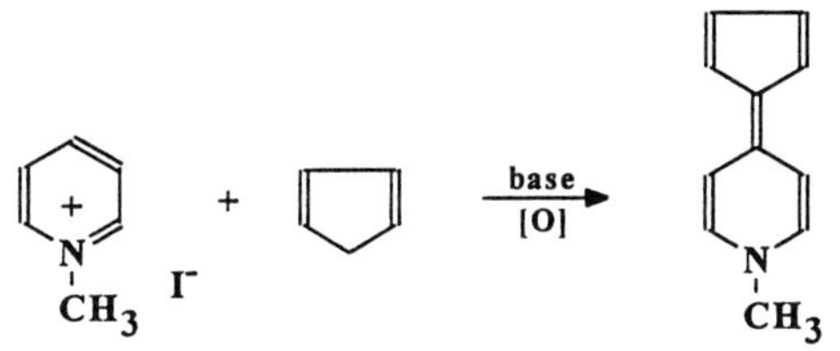

Scheme 99

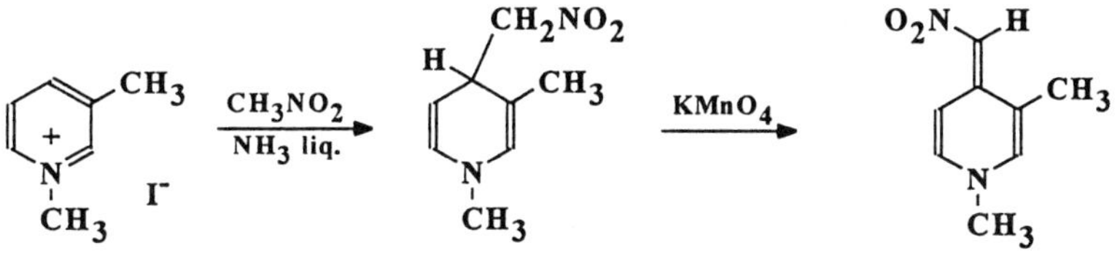

N-Alkylpyridinium [51JA3325; 56LA176; 70JCS(C)800; 81CCC503], quinolinium [56LA176; 51JA3325; 81CCC503], isoquinolinium [51JA3325], quinazolinium [71MI1], and acridinium [76H987] cations can also undergo smooth S_N^H substitutions, when reacting with carbanions generated from CH-active compounds, under oxidative conditions. For instance, base-induced carbanions derived from cyclopentadiene or indene by action of sodium methoxide react with N-methylpyridinium iodide into the S_N^H products **126** (Scheme 99); fluorene, however, proved to be unreactive under the same conditions and was recovered unchanged [70JCS(C)800]. Analogously, treatment of N-methylpyridinium salts with nitromethane in liquid ammonia in the presence of potassium permanganate gives 4-nitromethylenepyridines **127** (Scheme 99) [88OPPI591; 90JOC778].

Another example is given in Scheme 100. Since in the compounds **128** the substituent at C-4 contains an acid hydrogen, they are usually deprotonated in the presence of a base, yielding the quinoid anhydrobases **129** (Scheme 100).

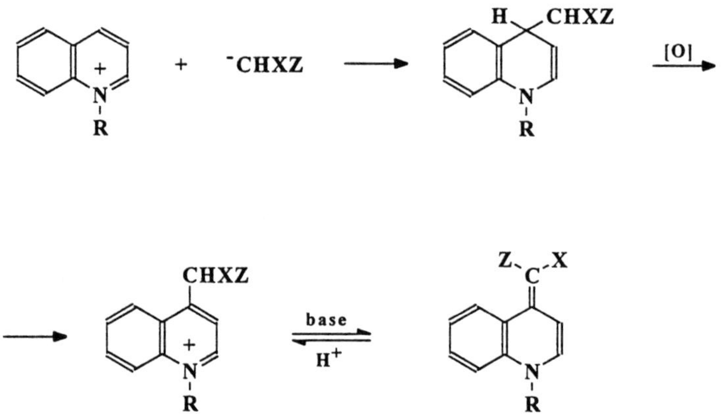

X, Z = COR, CN, NO$_2$, Ar; R = alkyl

Scheme 100

Alkylacridinium salts are prone, as can be expected, to undergo aminoarylation by action of arylamines. The reaction of N-methylacridinium iodide with arylamines enables the introduction of a variety of aminoaryl substituents (Scheme 101; Table XXIV) [76RCR454; 77MI1].

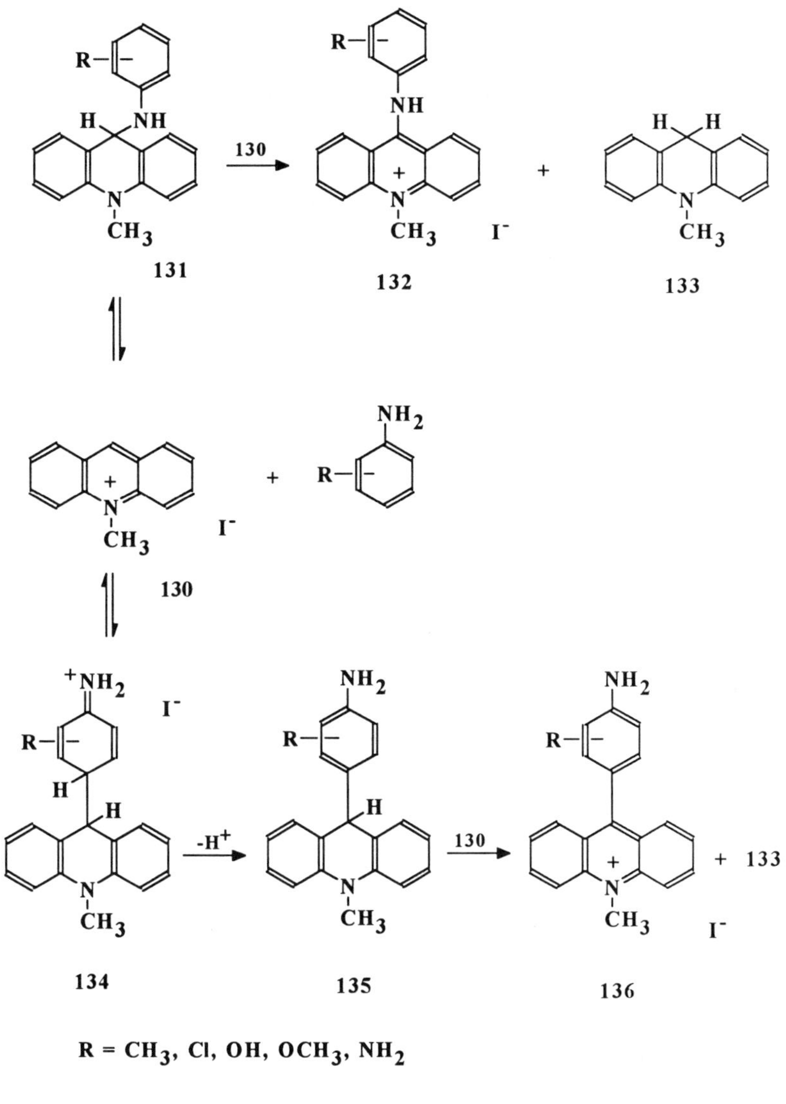

131 132 133

130

134 135 136

R = CH_3, Cl, OH, OCH_3, NH_2

Scheme 101

TABLE XXIV
Nucleophilic Substitution of Hydrogen
in the N-Methylacridinium Cation by Action of Arylamines[1]

Arylamine	Reaction Temp. (°C)	Reaction Time (hours)	Aryl in 9-substituted 1-methylacridinium salts	Yield (%)
Aniline	130	1.5	4-Aminophenyl	90
N-Methylaniline	120	2.0	4-N-Methylaminophenyl	94
N,N-Dimethylaniline	130	1.5	4-N,N-Dimethylaminophenyl	91
N,N-Diethylaniline	120	1.5	4-N,N-Diethylaminophenyl	75
o-Toluidine	130	2.0	4-Amino-3-methylphenyl	90
o-Anisidine	130	1.5	4-Amino-3-methoxyphenyl	91
o-Aminophenol	130	1.5	4-Amino-3-hydroxyphenyl	85
o-Phenylenediamine	130	2.0	3,4-Diaminophenyl	85
m-Toluidine	120	1.5	4-Amino-2-methylphenyl	65
m-Chloroaniline	140	3.0	4-Amino-2-chlorophenyl	63
m-Aminophenol	120	1.5	4-Amino-2-hydroxyphenyl	92
1-Naphthylamine	120	1.5	1-Aminonaphthyl-2	95

[1] In a melt with sulfur. Molar ratio between N-methylacridinium iodide, arylamine, and sulfur 1:2:3 [76RCR454; 77MI1]

The reaction of N-methylacridinium iodide with arylamines has been the subject of extensive kinetic and NMR studies [76KG1227; 78ZOR140; 79ZOR117; 79ZOR206]. The ambident character of arylamines allows the formation of two types of σ-adducts due to both N- and C-addition. UV and NMR studies of the reaction between the

N-methylacridinium ion **130** and primary arylamines have revealed the formation of very unstable N-adducts **131** under kinetically controlled (–50°C) conditions (Scheme 101) [79ZOR206]. At temperatures above 0°C the N-adducts **131** are converted into the thermodynamically more favored C-adducts **135**, apparently via the dissociation and readdition mechanism involving intermediate **134** (Scheme 101).

NMR studies were undertaken dealing with the kinetics of the decrease of the starting material N-methylacridinium cation **130**, the formation of the intermediate C-adduct **135** and of the final product **136**. The results of these studies, combined with the observation of a small kinetic isotope effect (about 2.2) for the aminophenylation reaction with 2,4,6-D_3-aniline and the absence of base catalysis, revealed that the abstraction of a proton from the arylamine moiety in **134** does not occur simultaneously with the addition process; neither does it occur simultaneously with the departure of the hydride ion (Scheme 101) [76KG1227; 78ZOR140; 79ZOR117].

It has also been shown that, when no oxidant is added, the N-methylacridinium cation **130** acts as oxidant of the intermediate σ^H-adducts **131** and **135**. 9-Arylaminoacridinium salts **132**, 9-aminoarylacridinium salts **136**, and 10-methyl-9,10-dihydroacridine **133** are formed due to these hydrogen transfers (Scheme 101) [78ZOR140; 77MI1; 82TL3677].

Indolo-dehydrogenation of N-alkylpyridinium [84KG1383], and N-alkylquinoxalinium [75KG1433] salts by interlinking C-3 of the indole with C-2 (C-6) of the azinium salts is another good example of nucleophilic aromatic substitution of hydrogen in these salts. The reaction occurs when heating the reagents in DMF (120–150°C) in the presence of air oxygen (Scheme 102) [76KG1146; 78ZOR431; 89H237]. Interestingly, the addition of indole to pyrido[2,3-*b*]pyrazinium salts [78ZOR431] does not take place in the quaternized ring, but in the pyrazine ring. It leads to a 1:2 adduct which by heating in air oxygen aromatizes but under loss of one indole molecule. It could not be established thus far which of the two indole groups is lost [78ZOR431].

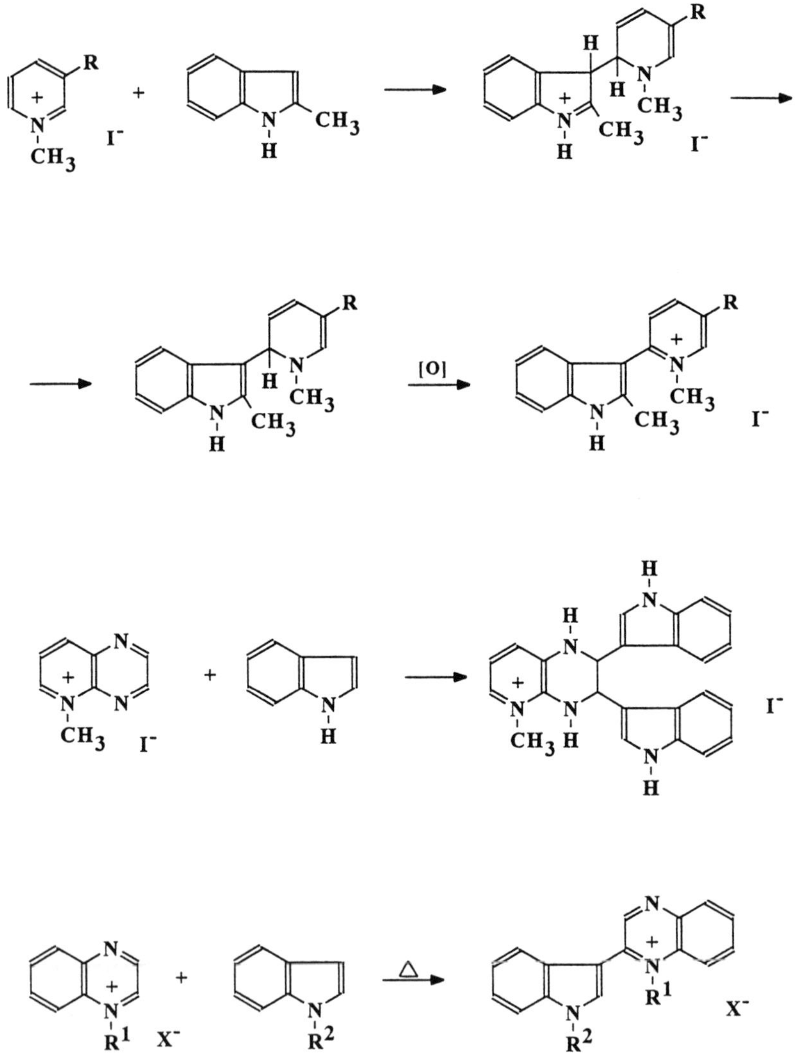

$R^1 = CH_3, C_2H_5; R^2 = H, CH_3; X = I, Ts$

Scheme 102

From the data discussed above it is already clear that the behavior of N-alkylazinium cations in S_N^H reactions is similar to that of their NH-analogues. Therefore, it is not surprising that also N-alkylquinolinium salts are able to undergo the hydroxymethylation reaction by free radicals (Scheme 103), although only a few papers on this subject have so far been published [86BCJ3905; 86BCJ3911] (see Chapter 3,II,A). The product obtained from the hydroxymethylation reaction on 1,4-dimethylquinolinium methylsulfate **137** is, however, not 2-hydroxymethyl-4-methylquinolinium salt **140**, but its dehydration product, the 1,2,4-trimethylquinolinium salt **141** [86BCJ3905; 86BCJ3911].

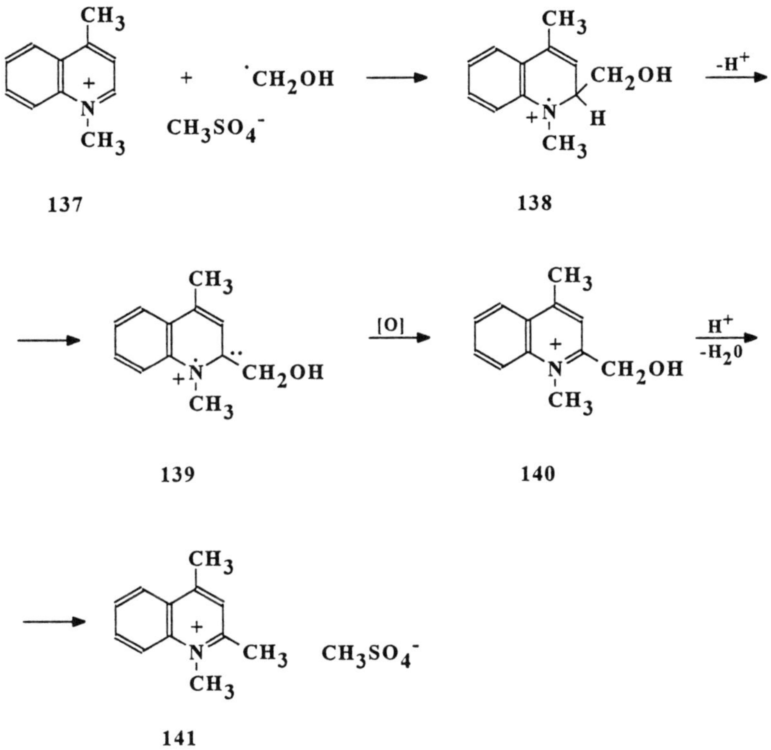

Scheme 103

There are examples known in which it is observed that hydrogen in a benzene ring being annelated to an N-alkylazinium salt is more easily substituted than halogen or another nucleofugal group. For instance, it is the hydrogen atom at C-2 of the 4-chloro-9-methylphenazinium salt **142** that is substituted by CH-active methylene compounds and not the chlorine atom (Scheme 104) [73MI1]. It is of interest that the reaction between the isomeric 2-chloro-9-methylphenazinium salt **143** and morpholine is very sensitive to the nature of solvent: in ethanol exclusively the S_N^H substitution at C-7 takes place, while in the more polar water solution only the $S_N(Ar)$ replacement of halogen at C-2 is observed (Scheme 104) [72KG1425].

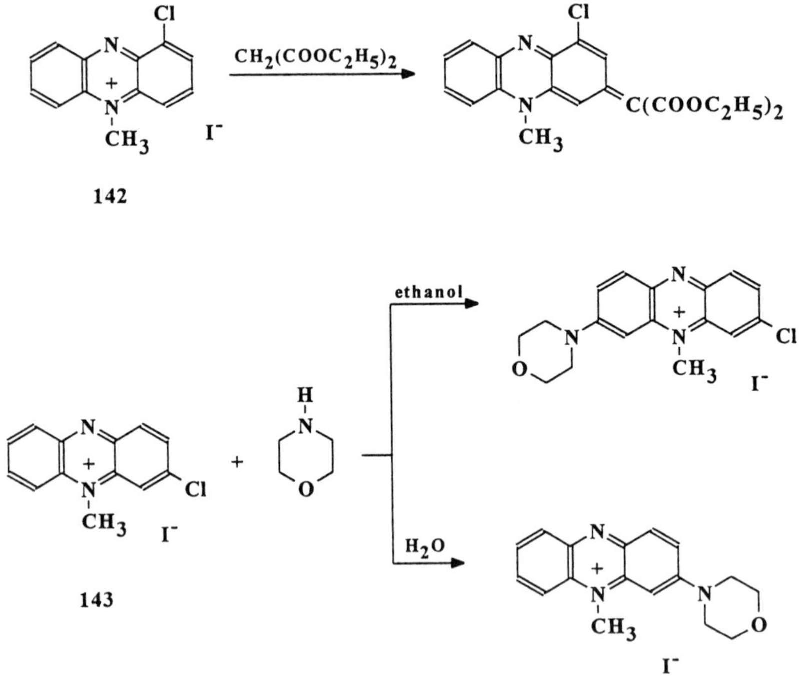

Scheme 104

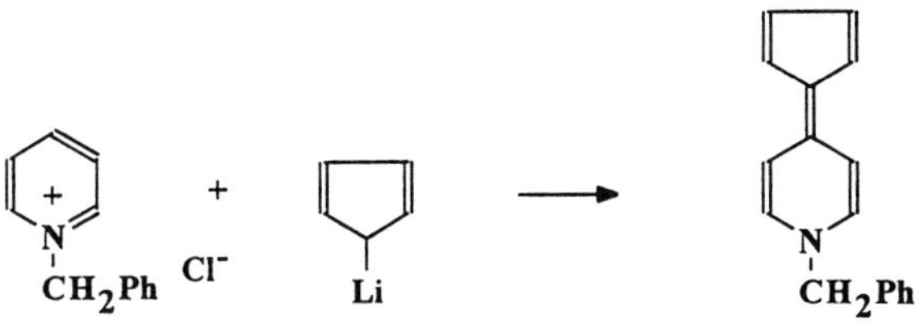

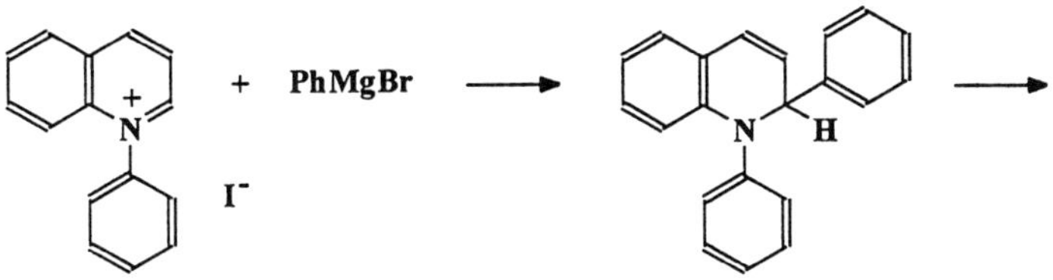

Scheme 105

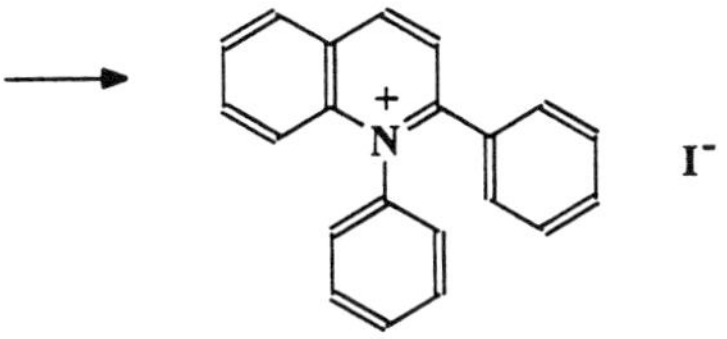

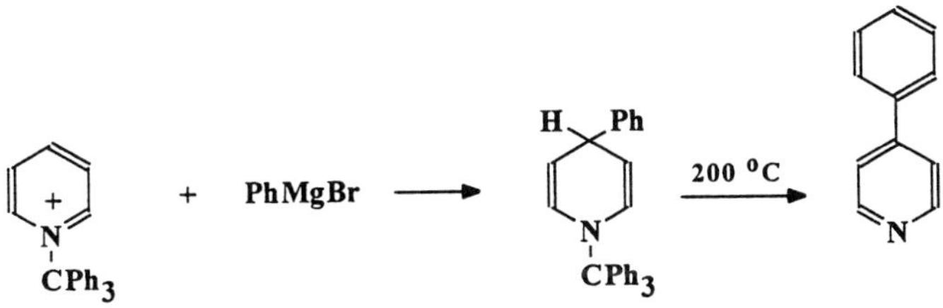

4. *Reactions with Organometallic Compounds.* Organolithium compounds were found to be effective reagents for the introduction of both alkyl and aryl residues into N-alkylazinium salts, as illustrated by the reaction of N-benzylpyridinium chloride with cyclopentadienyllithium (Scheme 105) [57JOC1370; 61MI1; 69TL433].

Unlike the organolithium compounds, the addition of alkyl and aryl Grignard reagents to N-methylpyridinium [71JOC772], N-methylquinolinium [66JA3376], and N-phenylquinolinium [78T363] salts has been shown to occur at α-position relative to the quaternary nitrogen. For instance, the 1,2-dihydroquinoline **145** is obtained from the N-phenylquinolinium salt **144**; oxidation *in situ* by iodine yields 1,2-diphenylquinolinium iodide **146** (Scheme 105) [78T363]. In the reaction of N-triphenylmethylpyridinium salt **147** with phenyl magnesium bromide the addition at C-2 is sterically hindered due to the presence of the bulky trityl group; therefore the reaction takes place at C-4 (Scheme 105). Another feature of the reaction is that the aromatization step proceeding at 200°C is accompanied by the removal of the trityl group to give 4-phenylpyridine (Scheme 105) [71JOC772].

5. *Reactions with S- and P-Nucleophiles.* Reactions of N-alkylpyridinium [80JCS(CC)1147] and N-alkylacridinium [77S862] salts with S- and P-nucleophiles, such as alkyl(aryl)thiolates and trialkylphosphites, result in the formation of the corresponding adducts **148** and **149** (Scheme 106); the dehydrogenation step has so far not been realized.

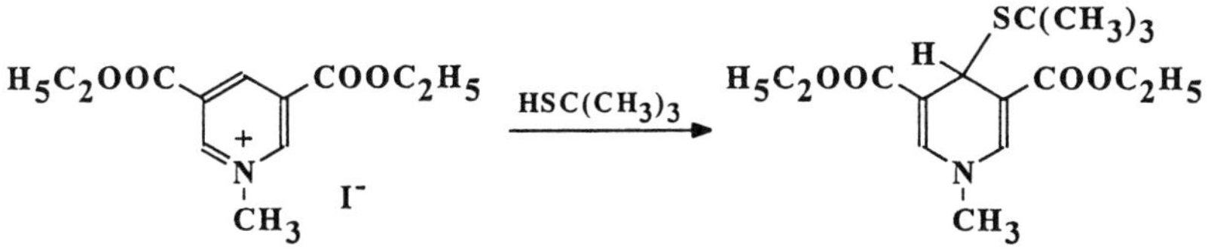

Chapter 3 S_N^H REACTIONS IN HETARENES

[Scheme showing reaction of diethyl 1-methylpyridinium-3,5-dicarboxylate iodide with HSC(CH$_3$)$_3$ to give compound **148**]

[Scheme showing reaction of 10-ethylacridinium iodide with P(OCH$_3$)$_3$ to give compound **149**]

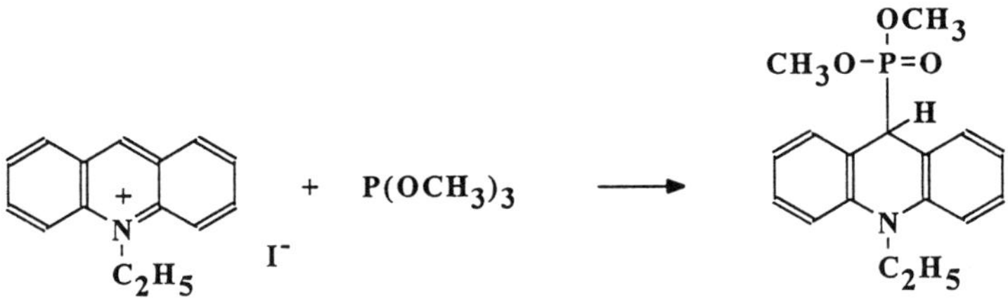

Scheme 106

Experimental Procedures

Example 1. A typical example for oxidative nitromethylation of N-alkylazinium salts. Preparation of 1,4-dihydro-1,3-dimethyl-4-nitromethylenepyridine

To a suspension of 570 mg (4 mmol) of 1,3-dimethylpyridinium chloride and 490 mg (8 mmol) nitromethane in 25–30 ml of liquid ammonia was added 0.65 g (4 mmol) of potassium permanganate. After stirring this suspension at –40 to –50°C for 4 hours, ammonia was evaporated and the residue extracted with chloroform. The extract was evaporated *in vacuo* and the residue obtained was purified by column chromatography, using silica gel and chloroform–methanol (4:1). Crystallization from benzene gave pure

1,4-dihydro-1,3-dimethyl-4-nitromethylenepyridine, melting point 229–230°C. Yield 530 mg (80%) [90JOC778; Copyright permission 1994 American Chemical Society].

Example 2. Preparation of 9-(3-indolyl)-10-methylacridinium iodide

A mixture of N-methylacridinium iodide (1.6 g, 5 mmol), indole (960 mg, 8 mmol) and 10 ml of DMSO was kept at room temperature for 3 days. The reaction mixture was filtered, diluted with 100 ml of ethanol and cooled with ice to deposit red crystals of 9-(3-indolyl)-10-methylacridinium iodide which were collected by filtration and recrystallized from ethanol. Melting point 256–258°C. Yield 820 mg (38%) [77MI1].

Example 3. A typical example of aminoaryl-dehydrogenation reaction. Preparation of 9-(4-N,N-dimethylaminophenyl)-10-methylacridinium iodide

A mixture of N-methylacridinium iodide (3.2 g, 10 mmol), N,N-dimethylaniline (2.14 g, 20 mmol), and sulfur (960 mg, 30 mmol) was stirred at 130°C for 1.5 hours. The mixture was then cooled, treated thoroughly with ether (5 x 50 ml) to eliminate an excess of dimethylaniline and sulfur. Recrystallization from ethanol–DMF (20:1) gave violet crystals of 9-(4-N,N-dimethylaminophenyl)-10-methylacridinium iodide, melting point 216°C. Yield 3.85 g (91%) [77MI1].

Example 4. Preparation of 2-(3-indolyl)-1-methylquinoxalinium iodide

A mixture of 1-methylquinoxalinium iodide (1.36 g, 5 mmol), indole (720 mg, 6 mmol), and 25 ml of ethanol was refluxed for 2 hours with bubbling air through the reaction mixture. The precipitate obtained on cooling to room temperature was filtered off and recrystallized from ethanol to yield 2-(3-indolyl)-1-methylquinoxalinium iodide, melting point above 300°C. Yield 1.18 g (60%) [77MI1].

Example 5. A typical experimental procedure for amination of N-alkylazinium cations. Preparation of 3-carbamoyl-1,6-dihydro-6-imino-1-methylpyridine

To a solution of 3-carbamoyl-1-methylpyridinium iodide (1.0 g, 4 mmol) in 40 ml of liquid ammonia at −33°C was added potassium permanganate (1.3 g, 8 mmol) in portions over a period of 10 minutes. The mixture was stirred for an additional 30 minutes. After evaporation of the liquid ammonia, 40–50 ml of water was added and the solution was continuously extracted with chloroform. Evaporation of the chloroform followed by column chromatography on cilica gel gave 3-carbamoyl-1,6-dihydro-6-imino-1-methylpyridine. Yield 0.41 g (75%), melting point of picrate 312–314°C. Base treatment gave 3-carboxy-1,6-dihydro-6-oxo-1-methylpyridine, melting point 239–240°C [84TL3763; 86JHC1015].

C. *N-Acylazinium Salts*

N-Acylazinium salts, *in situ* generated by adding acyl halides, arylsulfonyl chlorides, or anhydrides to azaaromatic substrates in dry aprotic solvents, are very accessible and reactive species [73UK1416; 74KG3; 79MI3; 80H1033; 88AHC199; 88SC1937]. The formation of the acylazinium salts is very rapid and usually prevails over possible interaction between acyl halide and the nucleophile, even in the case of organometallic reagents. The high reactivity of N-acylazinium salts allows the addition of relatively weak C-nucleophiles of aromatic nature (dialkylanilines, pyrroles, indoles, etc.), organometallic reagents and trialkylphosphites. Nucleophiles bearing NH_2, OH, or SH groups cannot usually be used because of possible transacylation [89MI3]; a rare example of using N-nucleophiles is the reaction of pyridine N-oxide with anilides of sulfinic acids in the presence of tosyl chloride, yielding 2-(N-tosyl-N-aryl)aminopyridines [82KG1278].

The characteristic feature of S_N^H reactions in the series of N-acylazinium cations is that the N-acyl-substituted dihydroazines derived from addition of nucleophiles are more stable and more reluctant to be aromatized due to the stabilizing electon-withdrawing effect of the acyl group [88AHC199; 89MI3]. The dehydrogenation step can be realized either through concerted elimination of the acyl residue and the sp^3-hydrogen, affording aldehydes (pathway a) or by action of an oxidant present in the reaction mixture (Scheme 107, pathway b) [72KG529].

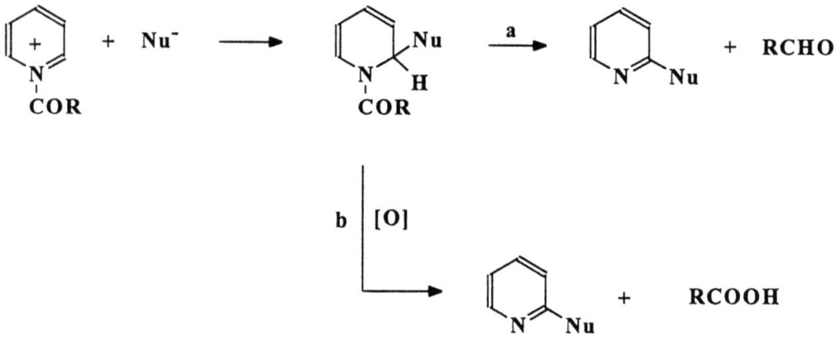

Scheme 107

1. Reactions with C-Nucleophiles The addition of the cyanide anion to N-acylazinium salts, well known as the Reissert reaction [05CB1603; 05CB1610; 05CB3415], has been the subject of a number of reviews [68AHC1; 79AHC187; 83JHC823; 84H2375; 88H2659]. A classical Reissert compound is considered to be 1-benzoyl-2-cyano-1,2-dihydroquinoline **150** which is converted on treatment with aqueous hydrochloric acid into benzaldehyde and quinoline-2-carboxylic acid (Scheme 108) [05CB1603]. Also isoquinoline [80H1033], phthalazine [78S206; 80H1033], naphthyridines [80H1033; 80JHC1211], as well as acridine [50CB10] are able to form the corresponding Reissert compounds.

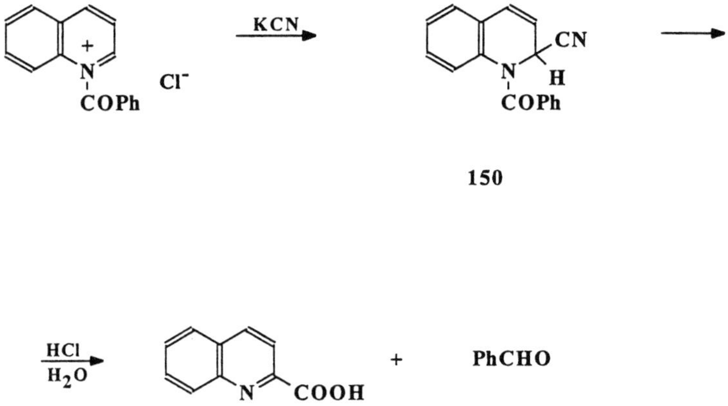

Scheme 108

As already mentioned, aromatic C-nucleophiles, such as N,N-disubstituted anilines, pyrroles, and indoles proved to react with N-acylazinium salts quite smoothly, thus providing a method for preparing a number of compounds which are otherwise difficult to obtain (Table XXV) [59AG310; 62CB1484; 64MI2; 67KG248, 70JHC1071; 70KG1291, 70KG1292; 70KG1515, 71KG82, 71KG648; 83USP4415578; 84H795; 89JCS(CC)727; 92LA813], as illustrated in Scheme 109.

It should be noted that the S_N^H reactions of N-acylazinium salts with aromatic nucleophiles can formally be rationalized as electrophilic substitution of hydrogen in an arene ring [70JHC1071; 70KG1292; 71KG648; 89MI3]. For instance, the reaction of N-acylpyridinium salts **151** with N,N-dimethylaniline, yielding 4-(4-N,N-dimethyl-aminophenyl)pyridine **152**, has been regared as a Friedel-Crafts type condensation; the

TABLE XXV
Yields of Products Obtained by Nucleophilic Substitution
of Hydrogen in N-Benzoylazinium Chlorides

Azine	Nucleophile	S_N^H Product	Yield (%)	References
Pyridine	1-Ethyl-1,2,3,4-tetrahydroquinoline	1-Ethyl-6-(pyridyl-4)-1,2,3,4-tetrahydroquinoline	85	67KG248
Pyridine	1-n-Butyl-1,2,3,4-tetrahydroquinoline	1-n-Butyl-6-(pyridyl-4)-1,2,3,4-tetrahydroquinoline	64	67KG248
Quinoline	1-Methyl-1,2,3,4-tetrahydroquinoline	1-Methyl-6-(quinolinyl-)-1,2,3,4-tetrahydroquinoline	46	70KG1515
Isoquinoline	N,N-Dimethylaniline	1-(4-N,N-Dimethylaminophenyl)isoquinoline	30	71KGS82
Acridine	N,N-Dibenzylaniline	9-(4-N,N-Dibenzylaminophenyl)acridine	47	70ZOR614
Acridine	Triethylphosphite	Acridinyl-9-phosphonic acid	40	71MI2
Acridine	Indole	9-(Indolyl-3)acridine	65	71KGS648
Acridine	Cyclopentanone	9-(Cyclopentanon-2-yl)acridine	31	71KGS643
Phenanthridine	Acetophenone	10-Phenacylphenanthridine	17	70KG1291

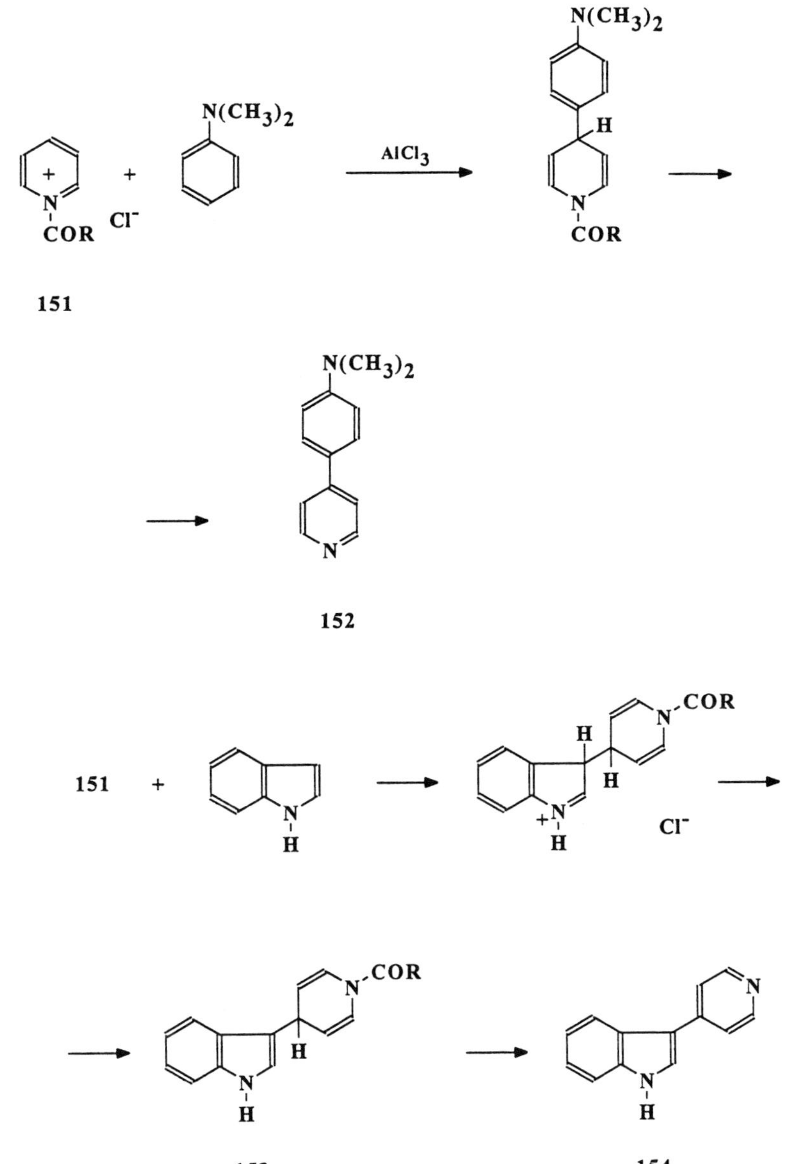
N(CH3)2
N(CH3)2
AlCl3
H
Cl-
COR
COR
151
N(CH3)2
152
151
COR
H
H
Cl-
H
H
COR
H
H
153
154

Scheme 109

S_N^H character of this reaction, featuring nucleophilic substitution of hydrogen in the pyridine ring was, however, ignored (Scheme 109) [64MI2; 89MI3].

The reaction of N-acylpyridinium salts **151** with indoles yields 4-(indolyl-3)-substituted pyridines **154** through simultaneous aromatization of the 1-acyl-1,4-dihydropyridine intermediates **153** (Scheme 109) [59AG310; 70JHC1071]. For comparison, when reacting the N-benzyloxycarbonyl salt of methyl nicotinate with N-metylindole the more stable 1,4-adduct, methyl 1-benzyloxycarbonyl-4-(1-methylindolyl-3)-1,4-dihydropyridine-3-carboxylate, was obtained [89JCS(CC)727].

Also, silyl enol ethers add to the 4-position of 1-alkoxycarbonylpyridinium salts to give 4-(2-oxoalkyl)pyridines after oxidation of the 1,4-dihydropyridine intermediates with sulfur [83TL5269; 85TL3267].

2. Reactions with Organometallic Reagents. Nucleophilic substitution of hydrogen in N-acylpyridinium salts by action of organometallic reagents has been extensively studied [70JA5442; 70AG518; 71KG1148; 76JOC3250; 82JOC4315; 82TL429; 83TL1801; 84H151; 84TL3297; 84TL3637; 84TL4867; 85JHC1419; 85JOC287; 85JOC4410; 85JOC5660; 85TL1027; 86H3199; 86TL211; 88AHC199; 90TL6287].

The regioselectivity found in these reactions proved to be dependent on both the nature of the reagent used and positions of the substituents in the ring [88AHC199]. This can be illustrated by the observation that phenylmagnesium bromide reacts with N-phenoxycarbonyl salt of methyl nicotinate **155** to give after oxidation with *ortho*-chloranil 6- and 4-phenyl-substituted nicotinates, **156** and **157**, in the ratio 6:1 (Scheme 110; Table XXVI) [84H151]. The reaction of 3,4-dimethyl-1-phenoxycarbonylpyridinium chloride with aryl Grignard reagents exhibits a higher site selectivity, showing the preference for the replacement of hydrogen at C-6 (α-position to the quaternary nitrogen) [76JOC3250]. Also, alkenyl and alkynyl Grignard reagents, as well as silver phenylacetylenide, add mainly or exclusively at C-2 (C-α) of 1-acyl-pyridinium salts [70JA5442; 70AG518; 82JOC4315; 83TL1801; 88AHC199].

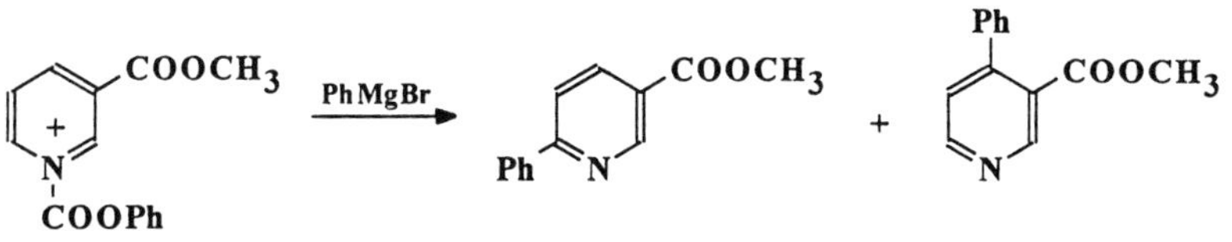

Scheme 110

Alkyl Grignard reagents are less regioselective than their aryl analogues and add to N-acylpyridinium salts at 2-(6-)- and 4-positions. After aromatization of the intermediary 1,2- and 1,4-dihydropyridines with chloranil comparable quantities of 2- and 4-substituted pyridines are obtained [88AHC199]. When the 4-position of the pyridine ring is blocked with a substituent, alkyl Grignard reagents add exclusively to the 2-position. Reacting 4-X-pyridines (X= Cl, Br) with alkyl Grignard reagents 2-alkyl-4-halopyridines are obtained, showing again (see Chapter 2,III,A) the interesting phenomenon: that the replacement of hydrogen is preferred to substitution of the nucleofugal chloride or bromide anion (Scheme 111; Table XXVI) [85JOC4410].

Regiospecific α-substitution in the pyridine ring with Grignard reagents is also achieved when the trimethylstannyl group is present at C-4. After acylation of 4-trimethylstannylpyridine **158**, and reacting the *in situ*-formed N-acylpyridinium salt **159** with Grignard reagent an 1,2-dihydropyridine intermediate **160** is obtained. Treatment of this intermediate with oxalic acid leads to aromatization and at the same time *ipso*-substitution of the stannyl group by hydrogen (Scheme 112) [84TL4867].

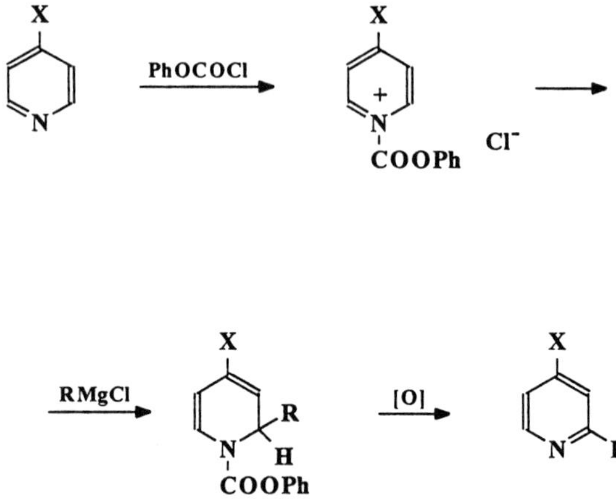

R = CH$_3$, C$_2$H$_5$, CH$_2$=CH, Ph, n-C$_3$H$_7$, n-C$_6$H$_{13}$,
X = Cl, Br

Scheme 111

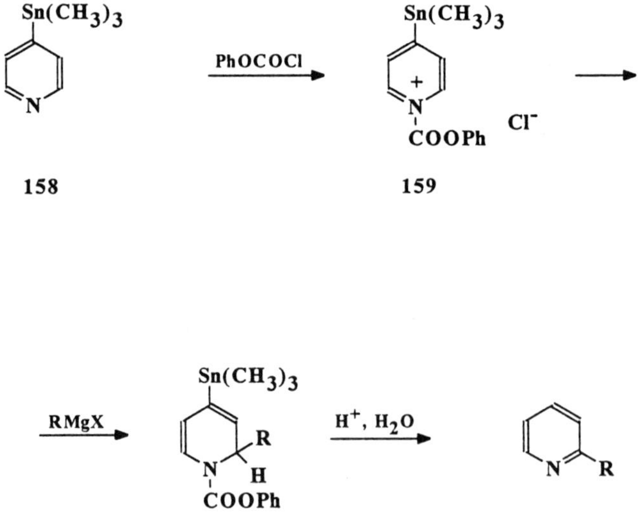

158, 159, 160

Scheme 112

TABLE XXVI

Yields of Products Obtained by Nucleophilic Substitution of Hydrogen in N-Phenoxycarbonyl Pyridinium Chlorides

Pyridine	Nucleophile	S_N^H Product	Yield (%)	References
3-Acetyl-pyridine	n-C$_3$H$_7$MgCl	3-Acetyl-4-n-propyl-pyridine	36	86H3199
3-Benzyloxy-pyridine	CH$_3$MgBr	3-Benzyloxy-4-methylpyridine	56	85JHC1419
Methyl nicotinate	C$_6$H$_5$MgCl	Methyl 6-phenyl-nicotinate	58	84H151
3-Bromo-pyridine	CH$_3$MgCl	3-Bromo-4-methyl-pyridine	51	84H339
3-Bromo-pyridine	n-C$_4$H$_9$MgCl	3-Bromo-4-n-butyl-pyridine	68	84H339
4-Chloro-pyridine	CH$_2$=CHMgBr	4-Chloro-2-vinyl-pyridine	54	85JOC4410
4-Chloro-pyridine	n-C$_2$H$_5$MgBr	4-Chloro-2-ethyl-pyridine	46	85JOC4410
4-Chloro-pyridine	n-C$_3$H$_7$MgCl	4-Chloro-2-n-propyl-pyridine	46	85JOC4410
4-Chloro-pyridine	C$_6$H$_{11}$MgCl	4-Chloro-2-cyclohexylpyridine	36	85JOC4410
4-Chloro-pyridine	C$_6$H$_5$MgCl	4-Chloro-2-phenyl-pyridine	55	85JOC4410

Organocopper compounds (RCuLi, RCu, RCu$_2$BF$_3$, RCuZnBr) are found to be very regioselective reagents and add exclusively to the 4-position of 1-acylpyridinium salts [82TL429; 88AHC199; 90TL6287]. It is interesting that the less regiospecific alkyl and aryl Grignard reagents change to high regioselectivity reagents when reacting in the presence of catalytic amounts of cuprous iodide. This method provides us with a convenient way for the synthesis of 4-alkyl- and 4-arylpyridines **162** through oxidation of the intermediary 1-phenoxycarbonyl-1,4-dihydropyridines **161** with hot sulfur or chloranil (Scheme 113) [82JOC4315; 88AHC199].

3-Halo and 3-acyl-substituted pyridines, as well as alkyl nicotinates react with Grignard reagents and cuprous iodide in a similar manner providing 3-halo-4-R-substituted pyridines (Scheme 113; Table XXVI) [83JHC1239; 84H339; 86H3199].

3. Reactions with P-Nucleophiles. N-Alkoxycarbonyl and N-aryloxycarbonyl-pyridinium salts were found to react smoothly with trialkylphosphites giving rise to 1,2- or 1,4-adducts [71MI2; 81TL4093; 82TL1709]. The reaction is mainly governed by steric effects of the substituents in the P-nucleophile as evidenced by the fact that tri(isopropyl)phosphite adds to the N-ethoxycarbonylpyridinium ion **163** at C-4, while N-[2,4,6-tri-*t*-butylphenoxycarbonyl]pyridinium chloride **164** gives with trimethyl-phosphite a 1:1 mixture of 2- and 4-adducts **165** and **166**, respectively. So, in spite of the presence of the bulky substituent at the ring nitrogen an addition at C-2 still takes place (Scheme 114) [81TL4093]. Position 2 proved to be the preferential site for nucleophilic attack of trimethylphosphite on the N-ethoxycarbonylquinolinium ion [82TL1709], as well as in the reaction of N-benzoylpyridinium chloride with triethylphosphite, yielding pyridine-2-phosphonic acid **167** (Scheme 114) [71MI2].

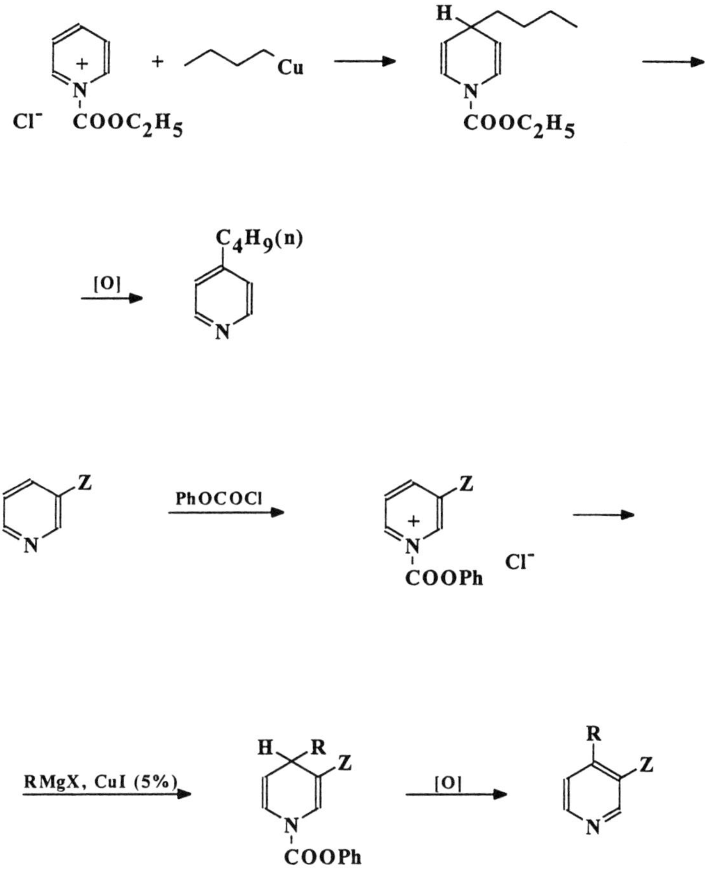

161 162

Z = H, Cl, Br, CH_3CO, $COOCH_3$, OCH_2Ph;
R = alkyl, aryl; X = Cl, Br

Scheme 113

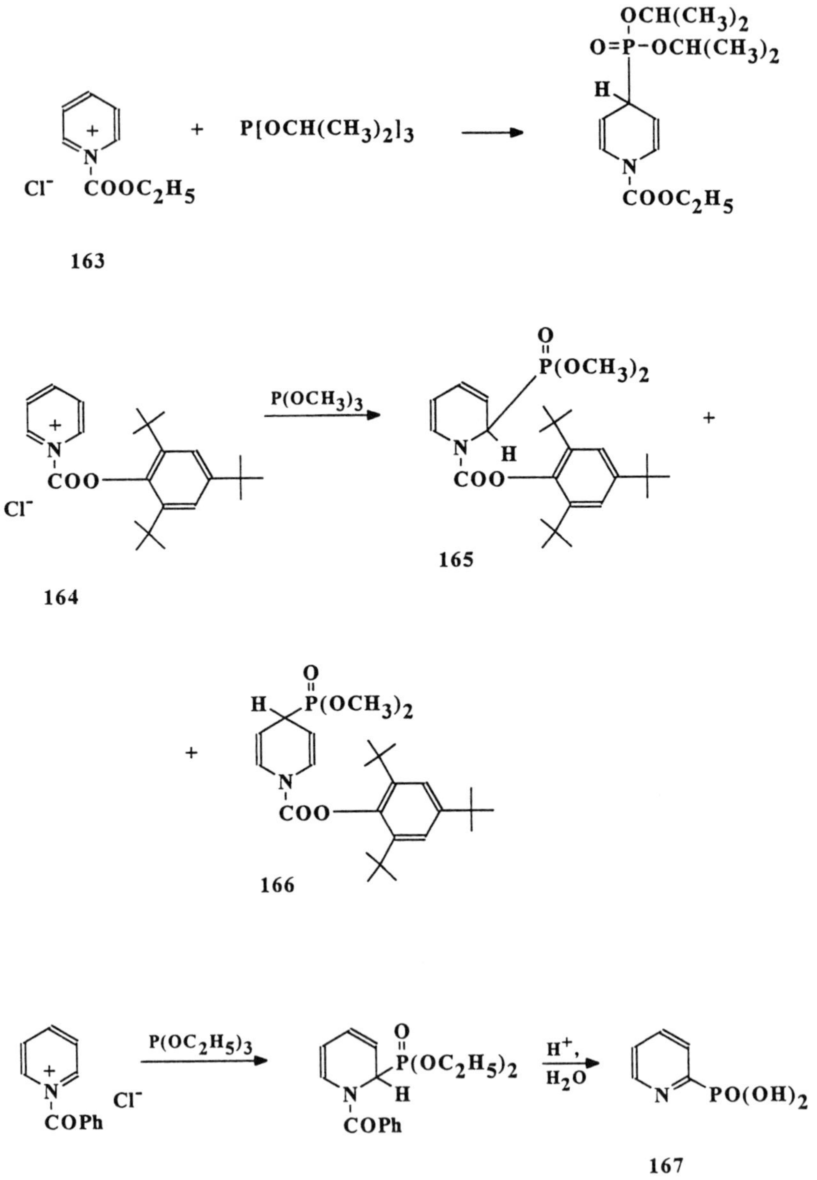

Cl⁻
COOC2H5
P[OCH(CH3)2]3
OCH(CH3)2
O=P-OCH(CH3)2
H
N
COOC2H5
163
P(OCH3)3
COO
P(OCH3)2
H
164
165
166
COPh
P(OC2H5)3
P(OC2H5)2
H+,
H2O
PO(OH)2
167

Scheme 114

Chapter 3 S_N^H REACTIONS IN HETARENES   173

Experimental Procedures

Example 1. Preparation of 9-(indolyl-3)acridine

A mixture of acridine (8.9 g, 50 mmol), indole (2.9 g, 25 mmol) and fresh-distilled benzoyl chloride (3.5 g, 25 mmol) was heated up to 70°C and kept at this temperature for 3 hours. The reaction mixture was then cooled to room temperature and the precipitate obtained was filtered off, washed with 10% aqueous ammonium hydroxide and water, dried in air and recrystallized from 1-pentanol to give 9-(indolyl-3)acridine, melting point 296–297°C. Yield 4.7 g (65%) [71KG648].

Example 2. Preparation of 1-(4-N,N-dimethylaminophenyl)isoquinoline

Step 1. A mixture of isoquinoline (6.45 g, 50 mmol), N,N-dimethylaniline (12.1 g, 100 mmol), and benzoyl chloride (7.03 g, 50 mmol) was heated up to 100°C and kept at this temperature for 8 hours. The reaction mixture was then neutralized with 2*N* sodium hydroxide and distilled in a stream of water vapor. A distillate obtained was cooled with ice water and the precipitate obtained was filtered off, washed with water, dried in air, and recrystallized from petroleum ether to yield 1-(4-N,N-dimethylaminophenyl)-2-benzoyl-1,2-dihydroisoquinoline, white crystals, melting point 113–114°C, yield 13.6 g (77%).

Step 2. A solution of 1-(4-N,N-dimethylaminophenyl)-2-benzoyl-1,2-dihydroiso-quinoline (2 g, 5.6 mmol) and 8 g of potassium hydroxide in 30 ml of 70% aqueous ethanol was refluxed during 5 hours. After addition of 30 ml of water ethanol was distilled off, and the suspension obtained was treated with 20 ml of benzene. The benzene extract was mixed with 10 ml of nitrobenzene and the resulting solution was refluxed for 3 hours. The reaction mixture was treated with 2N hydrochloric acid (3 x 20 ml) and collected acidic extracts were neutralized with 2N sodium hydroxide. A mixture of 1-(4-N,N-dimethylaminophenyl)isoquinoline and 1-(4-N,N-dimethylaminophenyl)-1,2-dihydroisoquinoline was obtained which was separated by column chromatography on Al_2O_3 with chloroform as eluent. Yield of 1-(4-N,N-dimethylaminophenyl)-isoquinoline 400 mg (30%), melting point 114–115°C [71KG82].

Example 3. Preparation of 3-bromo-4-phenylpyridine

In a 1-liter three-necked flask equipped with an overhead stirrer were placed 3-bromopyridine (10.6 ml, 0.11 mol), cuprous iodide (952 mg, 5 mmol), methyl sulfide (22 ml, 0.3 mol), and 250 ml of THF under nitrogen. The solution was cooled to –20°C and 12.9 ml (0.1 mol) of phenyl chloroformate was added via syringe with stirring. After 5 minutes, phenylmagnesium chloride (0.1 mol) in 50 ml of THF was added dropwise over 10 minutes. The mixture was stirred for 15 minutes at –20°C and then at room temperature for another 15 minutes followed by the addition of aqueous 20% ammonium chloride solution (75 ml). Ether (200 ml) was added and the organic layer was washed with 50 ml portions of 20% ammonium chloride–ammonium hydroxide (50: 50), water, 10% aqueous hydrochloric acid, and brine. After drying (magnesium sulfate), the solution was filtered and concentrated to yield 25.6 g of crude 3-bromo-1-phenoxycarbonyl-4-phenyl-1,4-dihydropyridine as a light yellow oil. To the crude dihydropyridine in 250 ml of dry toluene was added 29.5 g (0.12 mole) of *para*-chloranil. The mixture was heated at reflux for 10 hours under nitrogen and cooled to room temperature. An aqueous solution of 1N sodium hydroxide (100 ml) was added, stirred 15 minutes, and then the mixture was filtered through Celite. The dark organic layer was washed with 2 x 50 ml portions of 1N sodium hydroxide and water, then extracted with 10% hydrochloric acid (4 x 50 ml). The combined acid extracts were concentrated *in vacuo* to approximately 75 ml, cooled, made basic with 20% sodium hydroxide, and extracted with methylene chloride (3 x 75 ml). The combined organic layer was washed with brine, dried (potassium carbonate/Norite), filtered through a Florisil pad, and concentrated to yield 18.27 g of a red oil. Bulb-to-bulb distillation (105–140°C/0.3 mm) gave 15.8 g (68%) of a white solid. Recrystallization from cold hexanes provided 15.5 g (66%) of 3-bromo-4-phenylpyridine as white crystals, melting point 49–50°C [83JHC1239].

Example 4. Preparation of 4-chloro-2-ethylpyridine

To a cooled (–78°C) solution of 4-chloropyridine (360 mg, 3.17 mmol) in 10 ml of THF was added an ethereal solution of ethylmagnesium bromide (3.49 mmol) in one portion. Immediately following, phenyl chloroformate (0.41 ml, 3.17 mmol) was added dropwise. The mixture was stirred at –78°C for 10 minutes, allowed to come to room temperature, and quenched with aqueous 20% NH_4Cl solution (10 ml). Ether (10 ml) was added, and the organic layer was washed with 10-ml portions of water, 10% HCl, and brine. After drying ($MgSO_4$), the solution was concentrated to give 1.46 g of the crude dihydropyridine as a yellow oil, which was dissolved in 15 ml of dry toluene. To this solution *ortho*-chloranil (860 mg, 3.5 mmol) in 7 ml of glacial acetic acid was added dropwise at room temperature. The mixture was stirred at room temperature for 24 hours, cooled, and made basic with 10% NaOH. The mixture was stirred for 15 minutes and filtered through Celite. The dark organic layer was washed with water and then extracted with 3 x 10 ml of 10% HCl. The combined acid extracts were cooled, made basic with 20% NaOH, and extracted with methylene chloride (3 x 10 ml). The combined organic extracts were washed with brine, dried ($MgSO_4$), and concentrated to yield the crude product (260 mg) as a yellow oil. Purification by radial preparative layer chromatography (ethyl acetate–hexane) gave 207 mg (46%) of 4-chloro-2-ethylpyridine as a clear oil. The hydrochloride salt has melting point 193–194°C [85JOC4410; Copyright permission 1994 American Chemical Society].

Example 5. Preparation of methyl 6-phenylnicotinate

A solution of methyl nicotinate (411 mg, 3 mmol) in 6 ml of dry THF under nitrogen was cooled to −20°C (dry ice/CCl_4). Phenyl chloroformate (0.39 ml, 3.0 mmol) was added dropwise and the mixture was stirred at −20°C for 10 minutes. A solution of phenylmagnesium chloride (3.0 mmol) in 1.5 ml of THF was added dropwise. The mixture was stirred at −20°C for 15 minutes followed by the addition of aqueous 20% NH_4Cl solution (20 ml). Ether (50 ml) was added and the organic layer was washed with 20 ml portions of 10% HCl (twice), water, and brine. After drying ($MgSO_4$), the solution was concentrated to give the crude dihydropyridine as a viscous oil. To this oil in dry toluene (20 ml) was added 810 mg (3.3 mmol) of *ortho*-chloranil. The mixture was heated at reflux for 3 hours and cooled to room temperature. Ether (50 ml) and 25 ml of 1N NaOH were added, and after stirring for 10 minutes the mixture was filtered through Celite. The organic layer was washed with 20-mL portions of water and brine. After drying ($MgSO_4$), the solution was filtered and concentrated to yield 875 mg of a yellow solid. Purification by radial preparative chromatography (0.1% MeOH–CH_2Cl_2) gave 370 mg (58%) of methyl 6-phenylnicotinate as a white solid. The product was recrystallized from ethyl acetate to give an analytical sample, melting point 113–114°C [84H151].

D. *Azine N-Oxides, and N-Alkoxy- and N-Acyloxyazinium Salts*

The chemistry of heteroaromatic N-oxides has been extensively studied and well documented in the literature [67MI1; 71MI3; 73MI1; 79MI1; 82MI1; 85MI1]. In this branch of heterocyclic chemistry S_N^H reactions have been most intensively applied [67MI1; 79MI1; 86CCA89; 86H161].

The synthesis and chemical behavior of azine N-oxides in S_N^H reactions with C-, N-, O-, P-, S-nucleophiles, and halogenides were extensively discussed in excellent monographs by E. Ochiai [67MI1] and A.R. Katritzky and J.M. Lagowsky [71MI3]. These monographs contain a great deal of examples of nucleophilic replacement of ring hydrogen, however characteristic features of S_N^H reactions have not been analyzed. In more recent studies a lot of new interesting examples of S_N^H reactions in aromatic N-oxides and their derivatives have been found.

An interesting feature of S_N^H reactions with the title compounds is that usually no oxidant is required to perform the aromatization of the intermediate σ^H-adduct. This is due to the fact that in the unstable intermediate **169** formed after addition of the nucleophile the nucleofugal group OR is present on the ring nitrogen, which by elimination of HOR leads to the aromatic final product **170** (Scheme 115). This behavior is in quite a contrast to that observed with the N-alkylazinium salts, which usually require the addition of an oxidant for achieving S_N^H substitutions (see Chapter 3,II,B) [88T1].

The formation of the quaternary N-alkoxy or N-acyloxyazinium salts **168** by the attack of alkylating or acylating reagents on the azine N-oxide function proved to be a prerequisite step for the addition of Nu⁻ to the α- and/or γ-positions in the azine ring (Scheme 115). Elimination of ROH from the intermediate dihydroazine **169** leads to the deoxygenated product **170** [67MI1; 76MI1; 79MI1; 86CCA89; 86H161]. Due to the loss of the N-oxide function, these S_N^H reactions are usually classified as deoxygenative nucleophilic substitutions [67MI1; 86CCA89; 86H161].

In deoxygenative nucleophilic substitutions in the series of N-alkoxy and N-acyloxyazinium salts, the 1,2- or 1,4-elimination of ROH or RCOOH from the intermediate adduct appears to be concerted. Such *cine-* or *tele-*substitutions are, of course, much more thermodynamically favored than direct elimination of the hydride ion [67MI1; 88T1; 89MI3].

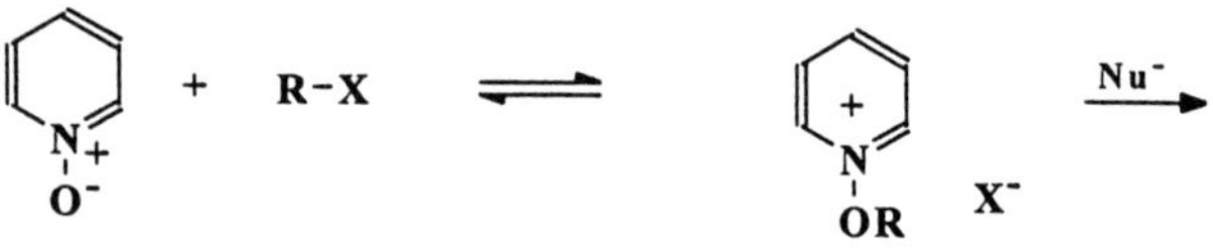

Scheme 115

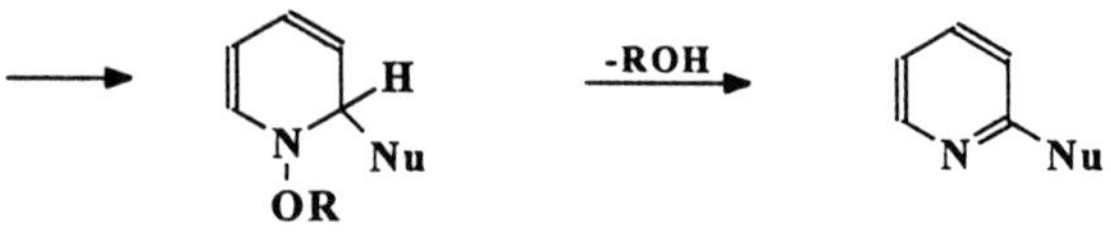

1. *Azine N-oxides*. The uncharged azine N-oxides are also able to form 1,2- or 1,4-adducts with participation of the N-oxide oxygen. Concerted deoxygenation and dehydrogenation results in the formation of α- or γ-substituted products. There are, however, interesting examples known of reactions with retention of the N-oxide function in the product [67MI1; 86CCA89; 89MI3]. For instance, quinoline N-oxide and 4-phenylphthalazine 2-oxide react with acetophenone under basic conditions to afford 2-phenacylquinoline N-oxide and 1-phenacyl-4-phenylphthalazine 2-oxide (Scheme 116) [78H371; 82H177; 86CCA89]. Also, 4-chloroquinoline N-oxide has been found to undergo the S_N^H reaction with methylketones to give 2-substituted 4-chloroquinoline N-oxides (Scheme 116; Table XXVII) [82H177; 84H1811; 86CCA89].

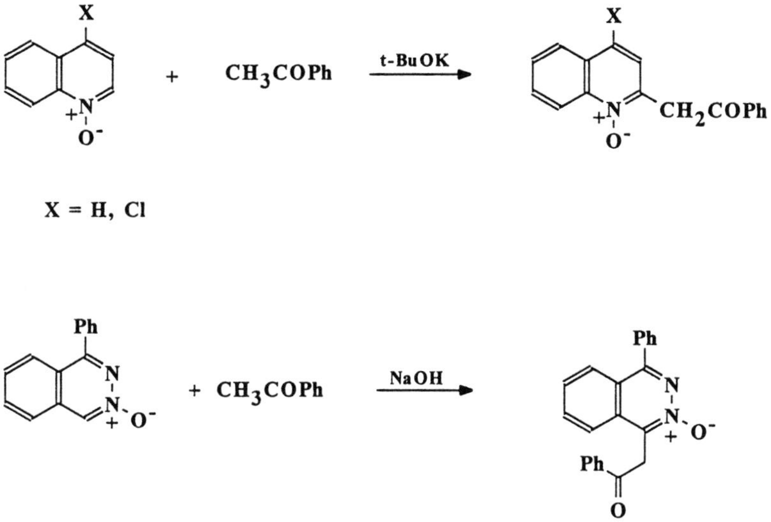

X = H, Cl

Scheme 116

TABLE XXVII
Nucleophilic Substitution of Hydrogen in Quinoline N-Oxides
by Action of Base-Induced Carbanions[1]

Quinoline N-Oxide	CH-Active Compound	S_N^H Product	Yield, %
Unsubstituted	Acetone	2-Acetonylquinoline N-oxide	37
	Pinacolone	2-Pinacolylquinoline N-oxide	81
	Acetophenone	2-Phenacylquinoline N-oxide	65
3-Bromo	Acetone	2-Acetonyl-3-bromoquinoline N-oxide	33
	Pinacolone	3-Bromo-2-pinacolylquinoline N-oxide	43
4-Chloro	Acetone	2-Acetonyl-4-chloroquinoline N-oxide[2]	80
	Pinacolone	4-Chloro-2-pinacolylquinoline N-oxide	
	Acetophenone	4-Chloro-2-phenacylquinoline N-oxide	65

[1] In t-butylamine at −10 °C in the presence of potassium t-butoxide.

[2] In liquid ammonia at −70 °C in the presence of potassium t-butoxide [82H177].

Experimental facts indicate that S_N^H reactions of azine N-oxides are accelerated by oxygen or added oxidants [86CCA89], although the role of oxidants, including that of air oxygen, in the nondeoxygenative S_N^H reactions of aromatic N-oxides is not quite clear and has not yet been thoroughly investigated. However, the fact that deliberate addition of oxidant into the reaction mixture enables retention of the N-oxide function in the S_N^H products is of great preparative value for the synthesis of functionalized N-oxides. Indeed, treatment of a great variety of 4-nitropyridazine 1-oxides **171** with a solution of potassium permanganate in liquid ammonia results in the formation of the corresponding 5-amino-4-nitropyridazine 1-oxides **172** in 50 to 75% yields (Scheme 117) [86JHC621].

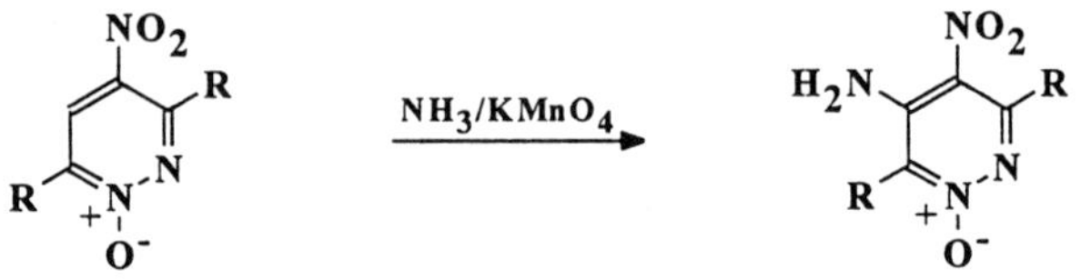

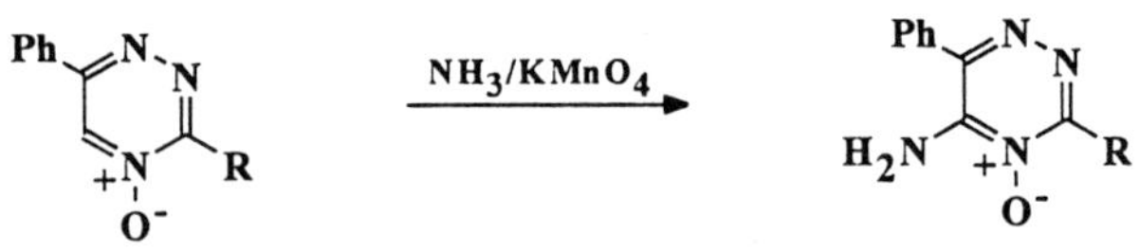

R = H, SCH$_3$, OCH$_3$, Ph

Scheme 117

In a similar manner 6-phenyl-1,2,4-triazine 4-oxides **173** are converted into the corresponding 5-amino derivatives **174** with retention of the N-oxide function (Scheme

117) [85S884]. Also oxidative cyanation of quinoline N-oxides **175** with a mixture of potassium cyanide and potassium ferricyanide has provided good yields of 2-cyanoquinoline N-oxides **176** (Scheme 118; Table XXVIII) [71CPB1809; 72JOC3588; 74JOC1836].

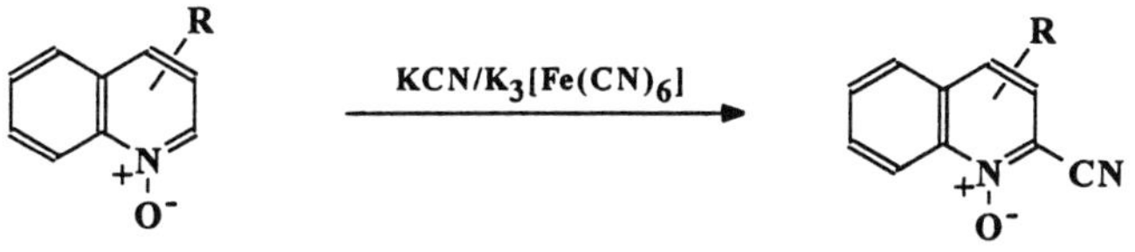

Scheme 118

R = H, CF$_3$, OCH$_3$

TABLE XXVIII
Oxidative Cyanation of Quinoline N-Oxides with Potassium Cyanide in the Presence of Potassium Ferricyanide

Quinoline N-oxide	S_N^H Product	Yield (%)	References
Unsubstituted	2-Cyanoquinoline N-oxide	85	74JOC1836
3-Trifluoromethyl	2-Cyano-3-trifluoromethylquinoline N-oxide	53	74JOC1836
4-Trifluoromethyl	2-Cyano-4-trifluoromethylquinoline N-oxide	47	74JOC1836
4-Methoxy	2-Cyano-4-methoxyquinoline N-oxide	17	74JOC1836
6-Nitro	2-Cyano-6-nitro-quinoline N-oxide	32	71CPB1809

Reactions of azine N-oxides with organometallic compounds are also synthetically very useful. Pyridine and quinoline N-oxides have been shown to undergo S_N^H

phenylation at C-2 by action of phenylmagnesium bromide in ether, however, with loss of the N-oxide function (Scheme 119) [54MI1]. The reaction involves interactions of the Grignard reagent with the N-oxide function, activating the pyridine ring. This reaction has been applied successfully for the synthesis of quinine and N-methylanabasine derivatives (Scheme 119) [67MI1, p. 252].

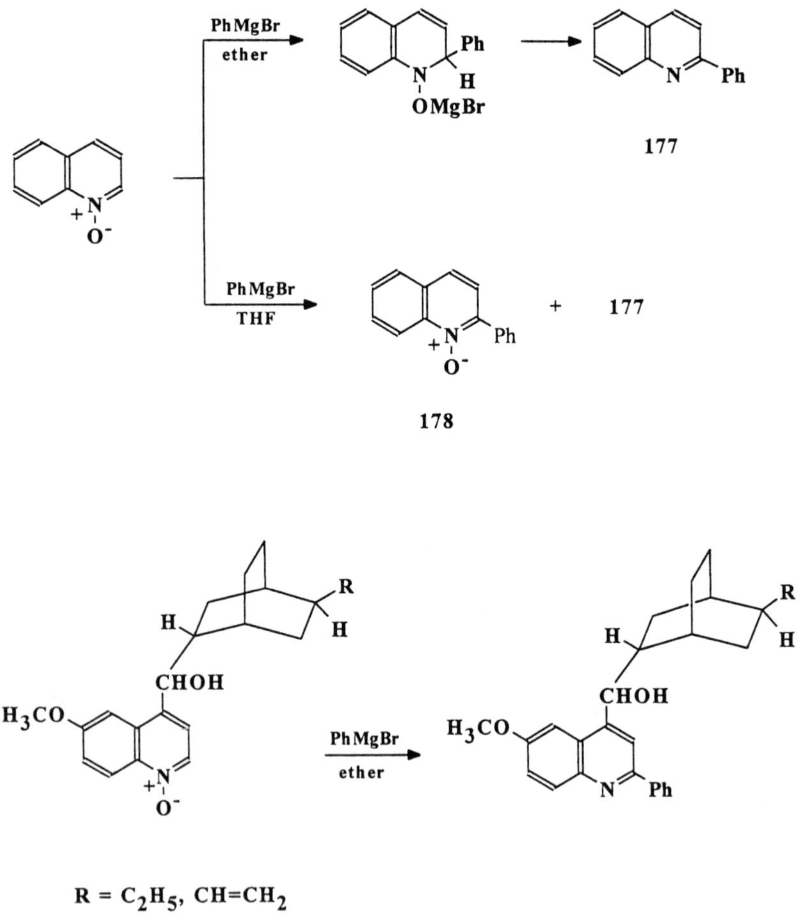

R = C_2H_5, CH=CH_2

Scheme 119

It is interesting to note that the reaction of quinoline N-oxide with phenymagnesium bromide in THF, instead of ether, yields 2-phenylquinoline **177** and its N-oxide **178** (Scheme 119). The solvent effect was explained on the basis that peroxides are more easily formed in THF than in ether to have some effect on the retention of the N-oxide group [65JOC910]. Also the reaction of 4-cyano-, 4-methoxy-, and 4-chloroquinoline N-oxide with phenylmagnesium bromide in THF has been reported to give 2-phenylquinolines as major products [80JHC293].

In azine N-oxides the vicarious nucleophilic substitution methodology was also successfully applied for structural modification of azines, with retention of the N-oxide function [87H229; 87H235]. When reacting quinoline N-oxide with phenoxyacetonitrile, chloromethyl phenyl sulfone, and methylthiomethyl *para*-tolyl sulfone, 2-substituted quinoline N-oxides were obtained (Scheme 120, Table XXIX) [87H229].

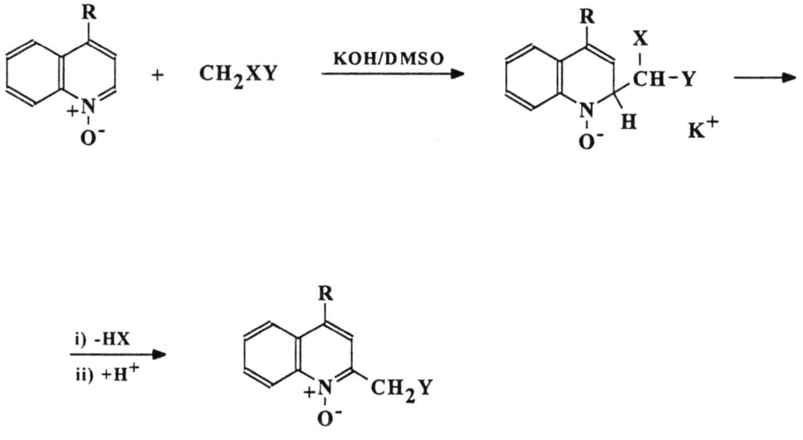

$X = Cl, CH_3S, PhO; Y = CN, SO_2Ph, SO_2C_6H_4\text{-}CH_3(p)$

Scheme 120

TABLE XXIX

Vicarious Nucleophilic Substitution in Quinoline N-Oxides[1]

Azine N-Oxide	Reagent[2]	S_N^H Product(s)	Yield (%)
Quinoline N-oxide	A	2-Cyanomethylquinoline N-oxide	49
4-Chloroquinoline N-oxide	A	4-Chloro-2-cyanomethylquinoline N-oxide	38
Quinoline N-oxide	B	2-Phenylsulfonylmethylquinoline N-oxide	51
4-Chloroquinoline N-oxide	A	4-Chloro-2-phenylsulfonylmethyl-quinoline N-oxide	88
Quinoline N-oxide	C	2-(*p*-Tolyl)sulfonylmethylquinoline N-oxide	24
		2-(Methylthio)(*p*-tolylsulfonyl)-methylquinoline N-oxide	16

[1] In tetahydrofuran in the presence of potassium *t*-butoxide [87H229].
[2] A: Phenoxyacetonitrile; B: Chloromethyl phenyl sulfone; C: Methylthiomethyl *p*-tolyl sulfone.

Also, the synthesis of a number of new methylene-substituted pyridazines, quinolines, isoquinolines, phthalazines, and quinoxalines has been developed using deoxygenative reactions of the corresponding N-oxides with CH-active malonodinitrile, indan-1,3-dione, and β-ketoesters [75M473].

2. N-Alkoxyazinium salts. N-Alkoxyazinium salts are easily accessible through O-alkylation of N-oxides with alkyl iodides, alkyl sulfates, or other alkylating agents, such as the very active trialkyloxonium tetrafluoroborates. Their enhanced reactivity toward nucleophilic reagents enables one to perform S_N^H reactions at milder conditions compared to those of uncharged azine N-oxides. A typical example is cyanation of the N-methoxypyridinium salt with potassium cyanide, which already occurs at room temperature and yields a mixture of 2- and 4-cyanopyridines (Scheme 121) [67MI1; 76MI1; 85MI1].

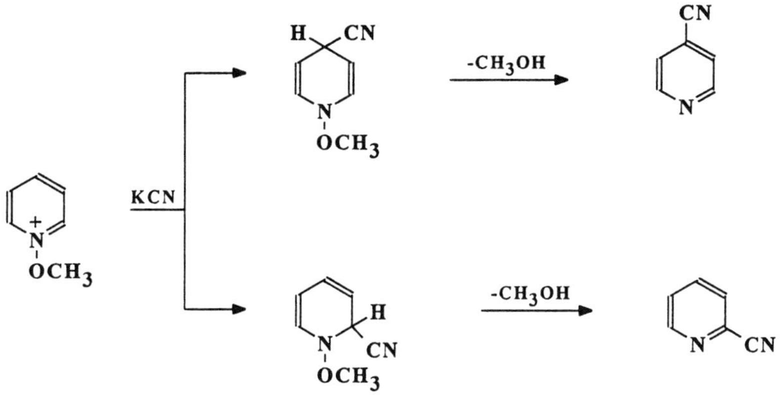

Scheme 121

The Minisci-type hydroxymethylation of N-alkoxypyridinium salts **179** in methanol with ammonium persulfate is another example of a deoxygenative nucleophilic substitution of hydrogen (Scheme 122; Table XXX) [89SC317].

Ammonium persulfate acts as a radical initiator and the first step of the reaction is supposed to be the addition of a nucleophilic hydroxymethyl radical to C-2 of the pyridine ring. It is suggested that radical cation σ^H-adducts **181** are formed which are then transformed into products **180** via elimination of a proton and the methoxyl radicals **183**. The latter can abstract a hydrogen atom from methanol, regenerating the carbon-centered hydroxymethyl radicals which can add again to the salts **179** (Scheme 122). The oxidative pathway involving the formation of 2-substituted N-methoxypyridinium salts **182** can only partly take place based on the consideration that the reaction requires less than equimolar amounts of ammonium persulfate (in contrast to the original Minisci procedure, see Chapter 3,II,A) [89SC317]. As discussed in Chapter 3,II,A the Minisci procedure, applied to NH-pyridinium salts under acidic conditions, usually gives a mixture of mono- and bis hydroxymethylpyridines. The advantage of using N-alkoxypyridinium salts **179** is that by this procedure only mono hydroxy-methylpyridines are obtained. The reason is that in the procedure with N-alkoxy-pyridinium salts no strong acidic conditions are created, so that after the mono hydroxymethyl pyridine is formed, no activation of the pyridine ring can take place for further substitution [89SC317].

TABLE XXX

Hydroxymethylation of Pyridine N-Oxides[1]

Pyridine N-oxide	S_N^H Product	Yield (%)
4-Chloro	4-Chloro-2-hydroxymethylpyridine	71
4-Methyl	2-Hydroxymethyl-4-methylpyridine	46
2-Methyl	4-Hydroxymethyl-2-methylpyridine	39
	2-Hydroxymethyl-6-methylpyridine	23
4-Cyano	2-Hydroxymethyl-4-cyanopyridine	40
2,6-Dimethyl	2,6-Dimethyl-4-hydroxymethylpyridine	45
3,4-Dimethyl	4,5-Dimethyl-2-hydroxymethylpyridine	13
	3,4-Dimethyl-2-hydroxymethylpyridine	8
4-Methoxy	2-Hydroxymethyl-4-methoxypyridine	16

[1] Reflux in methanol in the presence of ammonium persulfate [89SC317].

$$S_2O_8^{2-} + CH_3OH \longrightarrow {}^{\cdot}CH_2OH + H^+ + SO_4^{2-} + SO_4^{\cdot -}$$

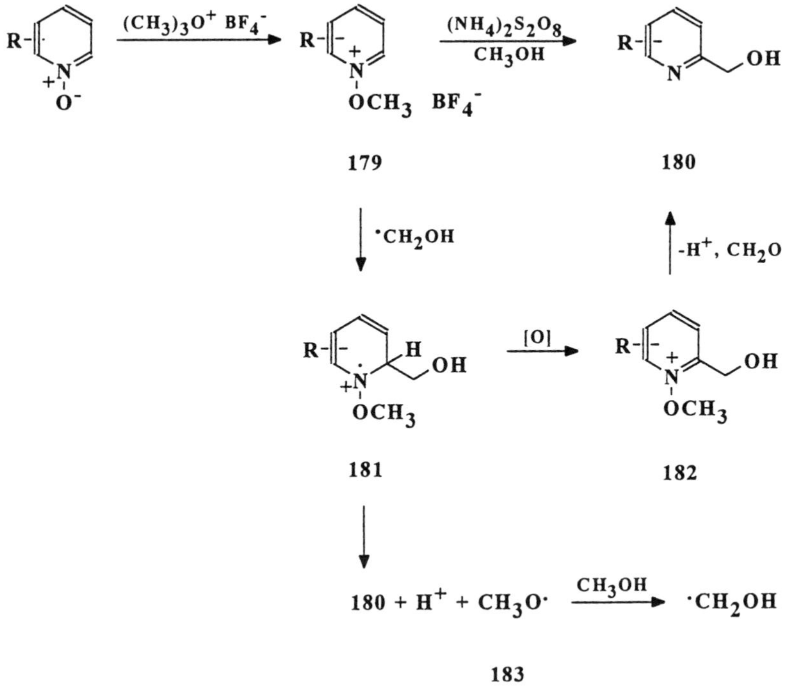

Scheme 122

3. *N-Acyloxyazinium salts.* N-Acyloxyazinium salts **184** are usually not stable enough to be isolated in a crystalline state. However, they can easily be generated *in situ* in solutions by means of a variety of acylating agents, such as acetic anhydride, tosyl chloride, benzoyl chloride, and the like. Reactions of pyridine and quinoline N-oxides with all kinds of O-, N-, S-, P-, and C-nucleophiles in the presence of acylating agents have been well studied and have been shown to provide a convenient way to 2-substituted pyridines and quinolines (Scheme 123; Table XXXI) [86CCA89].

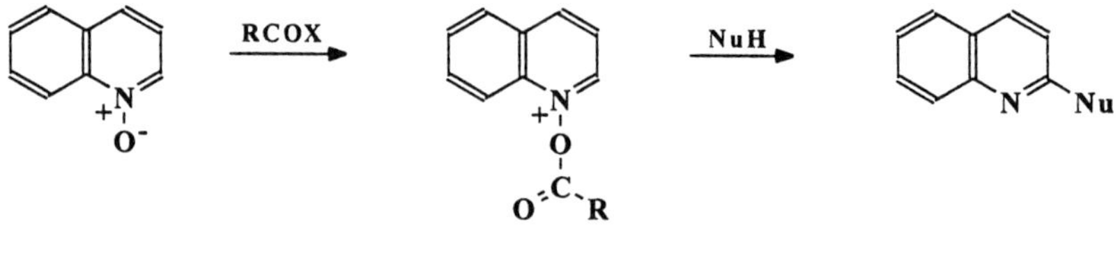

Nu = O-, N-, S-, P- and C-nucleophiles

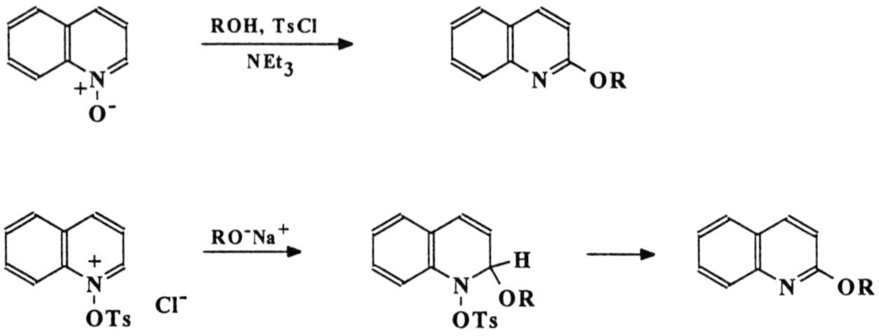

Scheme 123

Since most reactions have been well documented (Table XXXI), the following discussion is restricted to the illustration of some examples. The reaction of quinoline N-oxides with alcohols in the presence of ethyl chloroformate or tosyl chloride provides a simple preparative route to 2-alkoxyquinolines [79JHC1209; 90H1325]. A slightly modified procedure has been developed by using the isolated tosylate salt and sodium alkoxides as reagent (Scheme 123) [83ZOR663].

TABLE XXXI
Nucleophilic Substitution of Hydrogen in N-Acyloxyazinium Salts

Azine N-oxide	Nucleophile (Acylating agent)1	S_N^H Product	Yield (%)	References
Pyridine N-oxide	1-Morpholino-cyclohexene (A)	2-(Pyridyl-2)cyclohexanone	83	65CPB912
Pyridine N-oxide	Dimedone	4-(Pyridyl-2)dimedone	48	82CPB2326
4-Chloropyridine N-oxide	Ethyl cyanoacetate (B)	Ethyl cyano-(4-chloropyridyl-2)acetate	90	78CPB3504
Quinoline N-oxide	Methanol (C)	2-Methoxyquinoline	81	90H1325
Quinoline N-oxide	Ethanol (C)	2-Ethoxyquinoline	76	90H1325
Quinoline N-oxide	Indole (A)	2-(Indolyl-2)quinoline	67	67CPB363
Quinoline N-oxide	Barbituric acid (B)	5-(Quinolyl-2)-barbituric acid	86	81H1083
Quinoline N-oxide	N,N-Dimethylaniline (A)	2-(4-N,N-dimethylaminophenyl)-quinoline	59	64MI1
Quinoline N-oxide	Ammonium hydroxide (D)	2-Aminoquinoline	71	64MI1
Quinoline N-oxide	Triethylphosphite (C)	Diethyl (quinolyl-2)-phosphonate	95	86CCA89
3-Methoxyquinoline N-oxide	Acetic anhydride	4-Acetoxy-3-methoxyquinoline	65	91H1579

TABLE XXXI (Continued)

Azine N-oxide	Nucleophile (Acylating agent)[1]	S_N^H Product	Yield (%)	References
3-Fluoroquinoline N-oxide	POCl$_3$	2-Chloro-3-fluoro-quinoline	83	91H1579
3-Bromoquinoline N-oxide	Dimedone (B)	4-(3-Bromoquino-lyl-2)-dimedone	83	82CPB2326
Isoquinoline N-oxide	Ethanol (C)	1-Ethoxyisoquinoline	73	90H1325
Isoquinoline N-oxide	Barbituric acid (B)	5-(Isoquinolyl-1)-barbituric acid	82	81H1083

[1] A: Benzoyl chloride; B: Acetic anhydride; C: Ethyl chloroformate; D: Tosyl chloride.

Cyanation of the pyridine ring has effectively been achieved by reacting pyridine N-oxides with trimethylsilanecarbonitrile in the presence of dimethylcarbamoyl chloride [83JOC1375; 84H93; 84H2375]. Trimethylsilanecarbonitrile is an excellent source of the cyanide ion in organic solvents, which enables the reaction to proceed smoothly in methylene chloride. The reaction occurs regioselectively at C-2 (Scheme 124; Table XXXII) [83JOC1375], being in contrast to the cyanation of N-methyl and N-methoxy-pyridinium ions which proceed at both C-2 and C-4 (Sections II,B and II.D). The reaction when applied to a series of 3-substituted pyridine N-oxides results in all cases to direct substitution of hydrogen predominantly in the 2-position of the pyridine ring with some minor amounts of 6-cyanopyridines (Table XXXII) [84H93; 84H2375]. The lack of cyanation at the 4-position of the pyridine ring suggests "intramolecular delivery" of the cyanide ion from the intermediate **185**, in which silicium is coordinated to the carbamoyl moiety (Scheme 124) [84H93].

TABLE XXXII

Cyanation of Pyridine[1] and Pyrazine[2] N-Oxides with Trimethylsilane-carbonitrile in the Presence of Dimethylcarbamoyl Chloride[1]

Azine N-oxide	S_N^H Product	Yield (%)
Pyridine	2-Cyanopyridine	94
2-Methypyridine	6-Cyano-2-methylpyridine	90
3-Methylpyridine	2-Cyano-3-methylpyridine	85
	6-Cyano-3-methylpyridine	10
4-Methylpyridine	2-Cyano-4-methylpyridine	90
2-Methoxypyridine	6-Cyano-2-methoxypyridine	90
2-Chloropyridine	2-Chloro-6-cyanopyridine	90
3-Hydroxypyridine	2-Cyano-3-hydroxypyridine	86
3-Cyanopyridine	2,3-Dicyanopyridine	45
	2,6-Dicyanopyridine	37
3-Cyanopyrazine	2,5-Dicyanopyrazine	25
3-Aminopyrazine	2-Amino-3-cyanopyrazine	93
3-Methoxypyrazine	2-Cyano-3-methoxypyrazine	43
	5-Cyano-3-methoxypyrazine	19
3-Methoxycarbonyl-pyrazine	2-Cyano-3-methoxycarbonyl-pyrazine	28
	2-Cyano-5-methoxycarbonyl-pyrazine	32

[1]See ref. [84H2375]. [2]See ref. [91JCS(P1)2877].

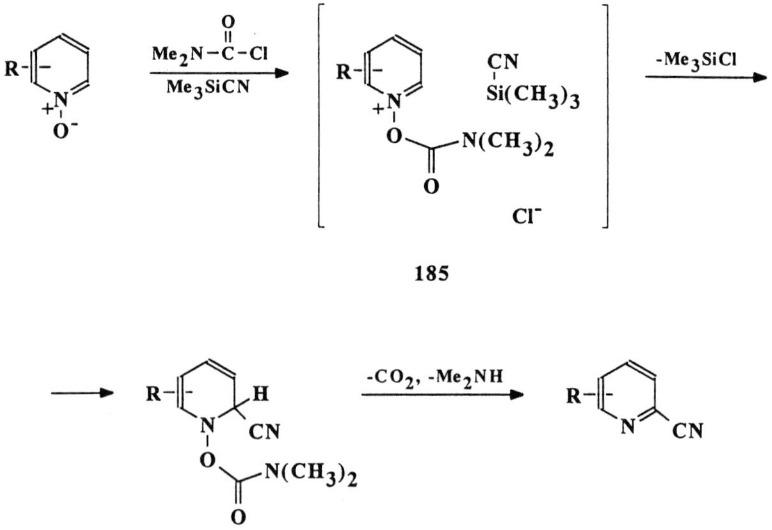

Scheme 124

R = H, CH₃, Cl, CN, OH

Also pyrazine N-oxides can easily be cyanated with trimethylsilylcyanide (Table XXXII). As can be seen from the results the regioselectivity is at positions 2 and 6. The reactivity order is found to be CN > NH$_2$ > OH > COOMe > Cl [91JCS(P1)2877]. In the case of 3-chloropyrazine N-oxide the reaction did not occur; the reaction requires a Lewis acid, such as zinc bromide to afford 3-chloro-2-cyanopyrazine, yield 76% [91JCS(P1)2877]. MO calculations concerning the reactivities and regiospecificities of the 3-substituted pyrazine N-oxides indicate that the cyanation directly occurs in the cases of 3-cyano- and 3-methoxycarbonyl pyrazine N-oxides, whereas in the case of the 3-amino- and 3-methoxypyrazine N-oxide the cyanation occurs after the rate-determining silylation [93BCJ1149].

Use of carbon nucleophiles such as enamines, arylamines, indoles, enol ethers, and others provides us with an excellent methodology to introduce a variety of carbon substituents into the pyridine ring (Table XXXI) [63CPB411; 63CPB1331; 64MI1; 74H167; 79MI1; 80JHC305; 80JHC1029; 82CPB1680; 82CPB1974; 86CCA89]. Of particular interest are those examples where nucleophilic substitution of hydrogen in the pyridine ring occurs in spite of the presence of a nucleofugal substituent. This is exemplified by the introduction of a cyclohexanone ring at position 4 of the quinoline system by the reaction of 2-chloroquinoline N-oxide with 1-morpholinocyclohexene (Scheme 125; Table XXXI) [66CPB762; 67CPB474].

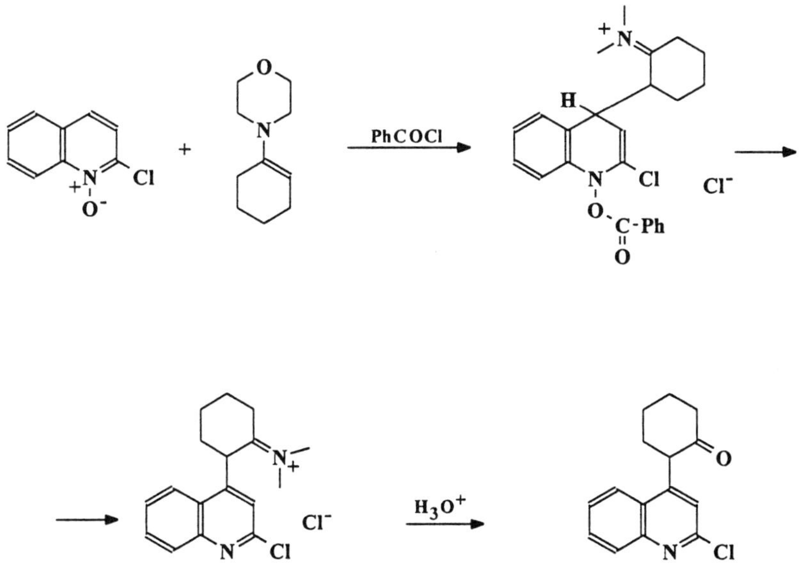

Scheme 125

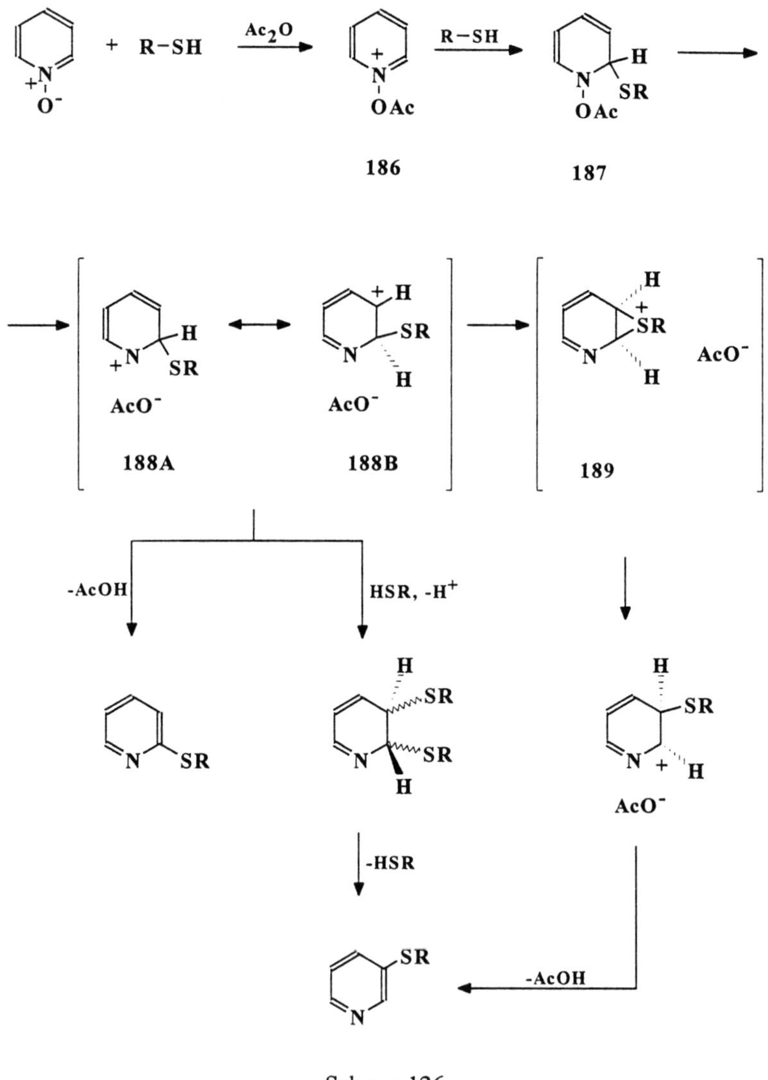

Scheme 126

In nearly all S_N^H reactions with pyridine N-oxides it is the α- or γ-position of the ring which is attacked by the nucleophiles. However, there are observed nucleophilic substitutions in which considerable quantities of β-substituted S_N^H products are formed [86H161; 93JCS(P1)15; 93JHC691]. For instance, pyridine N-oxides dissolved in acetic anhydride when reacting with thiols furnish α- and β-pyridyl sulfides (Scheme 126; Table XXXIII) [86H161; 91JHC1051]. A plausible hypothesis to account for both α- and β-substitution has been advanced involving common intermediates (Scheme 126). It has been suggested that the initially formed N-acetoxypyridinium cation **186** is attacked by thiols at C-2 to generate the 1,2-dihydropyridines **187**. Aromatization of these σ^H-adducts requires elimination of acetic acid. Since the sp^3-hydrogen is not particularly acidic, it is suggested that an ion pair **188** may be created in which the electron deficiency at the β-position in the pyridinium counter ion is enhanced [86H161]. The cationic character of C-3, symbolized by the resonance-hybrid structures **188A** and **188B**, stimulates a nucleophilic attack by the neighboring sulfide to form the episulfonium ion **189**. Intermediacy of episulfonium ions **189** explains the formation of β-pyridyl sulfides [91JHC1051]. It cannot be excluded that an intermolecular addition with the thiols takes place on the β-carbon, leading to the 2,3-dialkylthio-2,3-dihydropyridines, which by a 1,2-elimination yields the β-pyridyl sulfides (Scheme 126).

Reaction of N-benzoyloxypyridinium chloride with silver acetylide results in 2-phenylethynylpyridine; similar results have been obtained with substituted pyridine N-oxides, but the yields obtained are usually low (Scheme 127) [89CL773]. Replacement of benzoyl chloride by acetyl chloride lowers the reactivity, while no reaction at all occurred with acetyl anhydride. In all cases the substitution takes place at the 2- or 6-position; no formation of 4-phenylethynylpyridine was observed [89CL773]. The reaction is supposed to proceed with attack of the acetylide at C-2 (C-6) to form 1,2-dihydropyridine intermediate **190** followed by elimination of benzoic acid (Scheme 127) [89CL773].

TABLE XXXIII
Substitution of Hydrogen in Pyridine N-Oxides
by Thiols R-SH in Acetic Anhydride[1]

Pyridine N-oxide	Substituent R in thiol R-SH	S_N^H Product(s)	Yield (%)
Unsubstituted	Methyl	2-Mehtylthiopyridine	20
		3-Methylthiopyridine	18
Unsubstituted	t-Butyl	2-(t-Butylthio)pyridine	43
		3-(t-Butylthio)pyridine	19
Unsubstituted	1-Adamantyl	2-(1-Adamantylthio)pyridine	4
		3-(1-Adamantylthio)pyridine	40
2-Methyl	t-Butyl	6-(t-Butylthio)-2-methylpyridine	27
		5-(t-Butylthio)-2-methylpyridine	5
3-Methyl	t-Butyl	2-(t-Butylthio)-3-methylpyridine	30
		6-(t-Butylthio)-3-methylpyridine	12
		5-(t-Butylthio)-3-methylpyridine	24
3-Aminocarbonyl	1-Adamantyl	2-(1-Adamantylthio)nicotinamide	60
		6-(1-Adamantylthio)nicotinamidel	6
		5-(1-Adamantylthio)nicotinamide	1
4-Methyl	t-Butyl	2-(t-Butylthio)-4-methylpyridine	29
		3-(t-Butylthio)-4-methylpyridine	12
4-Phenyl	1-Adamantyl	2-(1-Adamantylthio)-4-phenyl-pyridine	40
		3-(1-Adamantylthio)-4-phenyl-pyridine	16

[1]See references [69JOC655; 86H161].

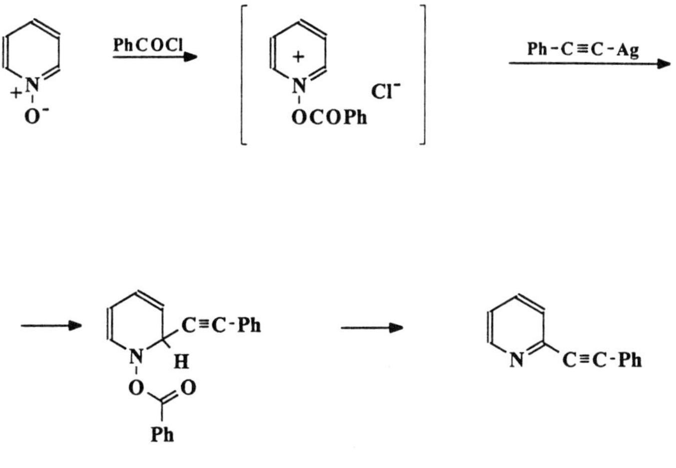

Scheme 127

190

Experimental Procedures

Example 1. Preparation of 4-(2-pyridyl)dimedone

A solution of pyridine N-oxide (380 mg, 4 mmol) in acetic anhydride (5 ml) was added all at once to a solution of dimedone (560 mg, 4 mmol) in acetic anhydride (8 ml), and the whole was heated at 90°C for 8 hours. The yellow reaction mixture was concentrated under reduced pressure, and the residue was recrystallized from methanol to give 420 mg (48%) of 4-(2-pyridyl)dimedone as pale yellow needles, melting point 162–163°C [82CPB2326].

Example 2. Preparation of ethyl cyano(4-chloropyridyl-2)acetate

A solution of 4-chloropyridine 1-oxide (1.3 g, 9.2 mmol), acetic anhydride (2.5 g, 25 mmol), and ethyl cyanoacetate (1.5 g, 13 mmol) in DMF was stirred at −15°C for 2 hours and then at room temperature overnight. The reaction mixture was poured into water, made alkaline with sodium carbonate, and extracted with methylene chloride. The extract was washed several times with NaCl solution and evaporated. The residue was chromatographed on alumina with benzene to give 2.0 g (90%) of ethyl cyano-(4-chloropyridyl-2)acetate, yellow needles, melting point 131–132°C [78CPB3504].

Example 3. A typical procedure for the hydroxymethylation of pyridine N-oxides

Pyridine-N-oxide (5 mmol) was dissolved in dry dichloromethane (15 m) and treated with trimethyloxonium tetrafluoroborate (5 mmol). After stirring at room temperature for 90 minutes the solvent was removed under reduced pressure and the resulting N-methoxypyridinium tetrafluoroborate was dissolved in methanol and the solution was heated under reflux. Ammonium persulfate (228 mg, 1 mmol) dissolved in water (1 ml) was added and after 30 minutes, a further portion of ammonium persulfate (114 mg, 0.5 mmol) in water (0.5 ml) was added. This solution was refluxed for a further 30 minutes. Then the mixture was evaporated to dryness. Chromatography on silica gel with chloroform–methanol (9:1) gave the corresponding 2-hydroxymethylpyridine (see Table XXX) [89SC317].

Example 4. Preparation of 2-ethoxyquinoline

To a stirred ice-cooled solution of quinoline N-oxide (720 mg, 5 mmol) and ethyl chloroformate (700 mg, 6.5 mmol) in ethanol (30 ml) triethylamine (1.01 g, 10 mmol)

was added. The reaction mixture was stirred at room temperature for an additional 12 hours, then concentrated under reduced pressure and the residue was extracted with 10% hydrochloric acid. The acidic solution was made alkaline with an aqueous solution of Na_2CO_3 and extracted with chloroform, and the residue from the extract was chromatographed on silica gel. Elution with hexane gave 2-ethoxyquinoline as a colorless oil. Yield 650 mg (76%). Picrate, yellow needles, melting point 148–149°C [90H1325].

Example 5. Preparation of 4-acetoxy-3-methoxyquinoline

A solution of 3-methoxyquinoline N-oxide (2.0 g, 11.4 mmol) in acetic anhydride (10 ml) was heated at reflux for 2 hours, and then excess of acetic anhydride was evaporated under reduced pressure and the residue was dissolved in chloroform. The chloroform solution was washed successively with aqueous $NaHCO_3$ and water. The residue from the $CHCl_3$ solution was recrystallized from chloroform–ether (1:1) to give 17.0 g (69%) of 4-acetoxy-3-methoxyquinoline as colorless needles, melting point 115–116°C [91H1579].

Example 6. Preparation of 2-phenylethynylpyridine

Benzoyl chloride (1.5 equiv.) was added dropwise to a solution of pyridine N-oxide (1.5 equiv.) in a solvent at room temperature to generate N-benzoyloxypyridinium salt as white precipitate. It is difficult to isolate the salt because of its hydroscopic character, so the reaction was carried out without isolation of the salt. After stirring for 0.5 hour, silver phenylacetylide (1.0 equiv.) was added and the reaction mixture was heated for 1 hour. Insoluble material was filtered off and the filtrate was washed (2 M NaOH), dried

dried (MgSO$_4$), and concentrated *in vacuo* to give 2-phenylethynylpyridine, which was isolated by column chromatography (Al$_2$O$_3$, hexane–benzene 1:1) [89CL773].

E. *N-Fluoroazinium Salts*

N-Fluoropyridinium salts form a unique and relatively new class of azinium cations which became known after Meinert's discovery of the parent unstable N-fluoropyridinium fluoride [65ZC64]. The salts can easily be prepared in good yields by reacting pyridines with a gas mixture of 10–15% of fluorine in nitrogen in the presence of a metal salt of a strong acid or Lewis acid. Preparation of stable N-fluoropyridinium salts, such as N-fluoropyridinium tetrafluoroborate [86TL3271], N-fluoropyridinium triflate [90OS129], and others [86TL3271; 91BCJ1081] caused a greater interest in these compounds and enabled researchers to elucidate their structure by ^{19}F NMR [91JFC369] and to study their behavior toward nucleophilic reagents.

N-Fluoropyridinium salts are reactive with a variety of N-, O-, P-, As-, and C-nucleophiles, and with halides (except iodide). The 2- or 4-substituted pyridines obtained in these reactions can formally be regarded as S_N^H products [83MI1; 87JA3789; 87TL2705; 88JOC1123; 89JOC1726; 90TL7379; 91JFC392; 91JOC6298; 91MC128; 93JHC329].

Treatment of N-fluoropyridinium tetrafluoroborates **191** with an excess of triethylamine at room temperature leads to a regiospecific substitution of hydrogen at 2-position of the pyridine ring by a fluorine atom (Scheme 128) [89JOC1726]. As seen in Table XXXIV, the reaction is successfully applied to prepare a host of fluoropyridines having electron-donating or electron-withdrawing substituents. The same results were obtained with other counteranions, such as SbF$_6^-$ or PF$_6^-$ [89JOC1726].

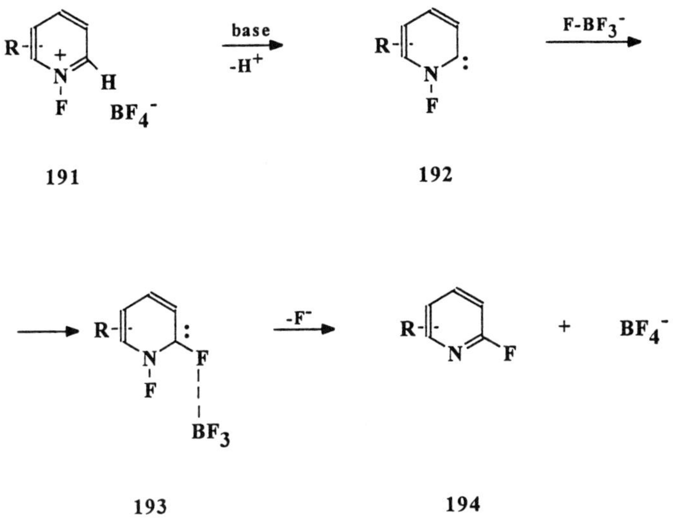

Chapter 3 S_N^H REACTIONS IN HETARENES

Scheme 128

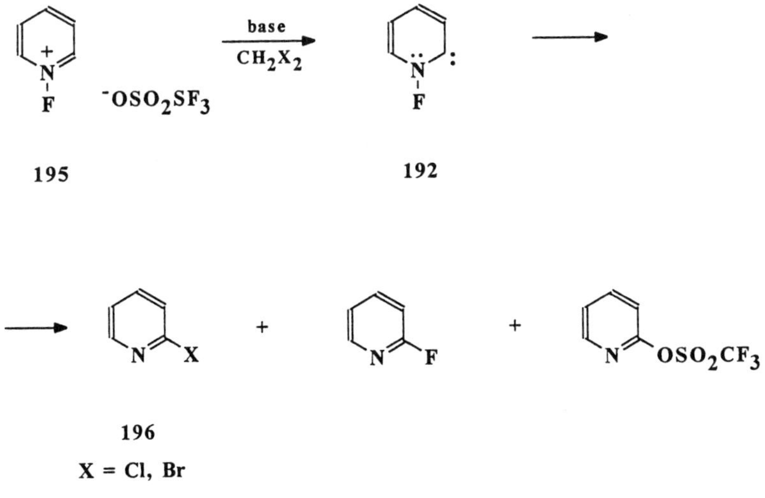

Scheme 129

TABLE XXXIV
Yields of 2-Fluoropyridines Obtained on Treatment of N-Fluoropyridinium Tetrafluoroborates with Triethylamine[1]

Substituent R in N-fluoropyridinium tetrafluoroborate	S_N^H Product	Yield (%)
4-Methyl	2-Fluoro-4-methylpyridine	80
3,5-Dimethyl	2-Fluoro-3,5-dimethylpyridine	87
4-t-Butyl	4-t-Butyl-2-fluoropyridine	91
2-Methoxy	2-Fluoro-6-methoxypyridine	75
4-Phenyl	2-Fluoro-4-phenylpyridine	40
4-(Methoxycarbonyl)	2-Fluoro-4-(methoxycarbonyl)pyridine	69
3-(Ethoxycarbonyl)	3-(Ethoxycarbonyl)-2-fluoropyridine	48
	5-(Ethoxycarbonyl)-2-fluoropyridine	7
2-Chloro	2-Chloro-6-fluoropyridine	72
3,5-Dichloro	3,5-Dichloro-2-fluoropyridine	62
3-Cyano	3-Cyano-2-fluoropyridine	51
4-Nitro	2-Fluoro-4-nitropyridine	21

[1] See ref. [89JOC1726].

To explain the observed regiospecific fluorination of pyridines at C-2 of N-fluoropyridinium salts **191**, singlet carbene **192** is postulated as intermediate, which reacts with a fluoride from either BF_4^-, SbF_6^- or PF_6^- to give adduct **193** [89JOC1726; 91JFC392]. By elimination of the fluoride ion the 2-fluoro compound **194** is formed with simultaneous regeneration of the counteranion (Scheme 128) [89JOC1726]. The behavior of N-fluoropyridinium triflate **195** to bases is somewhat different: treatment of the triflate with a 10-fold excess of triethylamine gave only a small yield of 2-fluoropyridine (21%) and a large amount of 2-pyridyl triflate (57%) [89JOC1726]. N-Fluoropyridinium triflate being treated with tetrabutylammonium fluoride (1 equiv.) in methylene chloride at room temperature for 5 minutes gave 2-chloropyridine **196** (28%) as main product together with some 2-fluoropyridine (14%) and 2-pyridyl triflate (4%) (Scheme 129) [89JOC1726]. The formation of **196** is explained by the reaction of the carbene intermediate **192** with methylene chloride. Using other solvents and different bases makes it possible to obtain a variety of 2-substituted pyridines (Table XXXV) [87TL2705].

Use of acetyl hypofluorite has been shown to open a new pathway for the synthesis of substituted pyridines, since this compound has been found to activate the pyridine ring toward attacks by weak nucleophiles, such as alcohols and halogen in CHX_2 (X= Cl and Br) [87JA3789; 88JOC1123; 89H249; 91JOC6298]. As one can see from the results presented in Table XXXVI, a variety of different S_N^H products, depending on the nature of both the pyridine derivative and the solvent, can be obtained in 60–90% yields.

The reaction of pyridine with acetyl hypofluorite in acetic acid at -75^oC for a short time forms 2-acetoxypyridine **199** (R= H) in 85% yield (Scheme 130) [87JA3789]. Similarly 4-methyl and 4-benzoyl pyridines [87JA3789], quinoline, and pyrazine [89H249] were transformed into 2-acetoxy-4-methylpyridine, 2-acetoxy-4-benzoylpyridine, 2-acetoxyquinoline, and 2-acetoxypyrazine in 60–90% yields (Table XXXVI) [87JA3789; 89H249]. The addition of radical inhibitors or initiators such as nitrobenzene or benzoyl peroxide did not affect the outcome of the reactions. Therefore, it seems unlikely that the S_N^H substitution proceeds via a radical pathway [87JA3789]. It has been suggested that the first step is an attack of acetyl hypofluorite on the ring nitrogen, producing ion pair **197**. A subsequent attack of the acetate on the pyridinium

TABLE XXXV
Yields of Products Obtained from the Reaction
of N-Fluoropyridinium Triflate with Bases[1]

Solvent	Base	S_N^H Product(s)	Yield (%)
Dichloromethane	Triethylamine	2-Chloropyridine	62
		2-Pyridyl trifluoromethanesulfonate	21
		2-Fluoropyridine	5
Dibromomethane	Triethylamine	2-Bromopyridine	60
		2-Pyridyl trifluoromethanesulfonate	27
		2-Fluoropyridine	4
Acetonitrile	Triethylamine	2-Acetamidopyridine	55
		2-Pyridyl trifluoromethanesulfonate	7
		2-Fluoropyridine	3
Benzene	Triethylamine	2-Phenylpyridine	33
		2-Pyridyl trifluoromethanesulfonate	42
		2-Fluoropyridine	15
Dichloromethane	Sodium hydroxide[2]	2-Hydroxypyridine	50
		2-Pyridyl trifluoromethanesulfonate	<1
		2-Fluoropyridine	<1
Methanol	Sodium methoxide[2]	2-Methoxypyridine	73
		2-Pyridyl trifluoromethanesulfonate	3
		2-Fluoropyridine	2

[1] See ref. [87TL2705]. [2] With 2 equiv. of base; in all other cases 1 equiv. of triethylamine was used.

TABLE XXXVI
Nucleophilic Halogenation, Alkoxylation, and Acetoxylation of
N-Fluoroazinium Acetates Derived from the Reaction of Azines
with Acetyl Hypofluorite

Azine	Solvent	S_N^H Product	Yield (%)	References
Pyridine	CH_3COOH	2-Acetoxypyridine	85	87JA3789
Pyridine	CH_2Cl_2	2-Chloropyridine	70	88JOC1123
		2-Acetoxypyridine	15	
Pyridine	CH_2Br_2	2-Bromopyidine	60	91JOC6298
		2-Acetoxypyridine	20	
Pyridine	$CHBr_3$	2-Bromopyridine	60	91JOC6298
		2-Acetoxypyridine	20	
Pyridine	CH_3OH	2-Methoxypyridine	70	91JOC6298
Pyridine	C_2H_5OH	2-Ethoxypyridine	70	91JOC6298
Pyridine	$n\text{-}C_4H_9OH$	2-n-Butoxypyridine	70	91JOC6298
4-Methyl-pyridine	CH_3COOH	2-Acetoxy-4-methyl-pyridine	75	87JA3789
4-Methyl-pyridine	CH_2Cl_2	2-Chloro-4-methyl-pyridine	60	91JOC6298
4-Methyl-pyridine	CH_2Br_2	2-Bromo-4-methyl-pyridine	55	91JOC6298
4-Benzoyl-pyridine	CH_3COOH	2-Acetoxy-4-benzoyl-pyridine	90	87JA3789

TABLE XXXVI (Continued)

Azine	Solvent	S_N^H Product	Yield (%)	References
3-Chloro-pyridine	CH$_2$Cl$_2$	2,3-Dichloropyridine	80	91JOC6298
3-Chloro-pyridine	CH$_2$Br$_2$	2-Bromo-3-chloro-pyridine	80	91JOC6298
3-Chloro-pyridine	CH$_3$OH	3-Chloro-2-methoxy-pyridine	75	91JOC6298
3-Fluoro-pyridine	CH$_2$Cl$_2$	2-Chloro-3-fluoro-pyridine	40	91JOC6298
4-Acetyl-pyridine	CH$_2$Br$_2$	2-Bromo-4-acetyl-pyridine	60	91JOC6298
3-Benzoyl-pyridine	CH$_2$Cl$_2$	2-Chloro-3-benzoyl-pyridine	80	91JOC6298
Methyl nicotinate	CH$_2$Cl$_2$	2-Chloro-3-methoxy-carbonylpyridine	60	91JOC6298
Quinoline	CH$_3$COOH	2-Acetoxyquinoline	90	87JA3789
Quinoline	CH$_3$OH	2-Methoxyquinoline	60	91JOC6298
4-Methyl-quinoline	CH$_3$OH	2-Methoxy-4-methyl-quinoline	60	91JOC6298

ion at C-2 results in a collapse of the ion pair. Elimination of HF from the S_N^H adduct **198** restores the aromaticity of the pyridine ring (Scheme 130) [87JA3789]. These reactions can be ascribed as a *cine*-substitution in which the entering group substitutes a hydrogen at the position *cine* to that of the leaving group (in this case the fluoride ion).

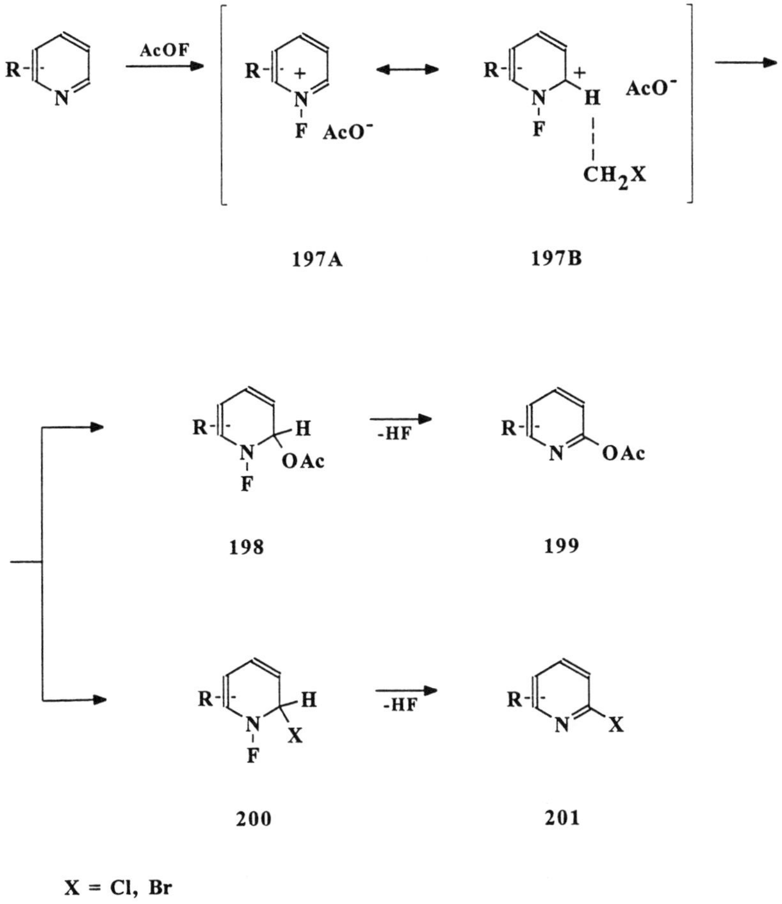

Scheme 130

X = Cl, Br

Reactions of pyridines with acetyl hypofluorite can also be carried out in solvents other than acetic acid, such as alcohols and methylene halides. As can be expected, the results proved to be strongly dependent on the nature of the solvent employed [88JOC1123; 89H249; 91JOC6298]. When a solution of acetyl hypofluorite was added at room temperature to pyridine dissolved in methylene chloride, an instantaneous reaction took place yielding as main product 2-chloropyridine (70%) and some 2-acetoxypyridine (15%) (Table XXXVI) [88JOC1123]. A plausible pathway for this reaction is also presented in Scheme 130. The carbocation **197B**, being a resonance form of the N-fluoropyridinium ion **197**, is probably responsible for holding a partially negatively charged chloro atom in the vicinity of C-2 of the N-fluoropyridinium salt. Addition of this chloro atom can then successfully compete with addition of the acetoxy residue of the acetyl hypofluorite [88JOC1123]. It should be noted that the formation of carbens as reactive species (as has been suggested in the base-induced conversion of N-fluoropyridinium tetrafluoroborate into 2-fluoropyridine) certainly has to be excluded in cases where the reactions with acetyl hypofluorite are carried out in acidic media [88JOC1123].

Methylene chloride also proved to be an efficient chlorinating agent for introducing a chloro atom in a host of other N-fluoropyridinium acetates (Table XXXVI) [88JOC1123; 91JOC6298]. Replacing methylene chloride with either methylene bromide or methyl bromide leads to substitution of the C-2 hydrogen for a bromo atom. Attempts to introduce an iodo substituent into the pyridine ring by employing CH_3I or CH_2I_2 have failed due to the fact that these solvents are easily oxidized by the acetyl hypofluorite [91JOC6298]. Among other solvents which are not rapidly oxidized, but still possess a nucleophilic moiety, primary alcohols seem to meet these requirements (Table XXXVI) [88JOC1123; 91JOC6298].

It is noteworthy that pyridine but also 3-chloro-, 3-fluoro-, 3-methoxycarbonyl-, 3-benzoyl-, 4-methyl-, or 4-acetylpyridines as well as quinoline and 4-methylquinoline react successfully with acetyl hypofluorite in all kinds of solvents (Table XXXVI). Pyridines with a low basicity, such as 2-cyano, 2-chloro, or 3,5-dichloropyridine do not interact with the electrophilic fluorine of the acetyl hypofluorite; only the starting material has eventually been isolated [91JOC6298]. It is also observed that with the 3-substituted pyridines the S_N^H reactions always take place at the *cine* position C-2. It supposes a intramolecular complexing of the reagent with the pyridine ring nitrogen and/or the C-3 substituent; thus the formation of an intimic ion pair may be involved in the reaction mechanism.

Reactions of C-nucleophiles with N-fluoropyridinium tetrafluoroborate are other examples of successful S_N^H substitutions in the pyridine ring (Table XXXVII) [83MI1; 90TL7379]. High yields of 2- and 4-substituted pyridines (even in water) are obtained. So, besides the *cine* substitution at C-2, the *tele* substitution at C-4 occurs [90TL7379]. Concerning the mechanism, it is assumed that the reactions of N-fluoropyridinium tetrafluoroborate with weakly basic carbanions proceed preferentially at C-2 or C-4 via the $S_N(AE)$ (addition-elimination) pathway (Scheme 131, path 3). In reactions with highly basic carbanions, C-2 substituted pyridines are formed from the solvent along with the C-substitution products through the possible intermediacy of carbens **202** (or cations **203**) which react with the solvent or a nucleophile (Scheme 131, path 1) [90TL7379].

The reaction of N-fluoropyridinium tetrafluoroborate with carbanions have recently been suggested to occur according to a $S_{RN}1$ process (Scheme 131, path 2), involving formation of the N-fluoropyridine radical **205** [90JA8563]. Recently published data on the high selectivity of the reaction of N-fluoropyridinium tetrafluoroborate **210** with sulfides **211**, which is regarded as a novel approach to the single-electron transfer

TABLE XXXVII
Yields of Products Obtained from the Reaction of N-Fluoropyridinium Tetrafluoroborate with Carbanions[1]

Reagent	Solvent	S_N^H product(s)	Yield[1] (%)
Sodium salt of diethylmalonate	EtOH	Di(ethoxycarbonyl)-(pyridyl-2)-methane	22
		2-Ethoxypyridine	41
Sodium salt of ethyl acetoacetate	EtOH	2-Ethyl-α-(Pyridyl-2)acetate	26
		2-Ethoxypyridine	38
Potassium salt of acetoacetone	CH_3CN	N-(Pyridyl-2)acetamide	24
		α-(Pyridyl-2)acetone	31
Potassium salt of trinitromethane	H_2O	2-Trinitromethylpyridine	7
		4-Trinitromethylpyridine	41
Potassium salt of trinitromethane	CH_3CN	2-Trinitromethylpyridine	21
		4-Trinitromethylpyridine	7
Sodium salt of ethyl nitroacetate	H_2O	$O_2N-C^-\text{-}COOEt$ on pyridinium Na^+	46
Sodium salt of ethyl dicyanoacetate	EtOH	pyridinium-$C(CN)_2COOEt$ Na^+	27
		2-Ethoxypyridine	37

[1]See ref. [90TL7379].

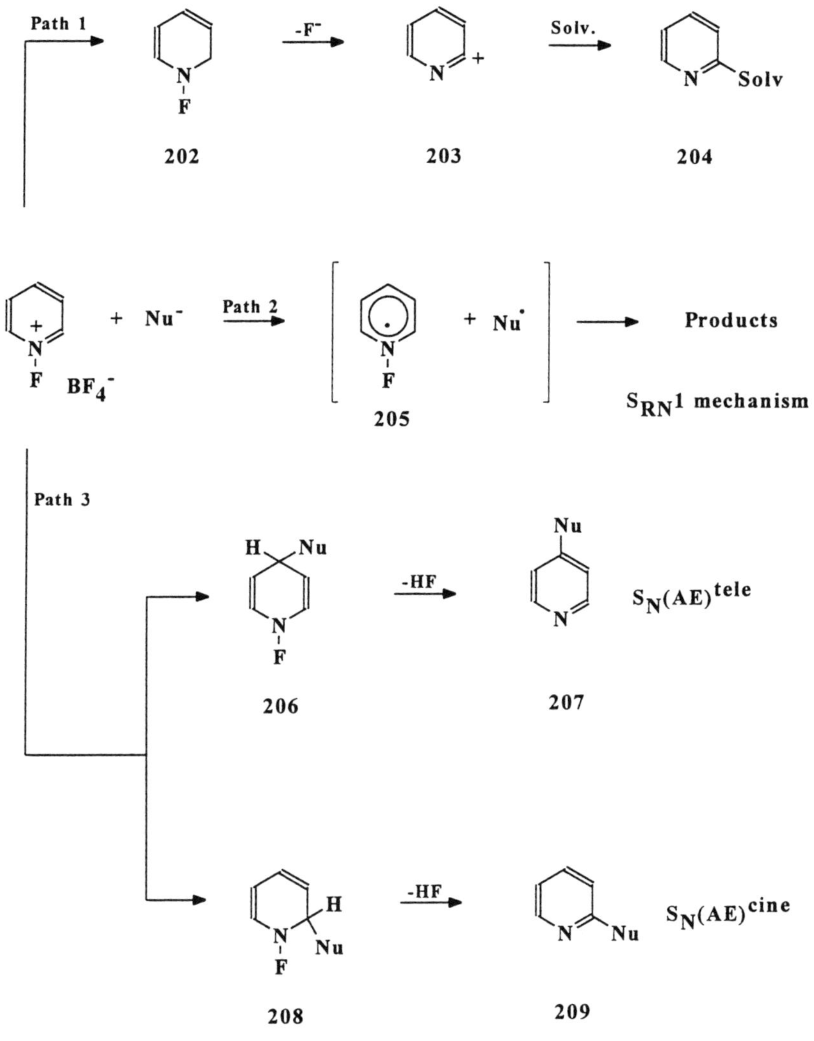

Chapter 3 S_N^H REACTIONS IN HETARENES   211

Scheme 131

from the sulfide **211** to the N-fluoropyridinium substrate **210**, giving the N-fluoro electrophilic radical **212** and the radical cation **213**. Loss of an α-proton from species **213** yields sulfide radicals **214**, which subsequently couple with **212** into the

1,2-dihydropyridines **215**. An alternative way to the dihydropyridines **215** is the nucleophilic addition of **214** to the N-fluoropyridinium ion **210**, yielding the radical cation species **217**, followed by a one-electron reduction of the latter by sulfides **211**. By elimination of hydrogen fluoride from **215** the products **216** are formed (Scheme 132) [93JHC329].

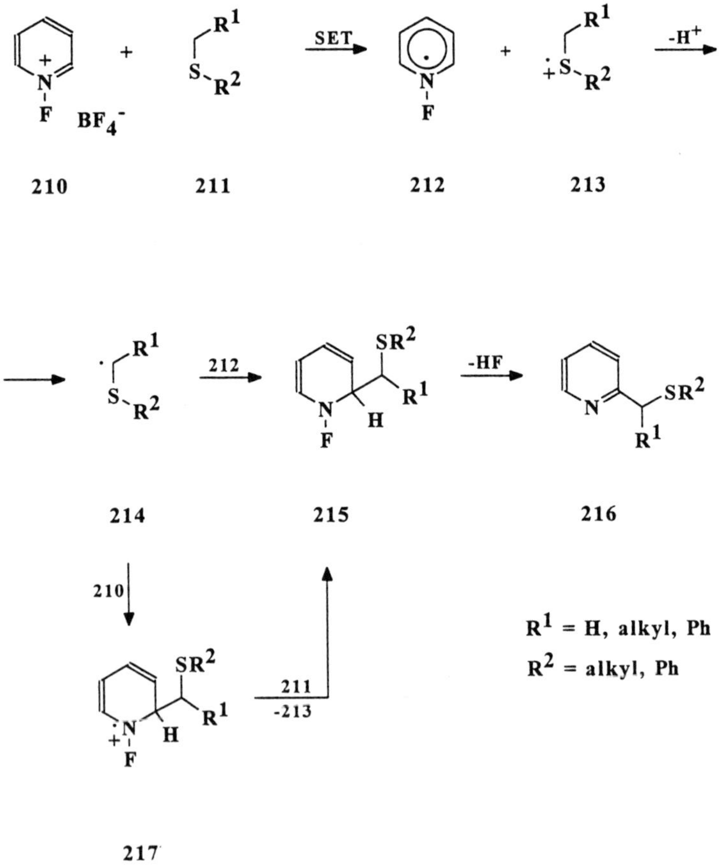

Scheme 132

Chapter 3 S_N^H REACTIONS IN HETARENES 213

The *cine*-(AE) aromatic substitution mechanism has been advanced for the S_N^H reactions of N-fluoropyridinium tetrafluoroborate with trialkyl/aryl phosphines and arsines, yielding 2-pyridyl-substituted phosphonium and arsenonium salts **218** (Scheme 133; Table XXXVIII) [91MC128]. In a similar reaction with trialkylphosphites the S_N^H products **219** (esters of 2-pyridylphosphonic acid) were obtained only in small yields (Scheme 133; Table XXXVIII).

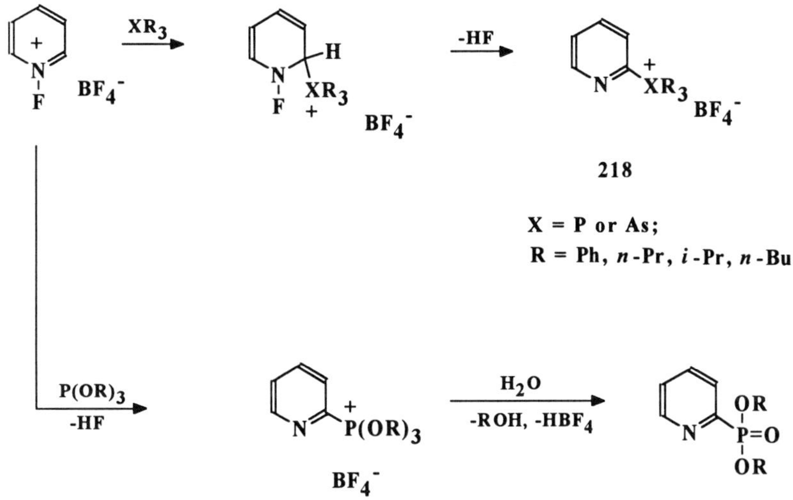

X = P or As;
R = Ph, *n*-Pr, *i*-Pr, *n*-Bu

R = Me, Et, *n*-Pr, *i*-Pr,

Scheme 133

TABLE XXXVIII
Reactions of N-Fluoropyridinium Tetrafluoroborate with Alkyl and Aryl Phosphines, Arsines and Trialkylphosphites

Reagent	S_N^H Product	Yield (%)
$P(CH_2CH_2CH_3)_3$	(Pyridyl-2)tripropylphosphonium tetrafluoroborate	78
$P[CH(CH_3)_2]_3$	(Pyridyl-2)tri-i-propylphosphonium tetrafluoroborate	75
$P(CH_2CH_2CH_2CH_3)_3$	(Pyridyl-2)tributylphosphonium tetrafluoroborate	84
$As(CH_2CH_2CH_2CH_3)_3$	(Pyridyl-2)tributylarsenonium tetrafluoroborate	86
PPh_3	(Pyridyl-2)triphenylphosphonium tetrafluoroborate	92
$AsPh_3$	(Pyridyl-2)triphenylarsenonium tetrafluoroborate	96
$P(OCH_3)_3$	Dimethyl ester of pyridine-2-phosphonic acid	6
$P(OCH_2CH_3)_3$	Diethyl ester of pyridine-2-phosphonic acid	5
$P(OCH_2CH_2CH_3)_3$	Di-n-propyl ester of pyridine-2-phosphonic acid	5
$P[OCH(CH_3)_2]_3$	Di-i-propyl ester of pyridine-2-phosphonic acid	4

Experimental Procedures

Example 1. Preparation of 2-fluoro-4-*t*-butylpyridine

Ground crystals of N-fluoro-4-*t*-butylpyridinium tetrafluoroborate [89JOC1726] (1.90 g, 7.89 mmol) were added by portions to 8 ml (79 mmol) of triethylamine under stirring at room temperature for 30 minutes. A vigorous exothermic reaction occurred immediately each time a portion of the N-fluoropyridinium salt was added. After all the salt was added, the solution was stirred for an additional 5 minutes. The gas chromatographic analysis of the reaction mixture showed that 4-*t*-butyl-2-fluoropyridine was produced in a 91% yield. The reaction mixture was acidified with 1% hydrochloric acid and extracted with pentane. The organic layer was dried with anhydrous magnesium sulfate and filtered, and the evaporation of the solvent gave 4-*t*-butyl-2-fluoropyridine as an oil (purity by GC analysis > 98%). Yield 0.87 g (72%) [89JOC1726; Copyright permission 1994 American Chemical Society].

Example 2. A typical experimental procedure for direct chlorination, bromination, and etherification of pyridines activated by acetyl hypofluorite

A gas mixture of 15% F_2 in N_2 was bubbled into a cold (−75°C) suspension of 8 g of AcONa in 400 ml of $CFCl_3$ and 50 ml of AcOH. The amount of AcOF thus obtained can be easily determined by reacting aliquots of the reaction mixture with aqueous KJ solution and titrating the liberated iodine. After the desired concentration of AcOF is achieved (usually around 0.1–0.15 *M*) the oxidizing solution was added in portions of 10–20 ml each to the pyridine derivative, which was dissolved in the appropriate solvent (room temperature). The reactions were usually carried out on scales of 30–40 mmol using 1.5- to 2-fold excess of AcOF, with conversions higher than 95%. They were usually monitored by GC, TLC, or NMR and in most cases were complete within a few minutes. The reaction was terminated by pouring it into 500 ml of thiosulfate solution, washing the organic layer with a NaHCO$_3$ solution followed by water until neutral, drying the organic layer over MgSO$_4$, and finally evaporating the solvent. The crude reaction mixture was usually subjected to vacuum flash chromatography using silica gel 60-H (Merck), with mixtures of EtOAc in petroleum ether serving as eluent. The halo and alkoxy products were eluted first, followed by the more polar 2-acetoxy derivatives [91JOC6298; Copyright permission 1994 American Chemical Society].

Example 3. A typical experimental procedure for the reactions of N-fluoropyridinium salts with C-nucleophiles

N-Fluoropyridinium tetrafluoroborate (1.84 g, 10 mmol) was added to a stirred solution of appropriate carbanion in absolute ethanol at −70°C or in absolute acetonitrile at −40°C (for examples see Table XXXVII). The mixture was kept at this temperature until the starting N-fluoropyridinium salt had been consumed completely (KJ/starch indicator); then the reaction mixture was filtered and concentrated. Products were isolated by column chromatography (eluent hexane–ether, 2:1) (Table XXXVII) [90TL7379].

F. *N-Aminoazinium Salts*

Of this class of compounds only a few representatives are investigated on their behavior toward nucleophiles. N-Acetylaminopyridinium salts **220** were shown to undergo the cyanation reaction with aqueous potassium cyanide, yielding the S_N^H products at C-2 or/and C-4. The parent salt (R= H) and its 3-methyl and 3-phenyl analogs are cyanated predominantly at C-4, yielding **221** while treatment of 3-bromo and 3-methoxy substituted N-acetylaminopyridinium salts with a large excess of potassium cyanide results in the 2-cyanation products **222**. In both cases the displacement of hydrogen is accompanied by formation of N-methylacetamide, which is eliminated from the ring nitrogen in the aromatization of the intermediate adduct (Scheme 134) [66CPB518].

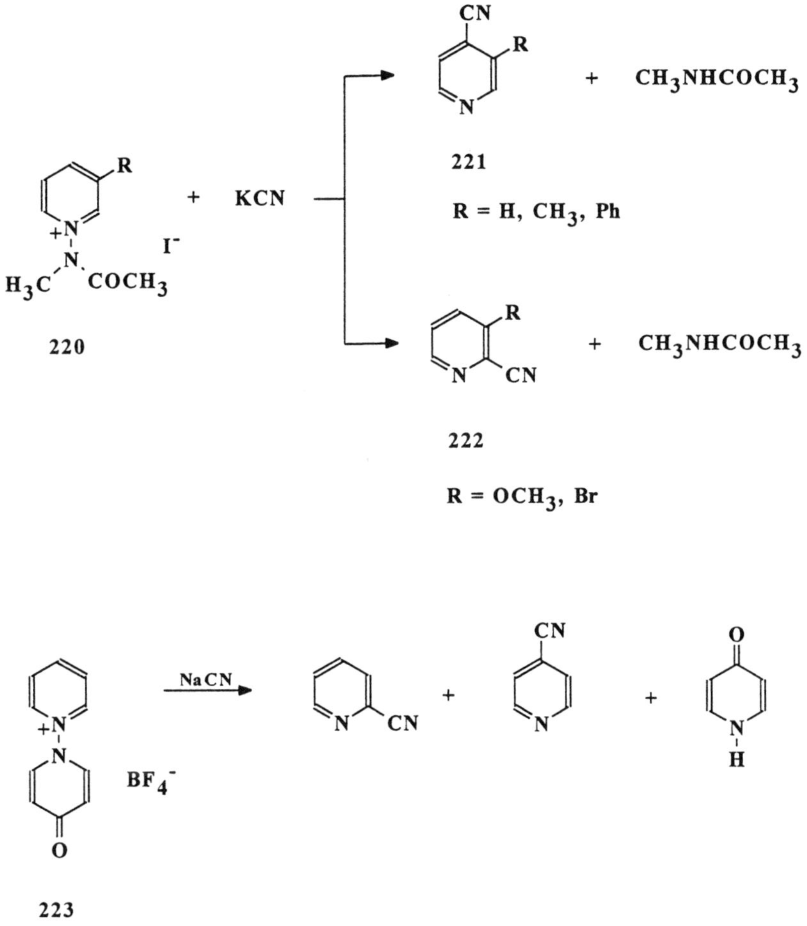

Scheme 134

In an analogous manner, treatment of N-pyridinio-4-pyridine **223** with aqueous sodium cyanide gives cyanation at C-2 and C-4; 4-pyridone is formed as a by-product (Scheme 134) [79JCS(P1)1698].

III. PYRYLIUM AND THIOPYRYLIUM SALTS

Pyrylium and thiopyrylium salts are known as highly π-deficient and reactive aromatics with a profound tendency to add nucleophilic reagents. On the other hand, these compounds are extremely vulnerable for ring opening and ring transformation reactions [73MI3]. In particular, when treated with primary alkyl, aryl, or hetarylamines 2,4,6-triphenylpyrylium salts are easily transformed into the corresponding N-pyridinium salts. In these 2,4,6-triphenylpyridinium salts the pyridine fragment is regarded as a leaving group for nucleophilic substitution reactions at the carbon atom bound to the quaternary nitrogen. This methodology has already found wide application in organic synthesis. It can be illustrated by the pyrylium-pyridinium ring transformation by action of benzylamine, followed by the reaction of N-benzyl-2,4,6-triphenylpyridinium perchlorate with piperidine, sodium phenolate, aniline, and other nucleophilic reagents, enabling one to obtain a variety of benzyl derivatives (Scheme 135) [79JCS(P1)418].

Due to possible ring transformations of pyrylium salts by action of amines, the scope of nucleophilic displacement of hydrogen in the series of pyrylium salts is limited compared to their azinium analogs. It concerns mainly the reactions of 2,6-disubstituted pyrylium and thiopyrylium cations with carbon nucleophiles, as follows from the examples discussed below.

1. *Reactions with C-Nucleophiles.* Cyanation of 2,6-diaryl pyrylium salts **224** with sodium cyanide is a typical example of the S_N^H reaction. It has been shown to give

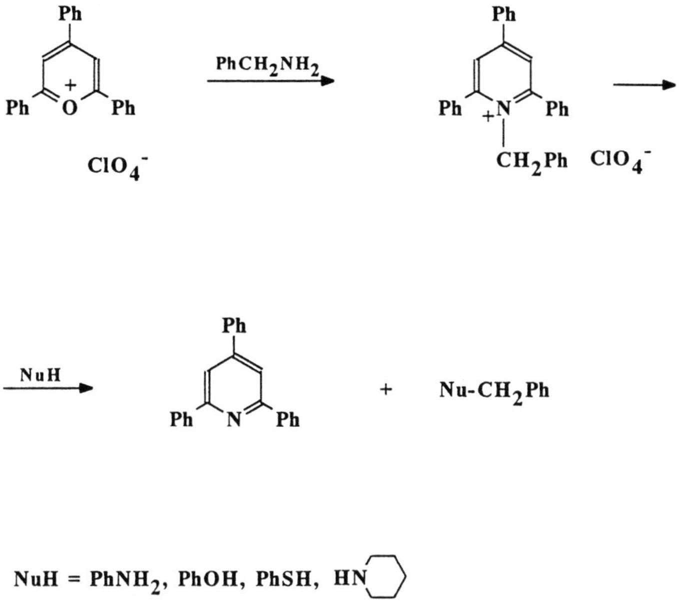

Scheme 135

quantitatively the pyran-4-carbonitriles **225** which can be hydrolyzed into the corresponding pyran-4-carboxylic acids **227**. The pyrans **225** and **227** can be oxidized with trityl perchlorate into 4-substituted 2,6-diphenylpyrylium perchlorates **226** and **228** in 44–63% yields (Scheme 136) [72KG1313].

As already discussed intensively in previous sections, N,N-dimethylaniline proved to be a very active nucleophile for nucleophilic displacements of hydrogen in different aromatic substrates. It reacts smoothly with 2,6-diphenylpyrylium [68MI2; 70KG733], 2-phenylbenzo[*b*]pyrylium [66HCA2046], and 2-phenylbenzo[*b*]thiopyrylium perchlorates, replacing hydrogen at C-4 with the aminophenyl group (Scheme 137; Table XXXIX). N-Alkyl-substituted pyrroles and indoles [68MI2; 70KG733], N-alkylpyrazoles [87KG182], as well as imidazo and pyrrolo[1,2-*a*]annelated benzimidazoles are also appropriate C-nucleophiles in replacement of H-4 in 2,6-diphenylpyrylium perchlorate [87KG309] (Scheme 138; Table XXXIX).

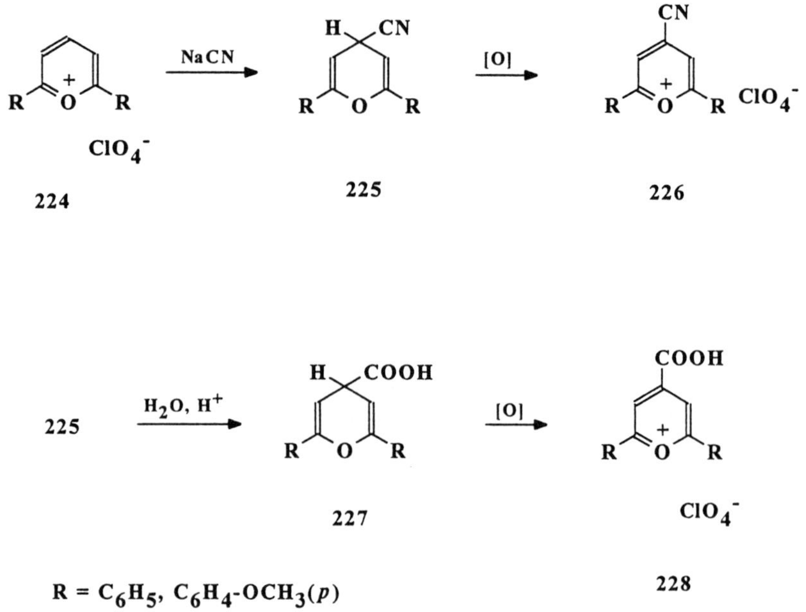

Scheme 136

R = C_6H_5, C_6H_4-$OCH_3(p)$

The C-C bond formation can also be reached with nucleophilic alkyl and aryl radicals generated by oxidative cleavage of carboxylic acids with ammonium persulfate. Methyl, ethyl, isopropyl, *t*-butyl, cyclohexyl, and adamantyl radicals were found to cause the Minisci-type substitution reaction on 2,6-disubstituted pyrylium salts, as shown in Scheme 139 [89MI4]. For a discussion on the mechanism of this radical reaction, see Section II.

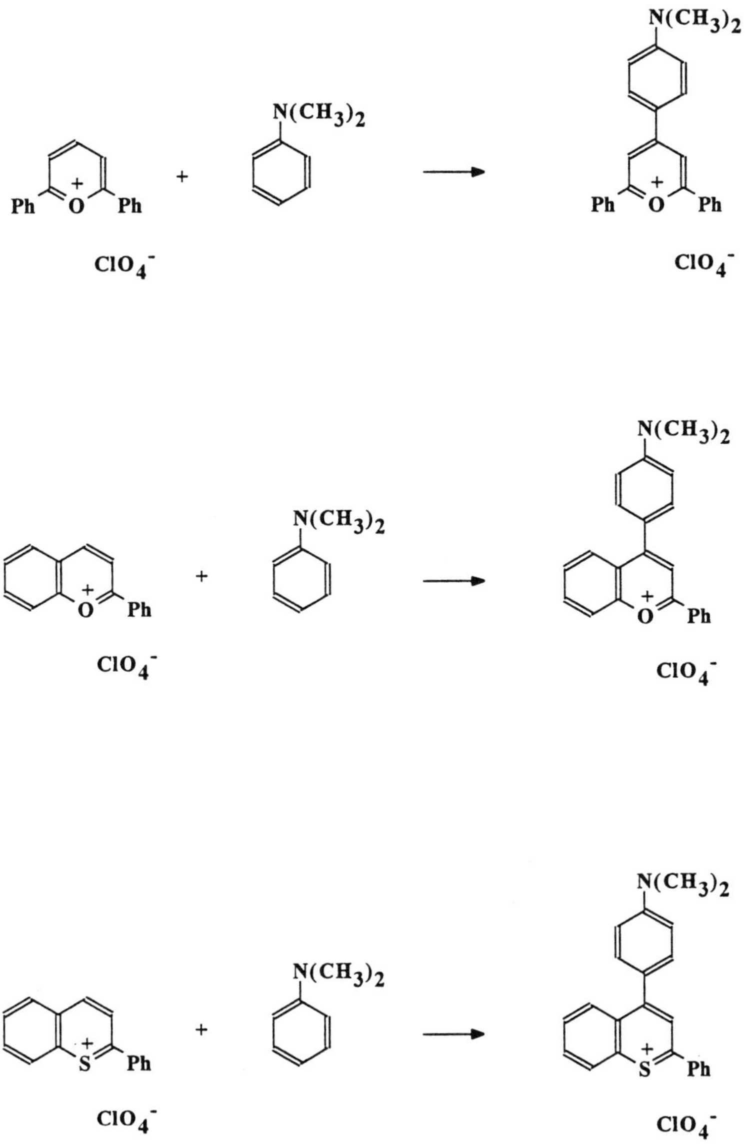
N(CH3)2
Ph
O
Ph
ClO4-
+
N(CH3)2
N(CH3)2
Ph
O
Ph
ClO4-
O
Ph
ClO4-
N(CH3)2
N(CH3)2
O
Ph
ClO4-
S
Ph
ClO4-
N(CH3)2
N(CH3)2
S
Ph
ClO4-

Scheme 137

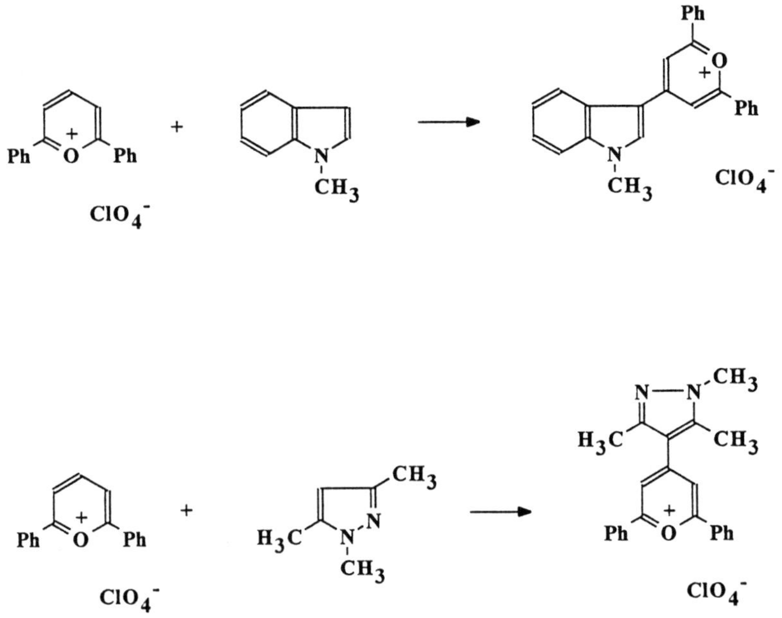

Scheme 138

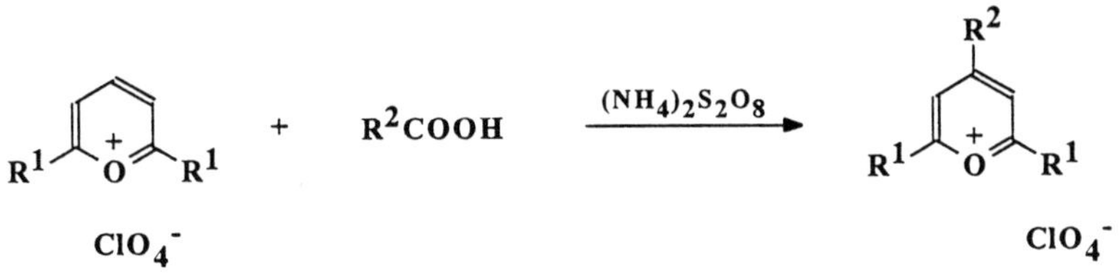

Scheme 139

R^1 = Ph, C(CH$_3$)$_3$; R^2 = CH$_3$, C$_2$H$_5$, cyclohexyl, adamantyl

TABLE XXXIX
Nucleophilic Substitution of Hydrogen at C-4
in 2,6-Diphenylpyrylium Perchlorate[1]

Nucleophile	S_N^H Product	Yield (%)
N,N-Dimethylaniline	4-(4-N,N-Dimethylaminophenyl)-2,6-diphenylpyrylium perchlorate	91
Resorcinol	4-(2,4-Dihydroxyphenyl)-2,6-diphenylpyrylium perchlorate	64
1,3-Dimethoxybenzene	4-(2,4-Dimethoxyphenyl)-2,6-diphenylpyrylium perchlorate	64
Phloroglucinol	4-(2,4,6-Trihydroxyphenyl)-2,6-diphenylpyrylium perchlorate	54
1-Methylpyrrole	4-(1-Methyl-2-pyrrolyl)-2,6-diphenylpyrylium perchlorate	85
1-Methylindole	4-(1-Methyl-3-indolyl)-2,6-diphenylpyrylium perchlorate	91
1-Methyl-1,2,3,4-tetrahydroquinoline	4-(1-Methyl-1,2,3,4-tetrahydro-5-quinolyl)-2,6-diphenylpyrylium perchlorate	78

[1]See ref. [70KG733].

2. Reaction with Organometallic Compounds. Addition of aryl or hetaryl lithiums at C-4 of 2,6-diphenylpyrylium perchlorate leads to the corresponding 4H-aryl(hetaryl) pyrans **229** followed by their dehydrogenation into 4-R substituted 2,6-diphenylpyrylium salts **230** (Scheme 140) [77MI4].

In a similar way, by reacting 3,5-dimethyl-2,6-diphenylpyrylium perchlorate **231** with lithium phenylacetylenide followed by oxidation one can introduce the phenylethynyl group at position 4 of the pyrylium salt **231** (Scheme 140) [84MI2].

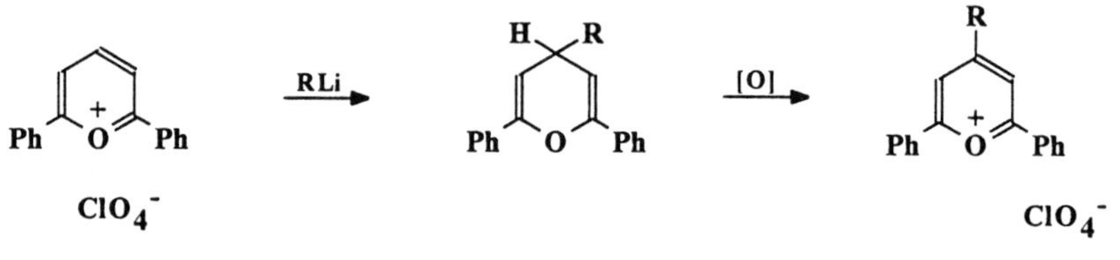

R = Ph, [thienyl], [benzothiazolyl]

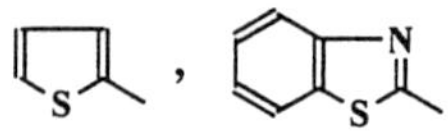

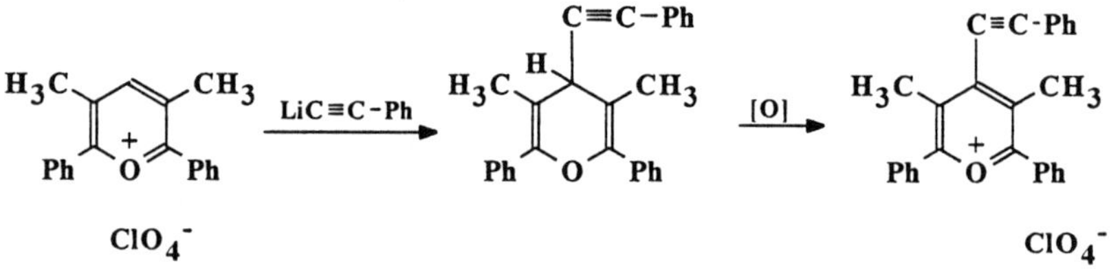

Scheme 140

When reacting 4-ethoxybenzo[*b*]pyrylium perchlorate **232** with phenylmagnesium bromide, 4-ethoxy-2-phenyl-2H-chromene **233** is formed in 50% yield; oxidation of **233** with perchloric acid in acetic anhydride gave 4-ethoxy-2-phenylflavilium perchlorate **234** in 80% yield (Scheme 141) [81ZOR880].

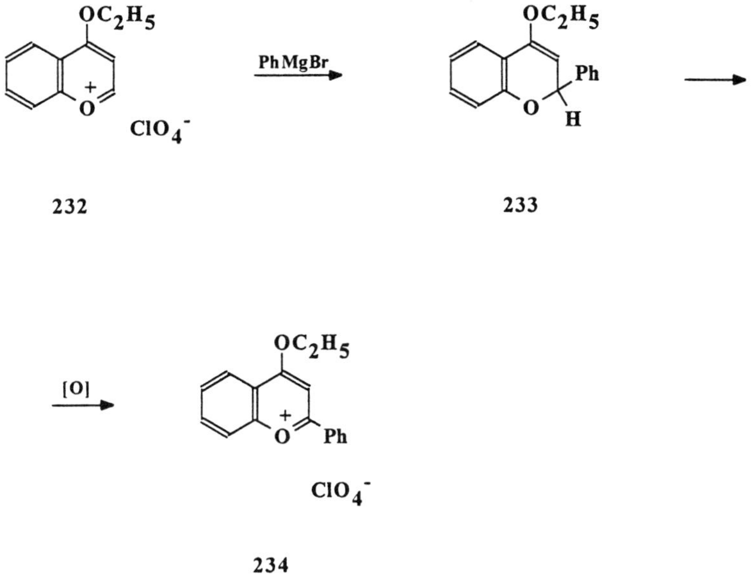

Scheme 141

3. Reactions with O- and P-Nucleophiles. A number of papers describe the adduct formation in the reactions of pyrylium and thiopyrylium salts with the methoxide ion [86JOC4385; 86JA3409; 88JOC1729], water [70KG1454] or phosphins [68MI3; 89ZOB1506]; however, effective synthetic methods based on the displacement of hydrogen with these nucleophiles have not yet been reported.

Experimental Procedures

Example 1. Cyanation of 2,6-diphenylpyrylium perchlorate

The Addition Step. A mixture of 2,6-diphenylpyrylium perchlorate (19.9 g, 60 mmol), sodium cyanide (4.9 g, 90 mmol), ether (400 ml) and water (400 ml) was vigorously stirred at room temperature till the pyrylium salt was fully dissolved. The ether layer was separated, washed with water and evaporated, yielding 15 g (97%) of 4-cyano- 2,6-diphenyl-4H-pyran, melting point 96–97°C.

The Aromatization Step. A mixture of 4-cyano-2,6-diphenyl-4H-pyran (1.7 g, 7 mmol), trityl perchlorate (2.7 g, 8 mmol), and acetic acid (15 ml) was heated at reflux for 5 minutes. The reaction mixture was cooled; the precipitate obtained was filtered off, washed with ether, and dried in air to give 1.1 g (44%) of 4-cyano-2,6-diphenylpyrylium perchlorate, melting point 296–297 °C [72KG1313].

Example 2. Preparation of 4-(1-methyl-3-indolyl)-2,6-diphenylpyrylium perchlorate

A mixture of 2,6-diphenylpyrylium perchlorate (3.3 g, 10 mmol), 1-methylindole (1.3 g, 10 mmol), and acetic acid (50 ml) was heated at reflux for 3 hours. The reaction mixture was cooled; the precipitate obtained was filtered off, washed with acetone and ether, and dried in air to give 1.0 g (43%) of 4-(1-methyl-3-indolyl)-2,6-diphenylpyrylium perchlorate, melting point 292°C [70KG733].

Example 3. Preparation of 4-(4-N,N-dimethylaminophenyl)-2,6-diphenylpyrylium perchlorate

A solution of 2,6-diphenylpyrylium perchlorate (3.3 g, 10 mmol) and N,N-dimethylaniline (700 mg, 5.8 mmol) in DMF (20 mL) was heated at reflux for 1 hour. The reaction mixture was cooled; the precipitate obtained was filtered off, washed with acetone and ether, and dried in air to give 2.05 g (91%) of 4-(4-N,N-dimethylaminophenyl)-2,6-diphenylpyrylium perchlorate, melting point 380°C (decomp.) [68MI2; 70KG733].

Example 4. Preparation of 4-(4-N,N-dimethylaminophenyl)-2-phenyl-benzo[b]pyrylium perchlorate

A solution of 2-phenylbenzo[b]pyrylium perchlorate (3.1 g, 10 mmol) and N,N-dimethylaniline (900 mg, 7.4 mmol) in DMF (20 ml) was heated at reflux for 10 minutes. The reaction mixture was cooled and diluted with ether (40 ml); the precipitate obtained was filtered off, washed with ether and dried in air to give 1.4 g (66%) of 4-(4-N,N-dimethylaminophenyl)-2-phenylbenzo[b]pyrylium perchlorate as violet crystals, melting point 250–251°C [68MI2].

Example 5. Synthesis of 4-(2,4-dimethoxyphenyl)-2,6-diphenylpyrylium perchlorate

A solution of 2,6-diphenylpyrylium perchlorate (3.3 g, 10 mmol) and 1,3-dimethoxy- benzene (1.4 g, 10 mmol) in dry DMF (20 ml) was heated at reflux for 2 hours. The reaction mixture was cooled, diluted with ether (60 ml), and was kept overnight in a refrigerator. A precipitate obtained was filtered off, washed with acetic acid and ether, and dried in air to give 1.5 g (64%) of 4-(2,4-dimethoxyphenyl)-2,6-diphenylpyrylium perchlorate as brown crystals, melting point 258–259°C [70KG733].

IV. FIVE-MEMBERED HETEROCYCLES

The Chichibabin amination of imidazoles, and their benzo and naphtho analogs has been the subject of intensive studies and presents a good example of S_N^H substitution reaction at C-2 in these systems [71MI4; 78RCR1042; 82CHE1221; 88AHC2]. For a successful amination reaction of these systems it has been established that the imidazole ring has to be condensed with an aromatic system, and that one of the imidazole nitrogens needs to bear an alkyl or aryl group [71MI4; 78RCR1042; 82CHE1221;

88AHC2]. Indeed, neither N-alkylimidazoles nor NH-imidazoles condensed with benzene, naphthalene, or phenanthrene are aminated with sodium amide in aprotic solvents even at elevated temperatures [71MI4; 73CHE88]. On the other hand, most of 1-alkyl or 1-aryl-substituted benzimidazoles and imidazo[4,5-*b*]pyridines undergo the heterogeneous Chichibabin amination reaction very smoothly to give the corresponding 2-amino compounds in good yields (Scheme 142; Table XL) [69CHE645; 70CHE987; 71CHE1036; 71MI4; 72CHE731; 73CHE88; 78RCR1042; 82CHE1221; 88AHC2].

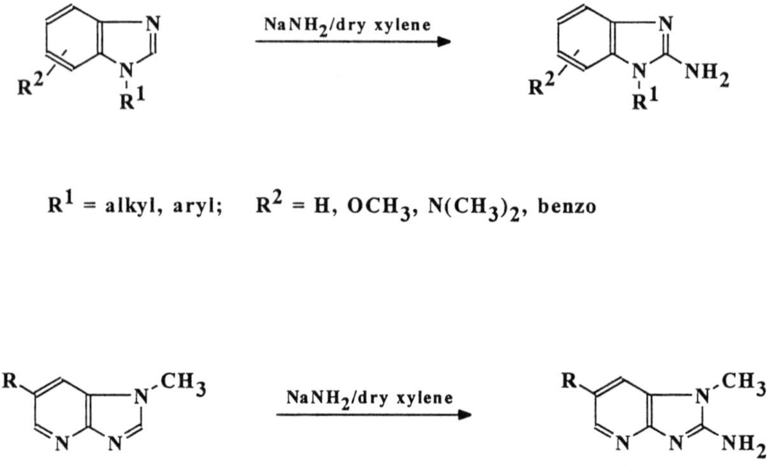

R^1 = alkyl, aryl; R^2 = H, OCH_3, $N(CH_3)_2$, benzo

R = CH_3, C_6H_5

Scheme 142

Electronic and steric effects of most alkyl substituents at N-1 in benzimidazoles have been found to be of only small influence, with the exception of 1-*t*-butylbenzimidazole which gave a poor yield of the 2-amino compound (Table XL) [88AHC2]. The presence of electron-donating substituents in the annelated benzene ring, such as the methoxy and

TABLE XL
The Chichibabin Amination of Condensed Imidazoles[1]

Starting benzimidazole	Yield (%) of the corresponding 2-amino compound	References
1-Methylbenzimidazole	80	78RCR1042
1-Ethylbenzimidazole	85	78RCR1042
1-n-Butylbenzimidazole	55	78RCR1042
1-t-Butylbenzimidazole	21	71CHE624
1-Cyclohexylbenzimidazole	70	88AHC2
1-Phenylbenzimidazole	68	88AHC2
1-(4-N,N-dimethylaminophenyl)-benzimidazole	71	70CHE987
1-Benzylbenzimidazole	67	78RCR1042
4-Methoxy-1-methylbenzimidazole	57	70CHE987
5-Methoxy-1-benzylbenzimidazole	85	70CHE987
6-Methoxy-1-ethylbenzimidazole	53	70CHE987
7-Methoxy-1-benzelbenzimidazole	64	70CHE987
5-Dimethylamino-1-methyl-benzimidazole	83	70CHE987
1,6-Dimethylimidazo[4,5-b]pyridine	50	93MI2
1-Methyl-6-phenylimidazo[4,5-b]-pyridine	90	93MI2
1-Methylnaphtho[2,3-d]imidazole	52	70CHE1075
1-Benzylnaphtho[2,3-d]imidazole	57	70CHE1075

[1] With sodium amide in inert aprotic solvents at elevated temperatures.

dimethylamino groups in different positions, and annelation of the imidazole ring with the naphthalene system have only a slight influence on the course of the amination reaction (Table XL) [88AHC2].

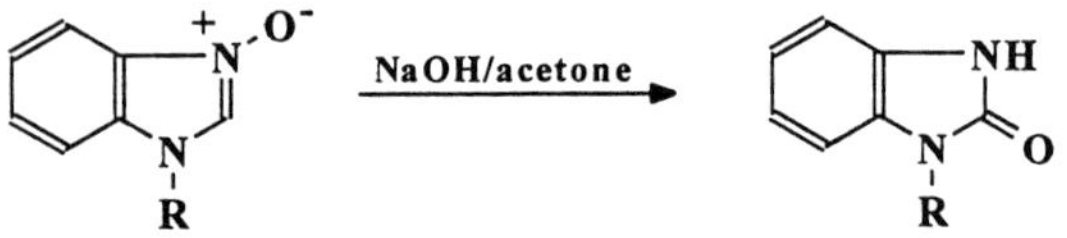

235 236

R = alkyl

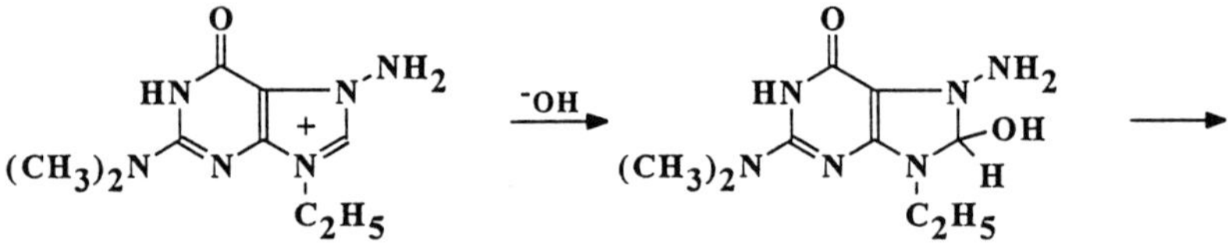

237 238

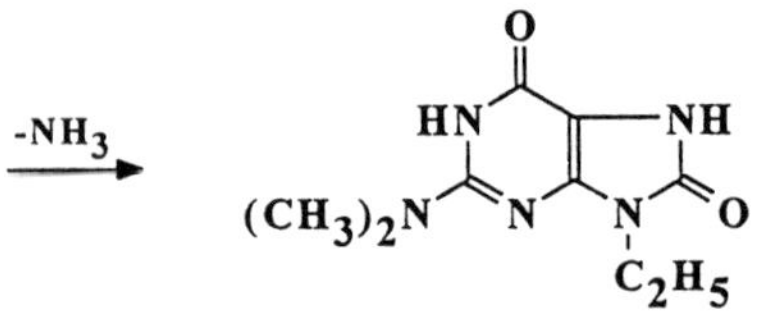

239

Scheme 143

Besides the Chichibabin amination, the hydroxylation of benzimidazoles is another example of the successful displacement of hydrogen at C-2 of the imidazole ring. When treated with sodium hydroxide under mild conditions, 1-alkylbenzimidazole N-oxides **235** are converted into 2-imidazolones **236** (Scheme 143) [64CPB783]. In a similar way the 7-aminoguanosine derivative **237** is converted into the 8-hydroxy compound **239** via intermediacy of the hydroxy adduct **238** (Scheme 143) [93MI1].

Some nitro derivatives of thiophen, furan and pyrrole [84BAP69], as well as nitro-imidazole [84BAP57] have also been found to undergo the nucleophilic substitution of the ring hydrogen very smoothly. For example, the reaction of 2-nitrothiophen **240** with the vicarious carbanion of chloromethyl phenyl sulfone in basic medium gives the 3-substituted 2-nitrothiophen (Scheme 144) [84BAP69]. Compound **240** was also nicely converted into 3-hydroxy-2-nitrothiophen through hydroxylation with *t*-butyl hydroperoxide in liquid ammonia in the presence of potassium *t*-butoxide (Scheme 144) [90JOC4979]

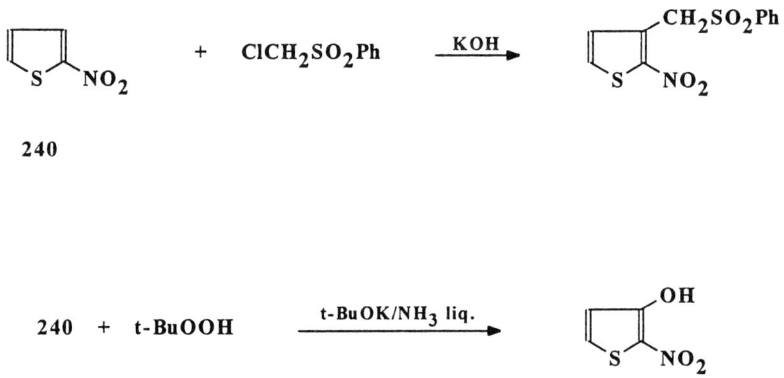

Scheme 144

1-Methyl-4-nitropyrazole reacts with the vicarious carbanion derived from chloromethyl *para*-tolyl sulfone to give a product in which the hydrogen at position 5 is substituted by the sulfone group (yield 87%) (Scheme 145) [89LA545]. The S_N(H-5) product was also obtained when reacting a similar vicarious reagent with 1,3-dimethyl-4-nitropyrazole (62%). The introduction of a second nitro group at position 3 has been found to decrease the yield of the S_N^H reaction dramatically (18%). The reason of this peculiar result is not evident [89LA545].

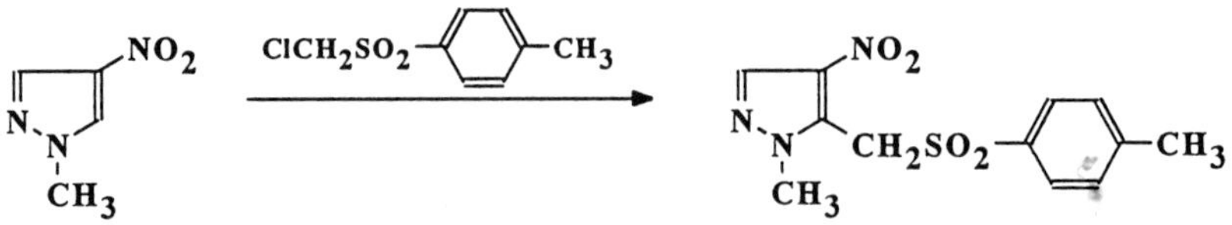

Scheme 145

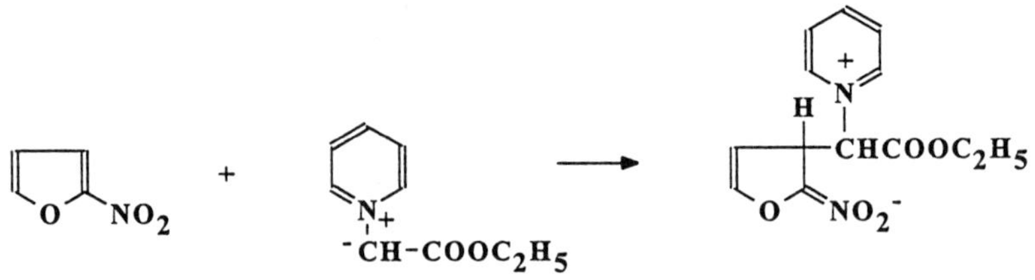

241

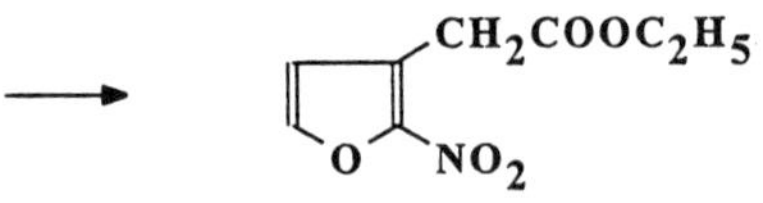

242

Scheme 146

Another example of vicarious substitution of hydrogen in the series of five-membered heteroaromatics is provided by the reaction of 2-nitrofuran with pyridinium methylide **241** (Scheme 146). The pyridine fragment present in the nucleophile acts as the leaving auxiliary group facilitating the displacement of hydrogen in 2-nitrofuran [79JHC1409].

It has been found that in 1-methoxy-1,2,3-triazolium salts **243** prepared by N-oxidation of 1,2,3-triazoles followed by methylation with trimethyloxonium tetrafluoroborate both carbon atoms in the ring are activated toward nucleophilic displacement reactions, although the C-4 position has been established to be less reactive than C-5 [81JCS(P1)503; 82JCS(P1)2749]. Indeed, by means of S_N^H reactions a variety of substituents can easily be introduced at C-5 of the 1,2,3-triazole ring (Scheme 147; Table XLI) [81JCS(P1)503]. When the C-5 position is blocked nucleophilic attack takes place at C-4. The nucleophilic addition at C-4 in 5-methyl-1-methoxy-2-phenyl-triazolium salt **244** is supposed to occur with a simultaneous loss of the methoxide ion. Subsequent aromatization through deprotonation results in fair to good yields of 4-fluoro, 4-cyano, 4-methylamino, and 4-methylthio-substituted triazoles (Scheme 147) [82JCS(P1)2749]. In the N$^+$-OR-substituted pyrazolium salts the C-4 position is less reactive in S_N^H reactions than C-3 and C-5 carbons [92H1129].

The 1,3,4-thiadiazolium salt **245** bearing the 4-nitrophenyl substituent at N-3 was reported to add a variety of nucleophiles, such as water, alcohols, triethylphosphite, or CH-active compounds, to give 2-substituted 1,3,4-thiadiazolines, as exemplified by the reaction of the salt **245** with acetone, yielding adduct **246** (Scheme 148). The rearomatization of adduct **246** which would result in the formation of the S_N^H product was not investigated [78LA98].

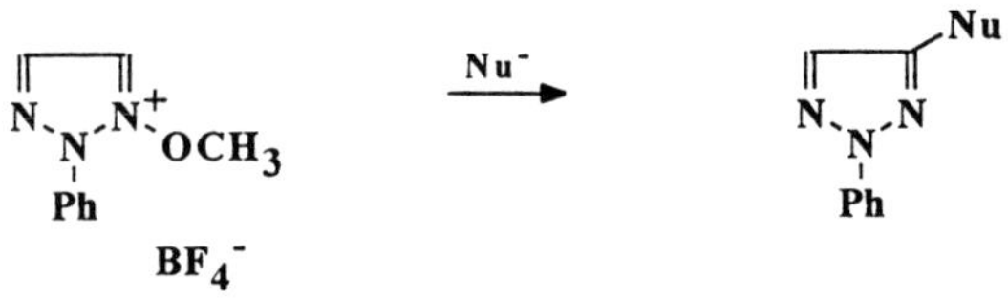

243

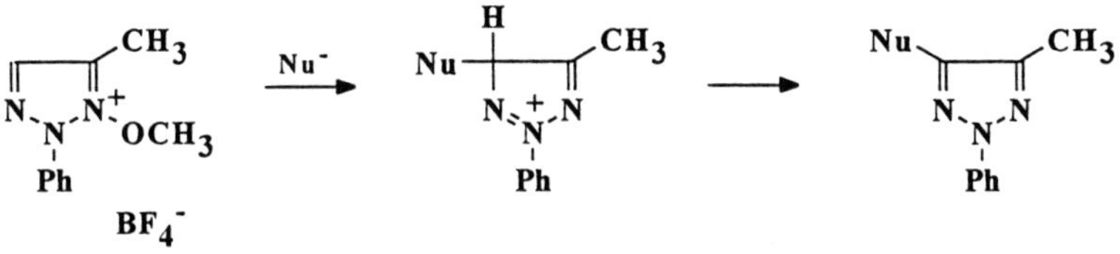

244

R = F, CN, NH$_2$, NHCH$_3$, N(CH$_3$)$_2$, N$_3$, SCH$_3$

Scheme 147

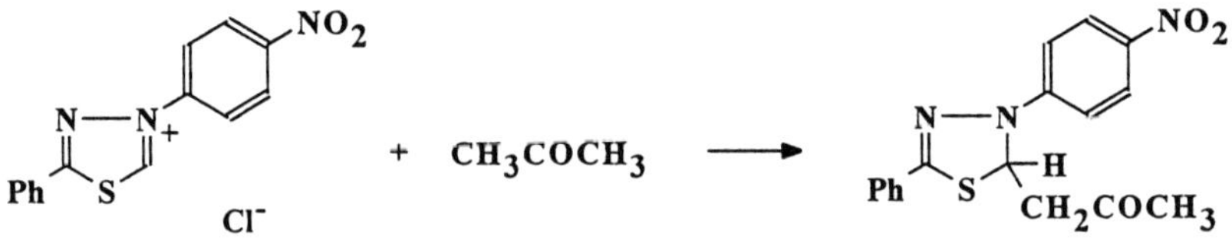

245   **246**

Scheme 148

TABLE XLI
Nucleophilic Substitution of Hydrogen
in 1-Methoxy-2-phenyl-1,2,3-triazolium Tetrafluoroborate

Nucleophile	S_N^H Product	Yield (%)
Potassium fluoride	4-Fluoro-2-phenyl-1,2,3-triazole	61
Potassium cyanide	4-Cyano-2-phenyl-1,2,3-triazole	97
Ammonium hydroxide	4-Amino-2-phenyl-1,2,3-triazole	79
Methylamine	4-Methylamino-2-phenyl-1,2,3-triazole	59
Dimethylamine	4-Dimethylamino-2-phenyl-1,2,3-triazole	90
Sodium azide	4-Azido-2-phenyl-1,2,3-triazole	73
Sodium hydroxide	4-Hydroxy-2-phenyl-1,2,3-triazole	83
Sodium methoxide	4-Methoxy-2-phenyl-1,2,3-triazole	94
Methyl mercaptane	4-Methylthio-2-phenyl-1,2,3-triazole	78

Experimental Procedures

Example 1. Preparation of 4-cyano-2-phenyl-1,2,3-triazole

1-Methoxy-2-phenyl-1,2,3-triazolium tetrafluoroborate (150 mg), potassium cyanide (75 mg), and acetonitrile (1.5 ml) were stirred for 3 days. Removal of the solvent, extraction with dichloromethane (4 x 10 ml), and removal of the dichloromethane afforded 96 mg (97%) of 4-cyano-2-phenyl-1,2,3-triazole, melting point 81–83°C [81JCS(P1)503].

Example 2. Preparation of 4-methyl-5-methylamino-2-phenyl-1,2,3-triazole

Methylamine, distilled from calcium sulfate, was condensed in a flask with 5-methyl-1-methoxy-2-phenyl-1,2,3-triazolium tetrafluoroborate (280 mg). Acetonitrile (1.8 ml) was added and the mixture kept at 20°C for 3 days. It was then evaporated to dryness and extracted with dichloromethane (3 x 5 ml), and the solvent was removed to give a residue which was subjected to flash chromatography with dichloromethane–hexane (1:2) to give a minute amount of 4-methoxy-5-methyl-2-phenyl-1,2,3-triazole and 4-methyl-2-phenyl-1,2,3-triazole (5 mg, 3%). Subsequent elution with ethylacetate–hexane (1:4) afforded 4-methyl-5-methylamino-2-phenyl-1,2,3-triazole as a colorless oil. Yield 110 mg (61%) [82JCS(P1)2749].

Example 3. Preparation of 1-methyl-4-nitro-5-(*para*-tosylmethyl)pyrazole

A fine powdered potassium hydroxide (2.0 g, 25 mmol) was added to a solution of 1-methyl-4-nitropyrazole (640 mg, 5.0 mmol) and chloromethyl *para*-tolyl sulfone (1.0 g, 5.0 mmol) in 8 ml of DMSO with vigorous stirring and cooling with ice. The reaction mixture was kept at 20°C for 1 hour, then poured in 100 ml of 1% aqueous solution of hydrochloric acid and extracted with chloroform (2 x 100 ml). The combined extracts were dried over sodium sulfate and evaporated *in vacuo*. The residue was recrystallized from ethanol to give 1.28 g (87%) of 1-methyl-4-nitro-5-(tosylmethyl)pyrazole, colorless prisms, melting point 168–169°C [89LA545].

V. ELECTROCHEMICAL S_N^H REACTIONS

As discussed in Chapter 2,IV, the electrochemical technique can be successfully applied for aromatization of some dihydroarene intermediates, instead of using an oxidizing agent. In the series of heteroaromatics, electrochemical oxidation of 1-acyl-substituted 1,2- and 1,4-dihydropyridines, derived from the addition of Grignard reagents to 1-acylpyridinium salts (see Section II.C) [82JOC4315], in methanol containing sodium methoxide as the electrolyte gave excellent results (Scheme 149; Table XLII) [90H2025]. It is understood that the 1-acylpyridinium salts **248** formed by the electrochemical oxidation of the N-ethoxycarbonyl 1,4-dihydropyridines **247** are deacylated in the presence of sodium methoxide into the free pyridine bases **249** (Scheme 149).

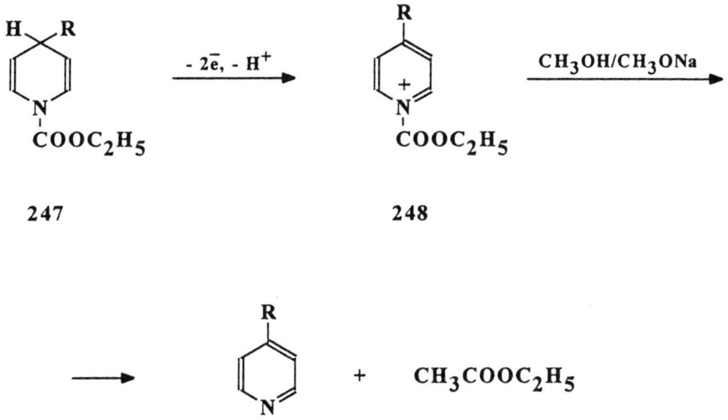

Scheme 149

Analogously 1-benzoyl-4-(3-indolyl)-1,4-dihydropyridine **250** [82KG361] as well as 9-substituted 10-methyl-9,10-dihydroacridines **251** [76MI3] were dehydrogenated on a rotating platinum disk electrode in dimethylformamide on an analytical scale (Scheme 150).

TABLE XLII
Anodic Oxydation of 1,2- and 1,4-Dihydropyridines Derived from the Addition of Grignard Reagents to 1-Acylpyridinium Salts[1]

Dihydropyridine	S_N^H Product	Yield (%)
1-Ethoxycarbonyl-4-phenyl-1,4-dihydropyridine	4-Phenylpyridine	65
1-Ethoxycarbonyl-4-*n*-butyl-1,4-dihydropyridine	4-*n*-Butylpyridine	79
1-Phenoxycarbonyl-4-*n*-butyl-1,4-dihydropyridine	4-*n*-Butylpyridine	60
1-Ethoxycarbonyl-2-methyl-4-phenyl-1,4-dihydropyridine	2-Methyl-4-phenylpyridine	60
1-Ethoxycarbonyl-3-methyl-4-phenyl-1,4-dihydropyridine	3-Methyl-4-phenylpyridine	90
1-Ethoxycarbonyl-2-phenyl-4-methyl-1,2-dihydropyridine	4-Methyl-2-phenylpyridine	38
1-Ethoxycarbonyl-2-*n*-butyl-4-methyl-1,2-dihydropyridine	4-Methyl-2-*n*-butylpyridine	31

[1]See ref. [90H2025].

Electrochemical oxidation has also been successfully used for the preparative cyanation of isomeric 2- and 3-methyl benzo[*b*]thiophens, **252** and **253**, in methanol containing sodium cyanide. In both cases S_N^H substitution products were obtained in 73–75% yields (Scheme 151) [92JCS(P1)333].

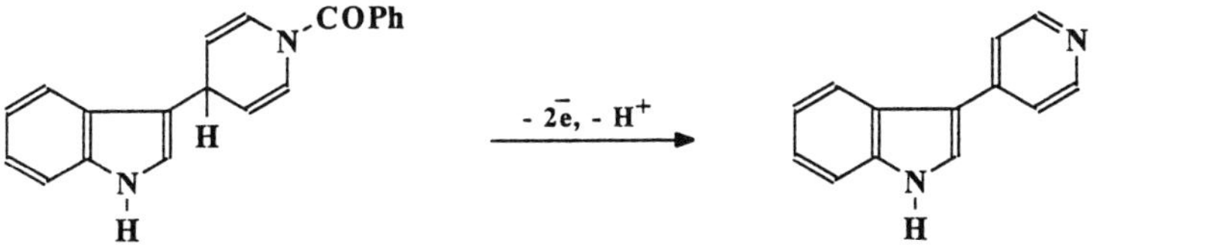

Chapter 3 S_N^H REACTIONS IN HETARENES

250

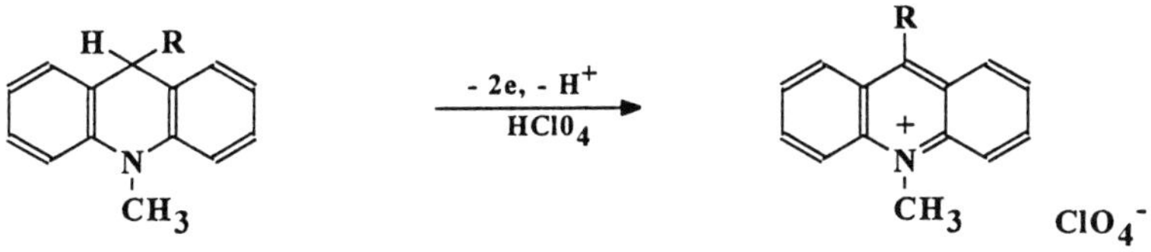

251

R = CH$_3$, C$_6$H$_5$, C$_6$H$_4$-N(CH$_3$)$_2$(*p*)

Scheme 150

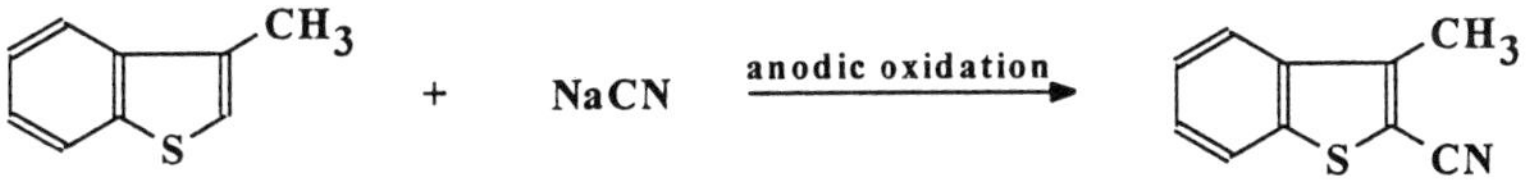

252

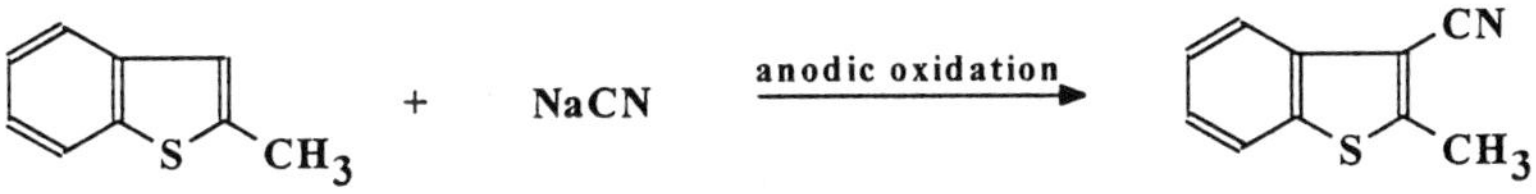

253

Scheme 151

Experimental Procedures

Example 1. General procedure for electrochemical oxidation of 1-acyldihydropyridines. Preparation of 4-*n*-butylpyridine

A solution of crude 4-*n*-butyl-1-(phenoxycarbonyl)-1,4-dihydropyridine (3.16 g, 12.3 mmol) in methanol (80 ml) was placed in an electrolysis cell equipped with carbon electrodes and a magnetic stirrer. Sodium methoxide (4.35 M in MeOH) was added dropwise as an electrolyte until a constant current (0.4 A) could be maintained with a voltage reading between 18V and 22V. The cell was cooled with an external water bath (18°C) while 6 F/mol of electricity was passed through the solution with stirring. The solvent was removed under reduced pressure at room temperature and ether was added (40 ml). The solution was extracted with 10% HCl (3 x 20 ml) and the organic layer was discarded. To the aqueous layer was added dichloromethane (2 x 20 ml). The combined organic extracts were washed with brine (20 ml), dried (K_2CO_3), and concentrated. The crude product was purified by radial preparative liquid chromatography (30% ethyl acetate–hexane–1% methanol) to give 981 mg (60%) of 4-*n*-butylpyridine as a clear yellow oil. Melting point of picrate 112–113°C [90H2025].

Example 2. Preparation of 3-cyano-2-methylbenzo[*b*]thiophen

Electrochemical cyanation of 2-methylbenzo[*b*]thiophen was carried out in a three-compartment cell at a platinum anode with a SCE (saturated calomel electrode) as the reference electrode. Oxidation of 2-methylbenzo[*b*]thiophen (300 mg, 2 mmol) in methanol (50 ml) containing sodium cyanide (1 g, 20 mmol) was performed at the peak potential of the substrate (1.5 V versus SCE). The reaction was terminated after passage of 2 F/mol of added 2-methylbenzo[*b*]thiophen. After completion of the oxidation, the of 2 F/mol of added 2-methylbenzo[b]thiophen. After completion of the oxidation, the anolyte was treated with brine and extracted with diethyl ether. The extract was concentrated, and column chromatography on silica gel with light petroleum (boiling point range 30–70°C) containing chloroform (15%) gave 3-cyano-2-methylbenzo[b]thiophen. Yield 260 mg (75%). Melting point 85–86°C [92JCS(P1)333].

VI. INTRAMOLECULAR S_N^H REACTIONS

In this Section we would like to discuss the examples of a successful use of the S_N^H reactions for the synthesis of condensed heterocyclic systems.

A recently developed methodology for the synthesis of condensed azine derivatives is based on nucleophilic substitution of two adjacent hydrogens in an azine ring [87KG1260; 88AHC301]. As known, a number of azine derivatives, especially pyrazines, 1,2,4-triazines, quinoxalines, pyrido[2,3-b]pyrazines, pteridines, and their quaternary salts are able to undergo the diaddition reactions [88AHC301; 88MI1]. When bifunctional nucleophilic reagents are used, these additions result in the formation of condensed systems through intramolecular *ortho*-cyclizations of intermediate monoadducts, as illustrated in a common form by the conversion of adducts **254,** formed by the addition of dinucleophiles to C-2 of N-methyl-1,4-diazinium salts, into condensed pyrazines **255** (Scheme 152) [88AHC301].

Oxidation of cycloadducts **255** by permanganate in acetone yields condensed pyrazine derivatives containing, of course, no ring junction hydrogen atoms. It is illustrated by conversion of 1,2,4-oxadiazino[5,6-b]quinoxalines **256**, obtained by the diaddition reaction of N-methylquinoxalinium iodide with the corresponding amidoximes [87KG1118] into the fully aromatics **257** [87KG1260]. Other condensed systems obtained by this method, **258** and **259**, are shown in Scheme 152.

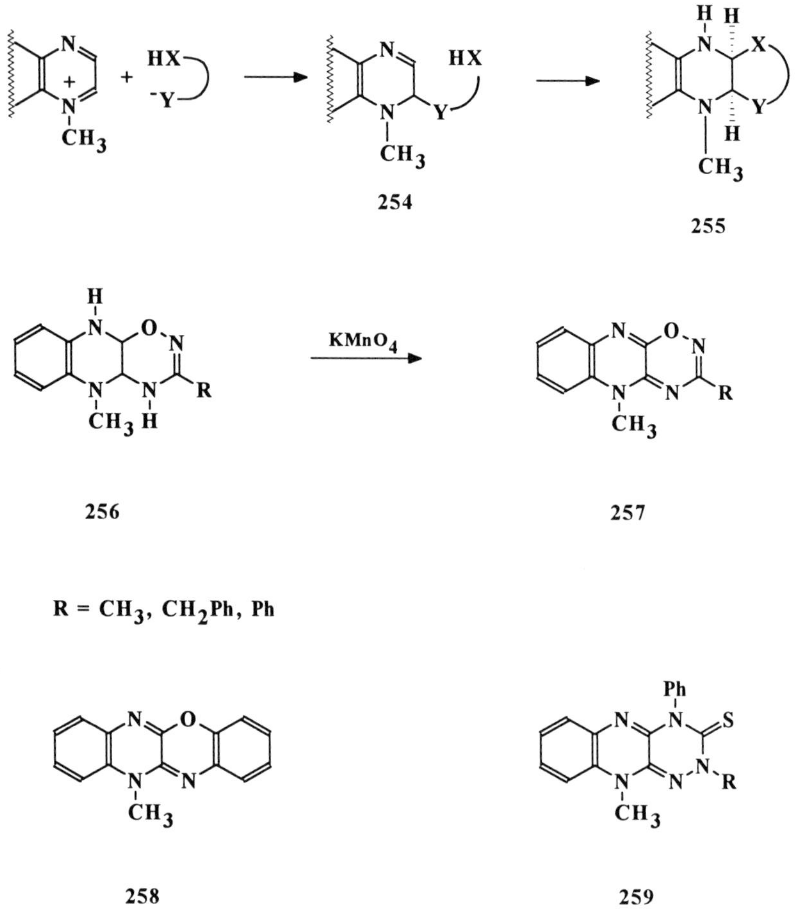

Scheme 152

R = CH$_3$, CH$_2$Ph, Ph

A new approach involving an intramolecular S_N^H process has recently been suggested for the structural modification of 6-fluoro-4-oxo-1,4-dihydroquinoline-3-carboxylic acids which are known as a new generation of the family of totally synthetic and highly effective "fluoroquinolone" antibacterials [92MC151]. It is based on the reaction of 1-amino-substituted 4-oxo-1,4-dihydroquinoline-3-carboxylates **260** with acetylacetone which proved to result in the formation of tricyclic 7-fluoro-4-oxopyrazolo[5,1-a]-quinoline-4-carboxylates **263** in 60–65% yields (Scheme 153).

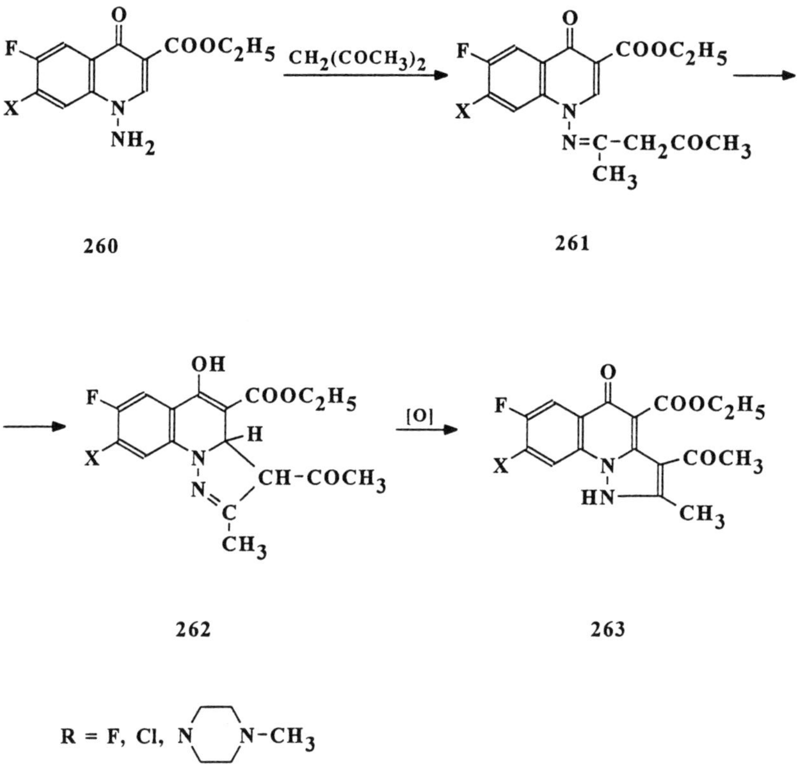

Scheme 153

The mechanism for the formation of **263** involves condensation of the N-amino compounds **260** with acetylacetone yielding the hydrazones **261** followed by the intramolecular nucleophilic attack at C-2. Oxidation with air oxygen leads to the described product (65%); only 4–5% yields of the compounds **263** were obtained in the absence of air oxygen [92MC153].

It is worth noting that only a few examples of the nucleophilic displacement of hydrogen at C-2 of the pyridine ring in the series of 4-oxoquinoline-3-carboxylic acids have been reported in the literature [88JHC1769; 89JHC1675]. The reaction under consideration seems to be a first example of using the intramolecular S_N^H reactions for the synthesis of tricyclic fluoroquinolones.

The synthesis of 2,3-fused quinolines from quinoline N-oxides bearing an appropriate substituent at C-3 was found to involve deoxygenative displacement of hydrogen at C-2 as the prerequisite step for subsequent intramolecular cyclizations [90H779; 92H1055]. Also the synthesis of 2,3-fused pyrazines has been shown to proceed via intramolecular displacement of hydrogen at C-2 of the pyrazine ring [92KG107].

Experimental Procedures

Example 1. Preparation of 2-benzyl-10-methyl-4-phenyl-2,3,4,10-tetrahydro-1,2,4-triazino[5,6-*b*]quinoxalin-3-thione (**259**, R= $CH_2C_6H_5$)

Potassium permanganate (1.6 g, 10 mmol) was added in small portions to a solution of 2-benzyl-10-methyl-4-phenyl-1,2,3,4,4a,5,10,10a-octahydro-1,2,4-triazino-[5,6-*b*]quinoxalin-3-thione (1 g, 2.5 mmol) [85KG960] in 10 ml of acetone under stirring and the reaction mixture was kept at room temperature for 2 hours. A precipitate of MnO_2 obtained was filtered off and washed thoroughly with acetone. Combined acetone filtrates were concentrated *in vacuo*, and the residue was recrystallized from acetone to give 700 mg (70%) of 2-benzyl-10-methyl-4-phenyl-2,3,4,10-tetrahydro-1,2,4-triazino-[5,6-*b*]quinoxalin-3-thione, melting point 223–225°C [87KG1260].

Example 2. Structural modification of "fluoroquinolone" antibiotics. Preparation of 3-acetyl-7,8-difluoro-4-ethoxycarbonyl-2-methyl-5-oxo-5,9b-dihydropyrazolo[5,1-*a*]quinoline (**263**, X= F)

A mixture of 134 mg (0.5 mmol) ethyl 1-amino-6,7-difluoro-4-oxo-1,4-dihydroquinoline-3-carboxylate and 500 mg (5.0 mmol) of acetylacetone was refluxed in acetic acid for 2 hours. Reaction mixture was then treated with water (1:1), cooled to room temperature and the precipitate obtained was filtered off. The solid was extracted with chloroform followed by a vacuum evaporation which gave 950 mg (65%) of 3-acetyl-7,8-difluoro-4-ethoxycarbonyl-2-methyl-5-oxo-5,9b-dihydropyrazolo[5,1-*a*]quinoline, melting point 170–172°C [92MC151].

Reactivity of Arenes and Heteroarenes and Mechanisms of the S_N^H Reactions

Some aspects of the reactivity of arenes and heteroarenes in the S_N^H reactions and the reaction pathways have already been mentioned in several sections of this book (especially the aromatization stage, see Introduction chapter), since they are a part of the common concept of S_N^H reactions.

In all S_N^H reactions two steps unevitably occur: the addition step of the nucleophile and the aromatization step involving elimination of hydrogen, although the sequence of these steps might be different as, for instance, in the aryne mechanism. We want to consider the following reaction pathways: (1) the well-documented $S_N^H(AE)$ mechanism referring to the reactions, occurring through the addition–elimination pathway; (2) the $S_N^H(EA)$ mechanism for the *cine*-substitution reactions; and (3) $S_N^H(ANRORC)$ mechanism which formally describes the displacement of the ring hydrogen through a nucleophilic addition, a ring opening, and a ring closure step.

I. THE ADDITION-ELIMINATION MECHANISM [$S_N^H(AE)$]

A. *The Addition Stage*

1. Formation of Molecular Complexes, Charge-Transfer Complexes and Radical Ions. It is now well recognized that the reactions of nitroarenes and other π-deficient aromatics with charged nucleophiles may involve the formation of charge-transfer complexes or radical ions [92JA7708]. Measurement of radical species in the reactions of π-deficient arenes with nucleophiles by means of electron spin resonance (ESR) spectroscopy [62JA4153; 64JA1807; 72MI1; 73MI4] or by trapping unstable radicals with *t*-nitrosobutane or nitroxides [75T969; 76BCJ3227; 78BCJ196] supported evidence for a radical mechanism in S_N^H reactions [64JA1807; 73MI4; 75T969; 76BCJ3227; 78BCJ196].

The classical "nucleophilic addition–elimination" schemes [51CRV273; 58MI1] are actually not capable in accounting for a number of experimental facts. For instance, the Janovsky reaction between *meta*-dinitrobenzene **264** and acetone in the presence of potassium hydroxide, yielding the anionic σ^H-adduct **265**, was shown to be accompanied by the formation of molecular oxygen, as measured by means of polarography [71MI5]. The reaction was suggested to involve electron transfer from the hydroxide ion to *meta*-dinitrobenzene, followed by recombination of the hydroxyl radicals into hydrogen peroxide. Decomposition of the latter accounts for the formation of oxygen (Scheme 154) [74MI3].

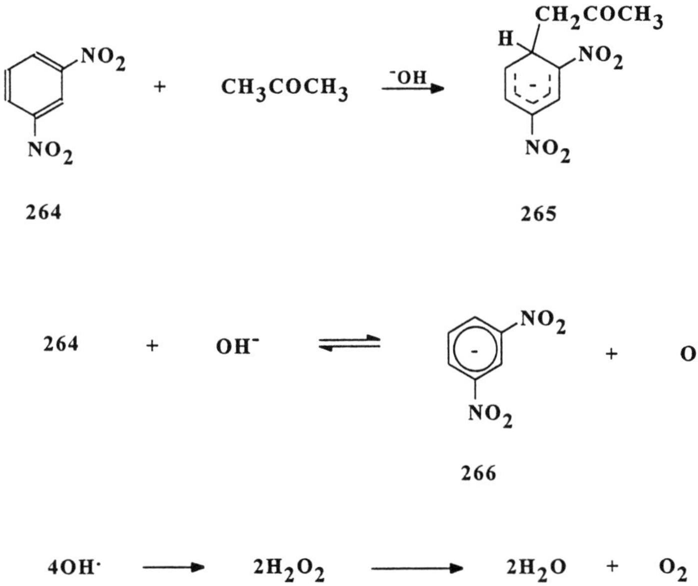

Scheme 154

There is also the observation that *meta*-dinitrobenzene as well as some other nitroaromatics undergo nucleophilic aromatic substitution of hydrogen by action of the hydroxide anion, yielding the corresponding phenols [64JA1807; 91MI2]. Taking into account that σ-adduct **267** can also act as an electron donor, a plausible mechanism for the formation of 2,4-dinitrophenol **270** has been suggested involving the intermediacy of radical and radical anion species **268** and **269** (Scheme 155) [64JA1807]. Similar electron exchanges between 1-halo-2,4-dinitrobenzenes and the hydroxide anion leading to the formation of the corresponding nitroarene anion radicals have been reported [87MI5].

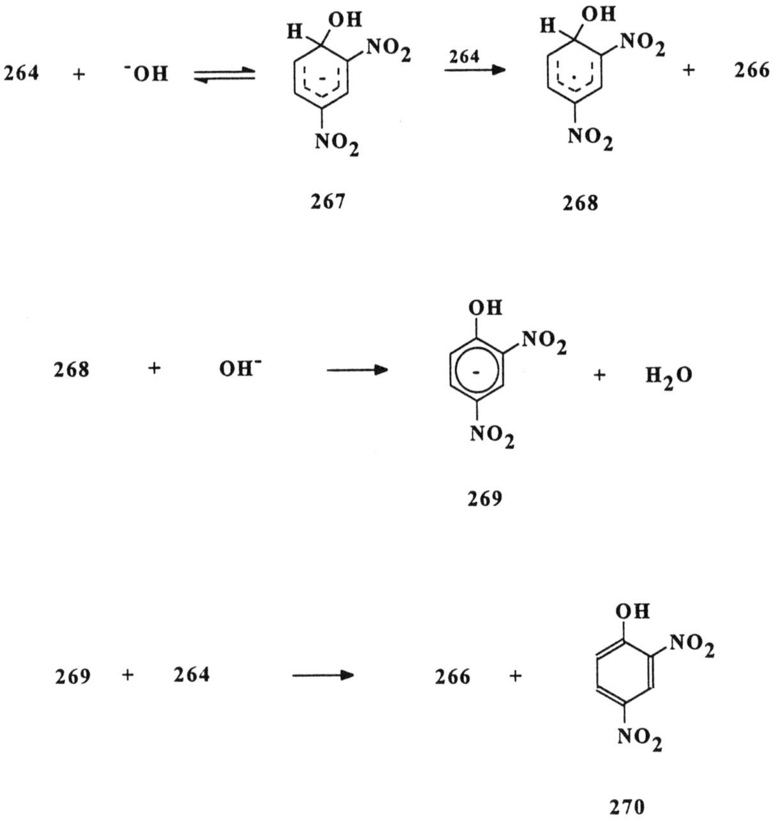

Scheme 155

Also the formation of molecular charge-transfer complex between *ortho*-dinitrobenzene and piperidine in hexane, as well as between 1-X-2,4-dinitrobenzenes (X= Cl, F) with various anilines [82JCR(S)258; 82MI4; 87JCS(P2)79] substantiates the single electron transfer (SET) mechanism.

Similar data have been obtained with π-deficient heterocycles. The initial interaction between azinium cations and nucleophiles often involves a fast reversible electron transfer from the nucleophile to the substrate, yielding charge-transfer complexes (CTC). For instance, in the reaction of acridinium iodide with arylamines, leading finally to the displacement of H-9 in the acridinium cation, the formation of CTC has been observed [70MI1; 77MI1]. Equilibrium constants for the formation of these electron transfer complexes have been determined [77MI1]. Judging from the ESR spectra of the reaction mixtures only a partial electron transfer in the reaction with anilines occurs; however, in the reaction of N-methylacridinium cation with N,N'-tetra-methyl-*para*-phenylenediamine a full SET has been observed [77KG690]. Based on UV and ESR spectra the "Wursters Blue" radical cation **272** could be identified and diacridanyl **271** could be isolated from the reaction mixture in 45% yield (Scheme 156) [77KG690]. This example shows that single electron transfer between a strong electron donor and the azinium cation can be a rather plausible elementary act in S_N^H substitutions.

Another example supporting the SET mechanism at the addition step is provided by the homolytic cleavage of the C-C bond in the σ^H-adduct **273** resulting from the addition of phenol to N-methylacridinium cation. When heating adduct **273** in an aprotic solvent and in an inert atmosphere the formation of N,N'-dimethyldiacridanyl **271** is observed (Scheme 157). Although acridine and phenol radicals have not been registered because of their instabilities, evidence for their presence in the reaction mixture has been obtained by adding diphenylpicrilhydrazine or 2,4,6-tri-*t*-butylphenol, giving stable, easily detectable radicals [79KG1736].

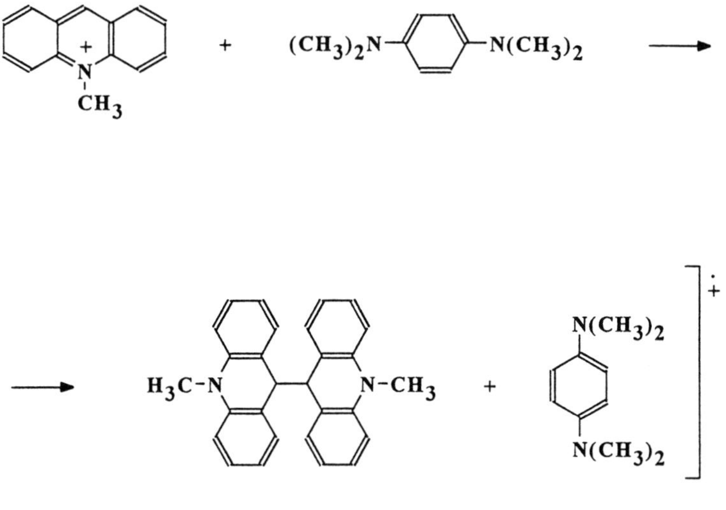

271 272

Scheme 156

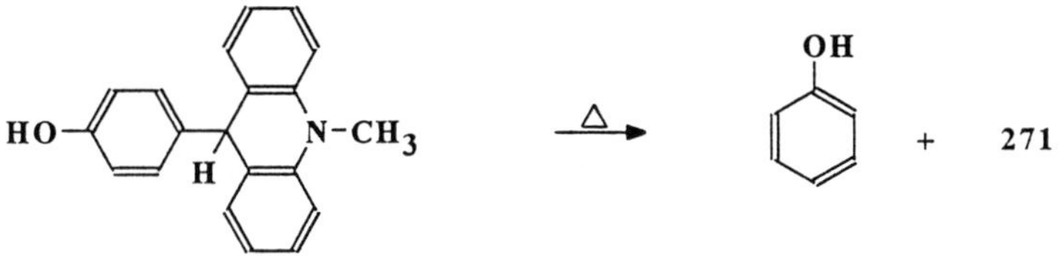

273

Scheme 157

This example seems to be the first case of the homolysis of σ^H-adducts arising as intermediates in nucleophilic substitution reactions. A similar dissociation of σ^H-adducts has later been found to occur in the S_N^H substitution at H-9 in the 10-methylacridinium ion by benzyl diethylamine (see Chapter 3,II,B, Scheme 97) [92BCJ55]. These facts can be considered as important evidences for the hidden radical nature of the addition step, since if there is no radical dissociation of σ-adducts, the principle of microscopic reversibility is disturbed.

A linear relationship is observed between the reactivity of a number of azinium cations, such as NH-pyrimidinium, NH-quinazolinium, NH-acridinium, and NH-imidazo-[4,5-*d*]pyrimidinium salts, to the addition of the bisulphite ion (Scheme 158), and the potentials of the one-electron polarographic reduction of these salts, what has also been regarded as an argument in favor of the SET mechanism [76JA5234].

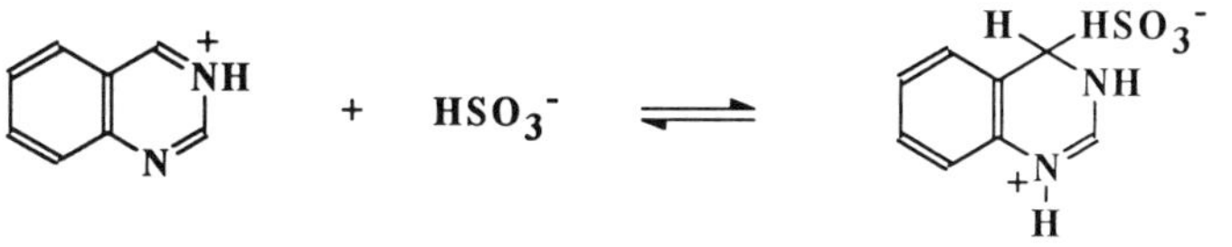

Scheme 158

Although radical or radical-chain mechanisms, operating evidently in S_N^H reactions of nitroarenes with nucleophiles, can rationalize the displacement of hydrogen in π-deficient heteroaromatics, there is a lack of experimental data substantiating the SET mechanism, as well as supporting the participation of radical ion species in the formation of σ^H-adducts. A certain limitation for the SET mechanism is also that the redox potentials of the reactive partners are supposed to fit each other [78RCR260; 84ACSA(B)439].

b. Molecular orbital (MO) calculations. Molecular orbital calculations. (*ab initio*, MNDO/3) for nitrobenzene, benzonitrile and other π-deficient arenes predict the preference of *ortho-* and *para*-position for the addition of nucleophiles, since the cyclohexadienide anions bearing electron-withdrawing substituents at C-2 or C-4 are thermodynamically more favored (Table XLIII) [80JA6430; 91MI2].

TABLE XLIII
Stabilization Energies (kJ/mol) for Cyclohexadienide Anions
Derived from Nucleophilic Addition Reaction on the Benzene Ring

Compound	Position of the sp^3-carbon in the benzene ring			
	C-1	C-2	C-3	C-4
Benzonitrile	54.1	125.8	69.6	148.6
Nitrobenzene	129.9	178.9	87.5	201.8

The presence of several electron acceptors at *ortho*- and/or *para*-position, as well as extension of the aromatic system through benzoannelation enhances the stability of a cyclohexadienide anion due to an increased capability in delocalization of the negative charge [49MI1; 68MI1; 91MI2]. It is consistent, for instance, with the experimental fact that benzenes can be aminated by hydroxylamine provided two nitro groups are present in the ring, while in the series of naphthalenes the presence of one nitro group is sufficient in order to perform this amino-dehydrogenation reaction [92MI3].

Attempts are reported to characterize the ability of azines and azinium cations to participate in the S_N^H reactions by various indices of quantum chemistry, such as π-charge on the carbon atoms [74KG1097; 78RCR1042], (σ+π) charge [69JA6381; 77KG690], energy of the lowest unoccupied molecular orbital, and so on [77MI1; 88T1]. However, from many studies it became clear that there is no straight dependence between the reactivity of azines in the S_N^H reactions and calculated positive charge values. For instance, the Chichibabin amination of a number of azines mentioned in Scheme 159 follows the reactivity order as indicated, which is, however, not in accordance with the order of the π-charge change (π-charges on the reactive carbon atoms are shown under the formulae of Scheme 159) [85MI4; 88T1].

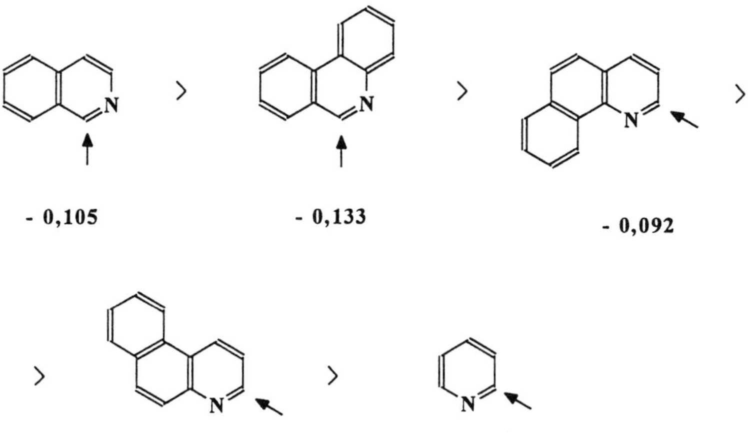

- 0,105 > - 0,133 > - 0,092 >

> - 0,095 > - 0,077

Scheme 159

These results are in agreement with FMO/calculation on the nucleophilic amination of nitroquinolines and isoquinolines with ammonia. Also with these compounds no correlation was found between the order of reactivity and the π-charge density in the different positions in these compounds (for the data, see Chapter 3,I,A). These aminations were shown to be controlled by the interaction between the HOMO of ammonia and the LUMO of the nitro compounds. It has recently been reported that amination of 3-nitropyridine with liquid ammonia seems to be directed by the π-charge on different positions; Colomb interaction is the main controlling factor [93ACS95].

When considering the reaction of isomeric diazanaphthalones **274–279** (Scheme 160) with N,N-dimethylaniline (see Chapter 3,I,A, Schemes 72 and 73) [77ZOR204], it has been found that the compounds **274–276** react smoothly with N,N-dimethylaniline to give S_N^H products (Chapter 3, Schemes 72 and 73), while diazanaphthalones **277–279** are unreactive under the same conditions [77ZOR204]. In order to relate chemical behavior of these compounds with reactivity indices some MO calculations were made. A good correspondence was found between the experimentally observed reactivity of the isomeric diazanaphthalones **274–279** and the so-called delocalization model of reactivity which takes into account the nature of nucleophiles. It has been shown that an attack by low active nucleophiles, for instance N,N-dimethylaniline, is favored for the diazanaphthalones **274–276**, and that for the isomeric compounds **277–279** no reactivity can be expected. This result is in full agreement with the experimental data [77ZOR204, 77MI1; 88T1].

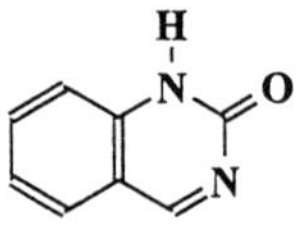

274

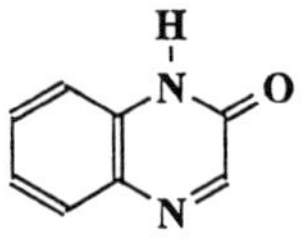

275

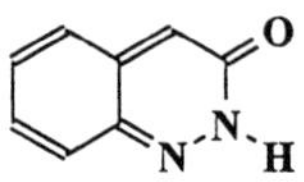

276

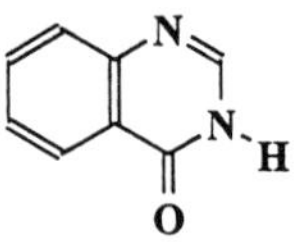

277

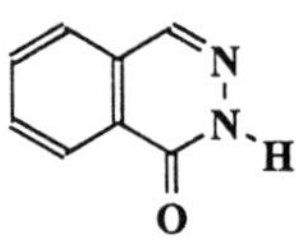

278

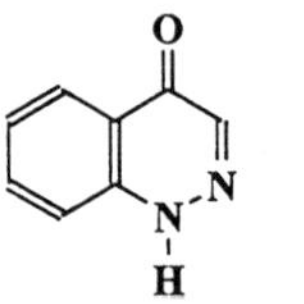

279

Scheme 160

From the data discussed above it follows that indices characterizing aromaticity of arenes and hetarenes are important criteria of their reactivity in the S_N^H reactions [77MI1; 85MI4; 88T1]. It is understood that introduction of any electron-withdrawing substituent, including a heteroatom, disturbs the symmetrical 6π-electronic system, thus reducing the aromatic character of the system. Moreover, these substituents or heteroatoms stabilize intermediate σ^H-adducts and the corresponding transition states. This is illustrated by the delocalization energies between nitroarenes and their anionic σ^H-complexes in the series nitrobenzene, 1,3-dinitrobenzene, and 1,3,5-trinitrobenzene differ from: −33 kJ/mol; 4 kJ/mol, and 42 kJ/mol [69AG136]. The fact that benzene is inert toward sodium amide while naphthalene affords a mixture of 1-amino and 1,5-diaminonaphthalene can certainly be accounted for in the lower aromaticity of naphthalene (see Chapter 2,I) [06CB3006; 06CB3023; 06CB3081]. Apparently due to the same reason anthracene and phenanthrene are smoothly alkylated by dimethyl sulfone and dimethylsulfoxide carbanions, while an aza-activated arene, pyridine, does not react [66JOC248; 66TL1123; 68TL4625].

3. Site Selectivity. An important aspect of the S_N^H reactions is the regioselectivity of a nucleophilic attack at the *ortho*- versus *para*-position relative to the nitro group in nitroarenes or at α- versus γ-position relative to the aza group in the series of azines. The directing effect of the nitro group at the *ortho*-position is usually more influental due to a stronger inductive effect, and also due to intramolecular hydrogen bonding in the reactions with primary and secondary amines, alcohols, and other hydrogen-bearing nucleophiles. Due to this "built-in solvation" nature of the zwitterionic intermediates the prevailing formation of *ortho*-nitro-substituted arenes is often observed [55JA5051; 71JPC3636]. For instance, the reaction of nitrobenzene with potassium hydroxide and oxygen, when carried out in dimethoxyethane at 60–70°C in the presence of 18-crown-6, gave 95% of *ortho*-nitrophenol and only 5% of its *para*-isomer (Scheme

161) [70GEP2006205]. Binding of the potassium ion in the crown ether makes it possible for the "free" hydroxide ion to participate in the built-in solvation of the *ortho* σ^H-adduct. On the other hand, the reaction of nitrobenzene with powdered potassium hydroxide in a steel autoclave at 75°C gave *para*-nitrophenol in 95% yield and only 5% of its *ortho*-isomer was obtained [82MI3], showing that other factors such as solvent, temperature, pressure, and catalyst are also of great importance and may result in opposite results.

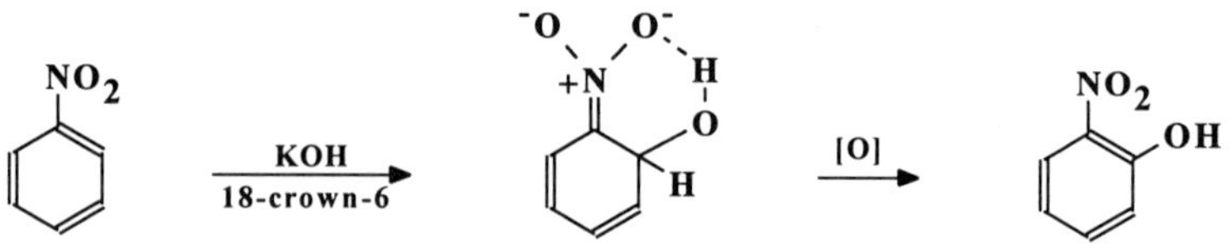

Scheme 161

Concurrent nucleophilic additions at unsubstituted carbon atoms is also observed in the series of mono nitro five-membered heterocycles. Several types of σ-adducts can be obtained, as shown in Sheme 162 for the reaction of 3-nitro five-membered systems with the methoxide ion. However, nucleophilic addition at mono nitro 2,4-disubstituted furans, pyrroles, thiophens, and selenophens has been shown to occur exclusively at α-position to the heteroatom, yielding σ-adducts **280**; no formation of σ-adduct **281** was observed (Scheme 162) [73TL1123; 76JOC2153; 78AJC2463; 78JOC4303]. This is probably due to the anomeric effect enhancing the stability of the C-α methoxy adducts.

Site selectivity in relation to temperature has been extensively studied in Chichibabin aminations of a number of azine derivatives [73JOC1947; 83AHC95; 85JHC353; 86CCA33]. This is illustrated by the following example. When quinoline is added to a solution of potassium amide in liquid ammonia at –60°C, the kinetically

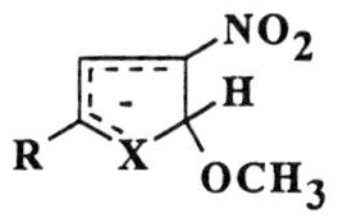

280 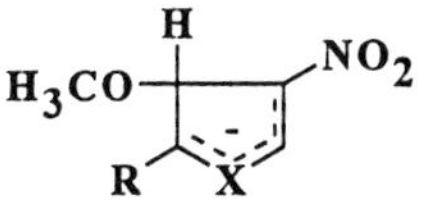 **281**

X = NH, O, S, Se;   R = CN, NO$_2$

Scheme 162

favored 2-aminodihydroquinolinide **282** is formed. However, when the same solution is warmed to about +10°C the thermodynamically favored 4-aminodihydroquinolinide **283** is almost exclusively present. At intermediate temperatures, mixtures of **282** and **281** are obtained (Scheme 163) [73JOC1947; 85JHC353]. From this example it is evident that site selectivity seems to play an important role at the stage of the addition reaction. It determines mainly the final composition of reaction products, although the difference in rates of the aromatization for isomeric σ-adducts also affects the reaction outcome. Indeed, when a mixture of adducts **282** and **283**, obtained at -40 °C in the ratio 3:1, was oxidized with potassium permanganate at the same temperature the corresponding 2- and 4-aminoquinolines, **284** and **285**, were obtained in 53% and 10% yields, respectively (Scheme 163) [85JHC353].

The problem of the regioselectivity in nucleophilic additions to pyridinium and related cations has attracted considerable interest. As discussed before (see Scheme 98, Chapter 3,II,B), the kinetic regioselectivity observed in the reactions of pyridinium and quinolinium cations with cyanide ion is in a sharp contrast with the thermodynamic one [65TL4615; 81ZOR418]. According to MO calculations the kinetic regioselectivity in nucleophilic additions to pyridinium ion is governed by the relative electron density at carbon atom under attack, independent of the hard or soft character of the nucleophile [92JOC4431]. Indeed, position 2 is the preferential site for nucleophilic addition in the kinetically controlled reactions of pyridinium and quinolinium cations with a vast majority of nucleophiles [76CL525; 76JOC1303; 79AHC1; 80H2015; 84T433; 86H181]. As far as the thermodynamic regioselectivity is concerned, many factors can contribute to the stability of σ-adducts, such as for instance, the anomeric effect operating in the reaction of pyridinium salts with methoxide ion, which leads to the thermodynamically controlled formation of 2-methoxy adducts [92JOC4431].

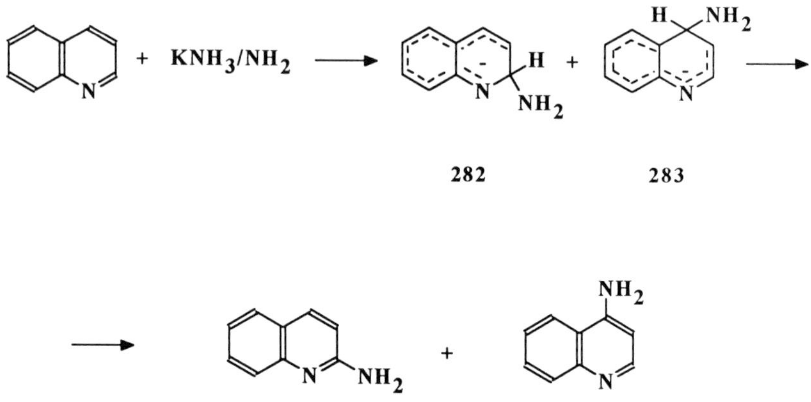

Scheme 163

4. *Structure and Stability of σ-Adducts.* σ-Adducts have already been postulated as intermediates in the nucleophilic substitution reactions, earlier than any other species. It is worth mentioning that the first example of σ-adduct formation refers to the field of the S_N^H reactions. In 1886 Janovsky discovered that treatment of a solution of *meta*-dinitrobenzene in acetone with sodium hydroxide gave a violet coloring of the reaction

mixture which was ascribed to the formation of anionic σ-complex **265** (see Scheme 154) [1886CB2155; 1891CB971]. Later on it was found that standing of the reaction mixture led to formation of 2,4-dinitrophenylacetone [50YZ444; 62JCS367].

The formation of σ-adducts is well documented in the literature [73MI3; 75MI1; 76RCR454; 78MI2; 79AHC1; 80H1033; 80H2015; 82CRV77; 82CRV427; 83AHC305; 84MI1; 85T237; 88AHC301; 88MI1; 89AHC73; 91MI1; 91MI2; 92H931].

In order to get insight into the structure and stereochemical features of these σ^H-adducts, a quite impressive number of studies have been performed (for reviews see [79AHC1; 82CRV77; 82CRV427; 83AHC305; 84MI1; 88MI1; 89AHC73; 91MI2; 92H931]). The NMR technique has proved to be a very successful tool for detecting these Meisenheimer type σ-adducts due to the fact the ring carbon to which the nucleophile adds undergoes a change in hybridization from sp^2 in the parent compound to sp^3 in the adduct. This results in pronounced upfield shifts for both the sp^3-carbon resonance and for a proton attached to it, as illustrated by ^1H NMR spectral data for some anionic σ-adducts of *s*-trinitrobenzene (Table XLIV) [91MI2].

Also, a considerable body of NMR spectral data has been accumulated in the literature for both anionic and neutral σ-adducts derived from reactions of pyridines [80H2015; 83AHC305; 84T433; 85T237], quinolines and isoquinolines [82MI2; 83AHC305], acridines [78ZOR140; 79ZOR206; 83AHC305], pyrimidines [82MI2; 83AHC305; 85T237], other diazines and benzodiazines [72CJC919; 72JA682; 73RTC708; 82MI2; 85T237; 88MI1], triazines [78RTC273; 85H2807; 85S884; 85T237; 86H239; 88MI1; 89AHC73; 92H931], pteridines [71JCS(B)2423; 75RTC45; 76OMR607; 82JHC1527; 86JHC477; 86JHC843; 87KG1385; 88MI1], naphthyridines [81JHC1349; 81JOC2134; 83AHC95], and *s*-tetrazines [81JOC3805; 85T237] with a variety of nucelophiles. They show that NMR spectral parameters of both anionic and neutral adducts are quite close to each other. In all the adducts studied a proton at the newly formed tetrahedral carbon resonates at a higher field ($\Delta\delta$= 3.5–5.5 ppm) relative

TABLE XLIV
^{1}H NMR Spectral Data for Some Anionic σ-Adducts
of s-Trinitrobenzene with Nucleophiles in DMSO

Nucleophile group attached to the sp^3-carbon	δ (ppm)	Δδ (ppm)
Hydroxy	6.1	3.1
Methoxy	6.1	3.1
Phenoxy	6.2	3.0
Amino	5.5	3.7
Phenylamino	6.2	3.0
Acetonyl	5.1	4.1
4-Hydroxyphenyl	5.6	3.6
Trichloromethyl	6.4	2.8

to the parent compound. An exception is found for the σ^H-aminoadduct of 3-phenyl-1,2,4,5-tetrazine, in which an upfield shift of 8.84 ppm (!) is observed [81JOC3805]. This unusual upfield shift is explained later in the text. Since ^{1}H chemical shifts for this resonance signal are usually rather sensitive to the nature of a nucleophile attached to the sp^3-carbon, they form a good diagnostic basis for structural elucidation of σ^H-adducts. This is especially important in those cases where reactions with nucleophiles are carried in water, alcohols, ammmonia, amines and other solvents which are able to add to aromatic substrates, yielding the corresponding solvent complexes. In such cases the addition stage is a rather complicated process, in which several concurrent addition and dissociation reactions are involved. For instance, the reaction of s-trinitrobenezene with aniline in methanol yields first the methoxy adduct **286** which is transformed slowly into the anilide complex **288** via the intermediacy of **287** (Scheme 164) [72CJC129; 77JA4429].

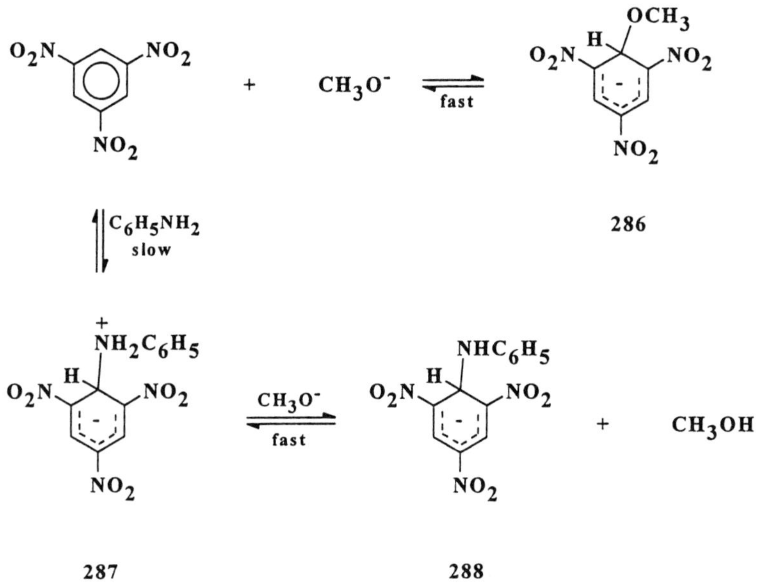

Scheme 164

Similar conversions of σ-adducts have been observed in the series of 1,4-diazinium salts, for instance, in the reaction of N-methylquinoxalinium ion with diethylmalonate in methanol in the presence of diethylamine [85KG669; 88MI1]. It results in the formation of the kinetically favored O-adduct **289** and N-adduct **290** which are gradually converted by action of diethyl malonate into the thermodynamically more stable C-adduct **291** (Scheme 165) [85KG669; 87MI1; 88MI1].

Also a few reports are published dealing with X-ray structural elucidation of σ^H-adducts derived from addition of nucleophiles to an unsubstituted carbon atom of aromatic or heteroaromatic ring [78RCR1061; 79AX(B)733; 86KG1544; 93ZOR622].

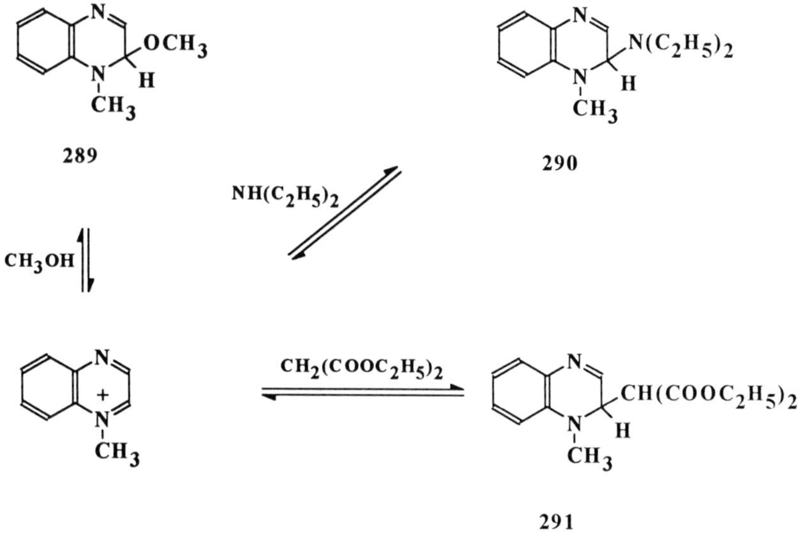

Scheme 165

The x-ray analysis of the trinitrobenzene–methoxide cyclohexadienide complex revealed that in this adduct there is a strong repulsion between the methoxy substituent and two nitro groups which forces the ring to be in a boat-like conformation [79AX(B)733]. Also in the Janovsky trinitrobenzene-acetone complex C-1 and C-4 atoms, being *cis*-oriented, are deviated from the plane formed by C-2, C-3, and C-5 atoms at 0.30 and 0.10 A, respectively (Figure 1) [78RCR1061]. The C(1)-C(2) and C(1)-C(6) bonds have the length of a normal ordinary C-C bond, while the C(2)-C(3) and C(5)-C(6) bonds are close to double C=C bond length, thus indicating at a considerable contribution of the quinoid structure **292**. Also a reduced length of the C(2)-N(1) bond (d= 1.38 A) shows that the negative charge is partly delocalized on this nitro group (Scheme 166; Figure 1). The capability of nitro groups to absorb the negative charge in anionic σ-adducts is in full agreement with MO calculations [72ACS2883], as well as with the data of ^{13}C NMR studies [76JOC3448].

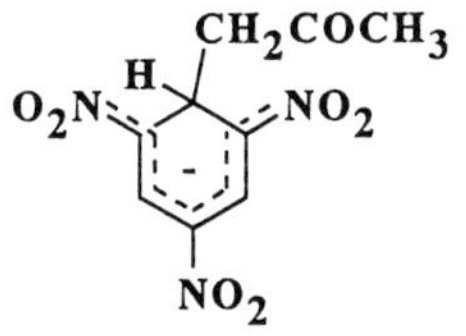

292

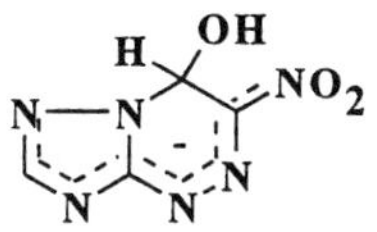

293

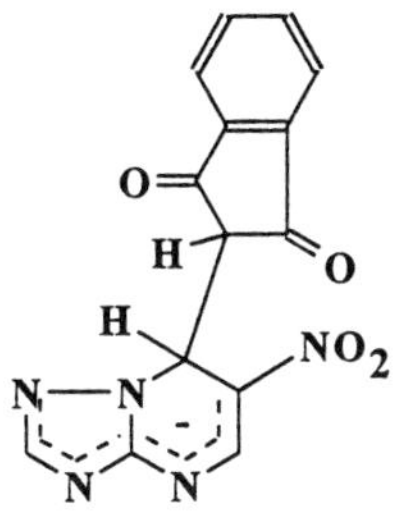

294

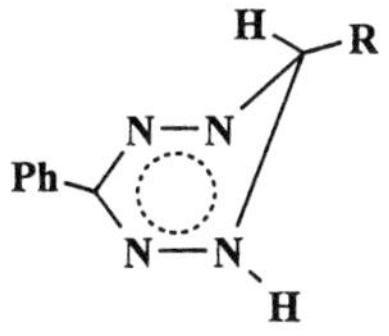

295a: R = NH$_2$
295b: R = C$_2$H$_5$

Scheme 166

However, localization of the negative charge is not the privilege of the nitro group. The X-ray data obtained for the adduct of 6-nitro-1,2,4-triazolo[5,1-c][1,2,4]triazine with hydroxide ion, that is, **293**, as well as for the adduct of 6-nitro-1,2,4-triazolo-[1,5-a]pyrimidine with 1,3-indandione carbanion, that is, **294** [93ZOR622], show that the negative charge is delocalized along the chain of atoms inside both rings,

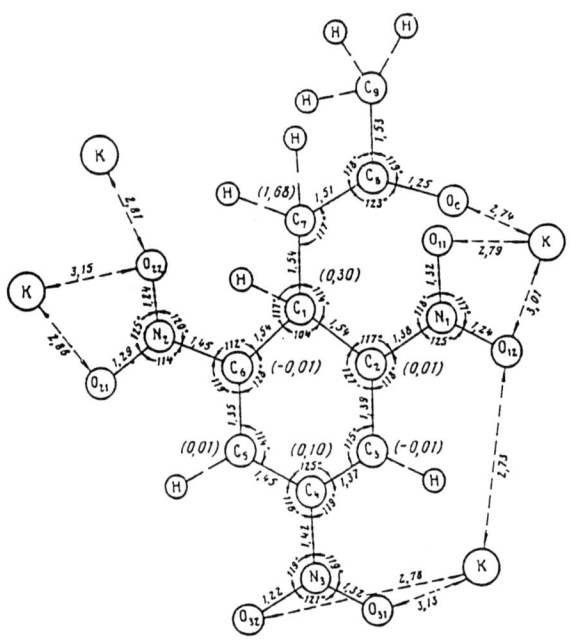

Figure 1. The X-ray structure of the Meisenheimer complex of 1,3,5-trinitrobenzene with acetonide anion (**292**).

while contribution of the nitro group appears to be small, as one can judge from the length of C(6)-N(5) bond in **294** (Figure 2; Scheme 166).

The unusual hydrogen upfield shift of 8.84 ppm observed on σ^H-aminoadduct formation of **295a**, obtained by addition of ammonia to 3-phenyl-1,2,4,5-tetrazine cannot be explained only by the change of hybridization of sp^2 to sp^3. It has been found that 6-amino-3-phenyl-1,6-dihydro-1,2,4,5-tetrazine **295a** can be characterized by a homotetrazole conformation; the amino group being large is in the *exo*-position and the

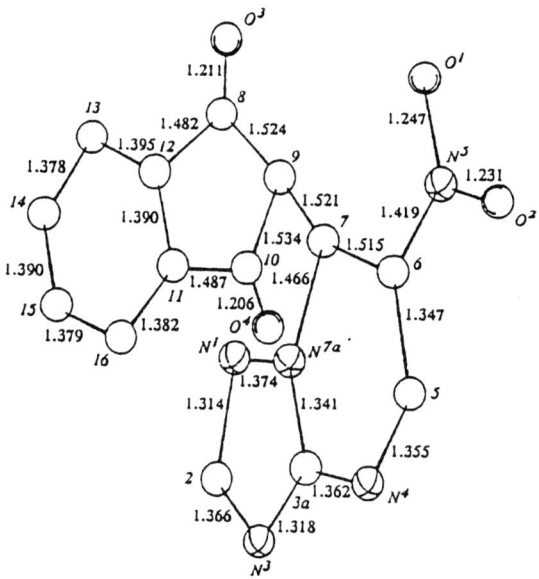

Figure 2. The X-ray structure of 6-nitro-7-(benzo[d]cyclopenta-1,3-dion-2-yl)-4,7-dihydro-1,2,4-triazolo[1,5-a]pyrimidine (**294**).

hydrogen is oriented above the plane of the ring and thus appears at a high field in the ^{1}H NMR spectrum [85T237].

An X-ray study of 6-ethyl-3-phenyl-1,6-dihydro-1,2,4,5-tetrazine (**295b**) confirms that the dihydro-1,2,4,5-tetrazines are boat shaped with both C-3 and C-6 upward (Figure 3). The sp^3-carbon atom is pointing upward with a dihedral angle of 49.3° between N(1)-C(6)-N(5) and the N(1)-N(2)-N(4)-N(5) planes; C-3 carrying the phenyl group is tilted 26.7° (see Figure 3). As predicted by ^{1}H NMR spectroscopy, H-6 is indeed above the puckered tetrazole ring and the ethyl group is in the *exo* position.

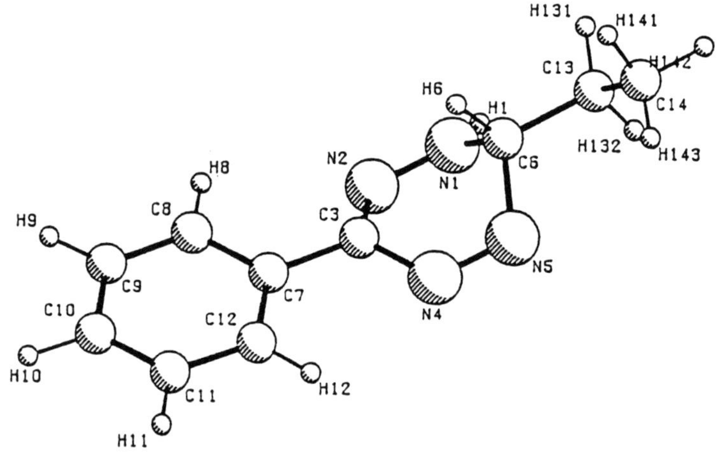

Figure 3. The X-ray structure of 6-ethyl-3-phenyl-1,6-dihydro-1,2,4,5-tetrazine (**295b**).

Besides formation of anionic σ-complexes, derived from reactions of nitroarenes with a vast majority of nucleophiles, there are examples of the addition reactions on the benzene ring, resulting in the formation of formally neutral species, which seem to be more stable than anionic σ-adducts. It is observed, for instance, in reactions of nitrobenzene with O-silylated enols [85JA5473], as well as with Grignard reagents [78HCA449; 82JOC5227] (Scheme 167, see also Chapter 2,III,A).

Since quantitative aspects of reactivity of nitroarenes, rate, and equilibrium constants for the formation of anionic σ-adducts with all kinds of nucleophiles are discussed in detail in excellent reviews and monographs [82CRV77; 84MI1; 91MI2], we wish to present just a few examples illustrating the ability of some nitroaromatics to add the methoxide anion and the stability of the formed σ-adducts (Table XLV).

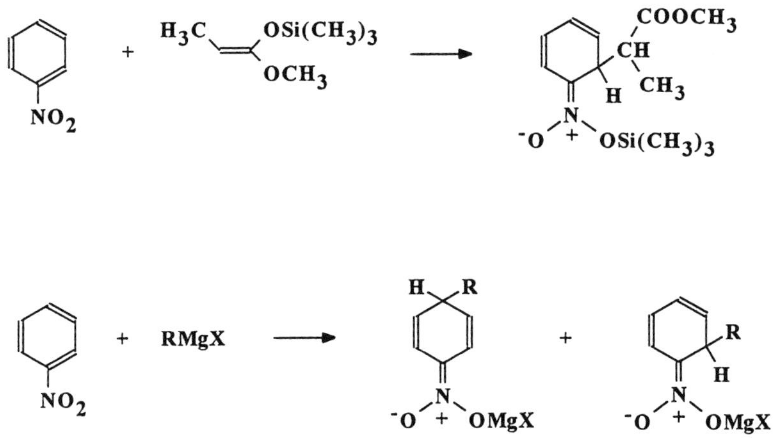

Scheme 167

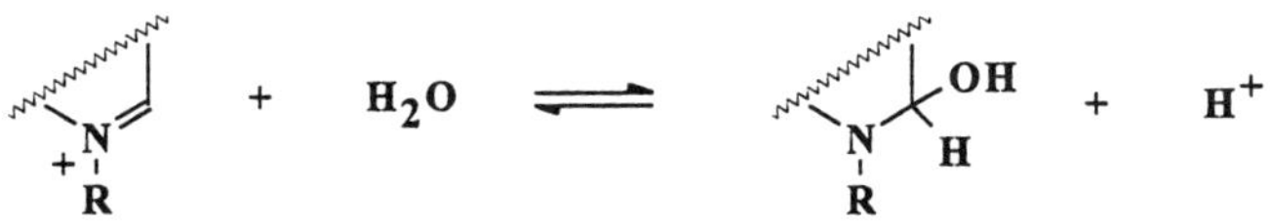

$$Q^+ + H_2O \rightleftharpoons QOH + H^+$$

$$K_{R^+} = \frac{[H^+][QOH]}{[Q^+]}; \qquad pK_{R^+} = -\log K_{R^+},$$

Q is quaternary N-alkylazinium salt, R is alkyl group

Scheme 168

TABLE XLV
Rate and Equilibrium Constants of Anionic σ-Adducts of Some Nitro- and Nitroheteroarenes (in Methanol at 25°C)

Complex	k_1 (l·mol^{-1}·s^{-1})	k_{-1} (s^{-1})	K_c (l·mol^{-1})	References
1,3,5-trinitrobenzene OCH$_3$ adduct	7050	305	23.1	70JA4682
1,4-dinitronaphthalene OCH$_3$ adduct	11.2	124	0.09	75JCS(P2)1568
3,5-dinitropyridine OCH$_3$ adduct	2460	35.5	69.5	72BSF4549
nitrofurazan OH adduct	4500	$< 9 \cdot 10^{-3}$	$> 5 \cdot 10^5$	78JOC4303

Also in the series of heteroaromatics rate and equilibrium constants for the formation of anionic σ-adducts have been extensively discussed [83AHC305]. One of the characteristic values describing in quantitative terms the tendency of azinium cations to interact with the hydroxide ion is the pK_{R^+} value (Scheme 168; Table XLVI)

[79AHC1; 86KG1380; 88AHC301]. The variation of pK_{R^+} values in the series of N-alkylazinium cations confirms that the stabilities of the hydroxy adducts increase by: (1) introduction of electron-withdrawing substituents; (2) benzo or heteroaromatic annelation; and (3) aza-activation (Table XLVI).

TABLE XLVI

pK_{R^+} Values for Some N-Alkylazinium Salts

N-Alkylazinium cation	pK_{R^+}	References
1-Methylpyridinium	>16.0	79AHC1
1-Methylquinolinium	16.5	79AHC1
2-Methylisoquinolinium	15.3	79AHC1
3-Aminocarbonyl-1-methylpyrazinium	8.04	79AHC1
3-Methoxycarbonyl-1-methyl-pyrazinium	6.37	88AHC301
1-Methylquinoxalinium	8.62	86KG1380
1-Methylbenzo[g]quinoxalinium	5.73	86KG1380
6-(N,N-Dimethylamino)-4-methyl-pyrido[2,3-b]pyrazinium	12.5	86KG1380
2-(N,N-Dimethylamino)-4-methyl-8-methylpteridinium	7.24	88AHC301
9-Methylacridinium	5.50	79AHC1

B. *The Aromatization Stage*

The aromatization of σ-adducts involving a C-H bond fission is certainly a key step in the nucleophilic aromatic displacement of hydrogen. The specific nature of the hydride ion as a leaving group, not being stable in the form of an anion, is the main feature of the S_N^H reactions. The tendency of σ-adducts to gain aromaticity by elimination of the hydride ion contradicts its nature, therefore elimination of the sp^3-hydrogen from the σ^H-adducts readily occurs only in those cases where the electron density in these adducts is redistributed in such a way that the abstraction of proton or hydrogen atom becomes possible. As already discussed in Chapter 1, it can be reached: (1) by adding an oxidant or by use of electrochemical technique to cause oxidative aromatization; (2) by introduction of an auxiliary group in reagent or substrate to create conditions for *auto-* aromatization of the intermediate σ^H-adducts.

From a great deal of substitution reactions described in this book it is clear that oxidative aromatization is a rather effective way to perform the S_N^H reactions. There are many examples where the S_N^H reactions do not take place in the absence of an oxidant or result in the formation of other products [76RCR454; 85MI4]. For instance, the reaction of 3-amino-1,2,4-triazine **296** with potassium amide in liquid ammonia at –40°C does not give any reaction product; also no adduct formation at C-5 could be observed by NMR spectroscopy. However, when the same reaction is carried out in the presence of potassium permanganate 3,5-diamino-1,2,4-triazine **299** is formed in a high yield (Scheme 169) [85S884]. This is probably due to the fact that in the KNH$_2$/NH$_3$ solution 3-amino-1,2,4-triazine is deprotonated by the strong base, and therefore, the equilibrium between the anion **297** and its σ^H-adduct **298** lies far to the left. On oxidation the equilibrium shifts in the direction of the σ^H-adduct **298**, being oxidized into the diamino compound **299** (Scheme 169). Also, it is a remarkable that the reaction of acridine with CH-active methylheterocycles mentioned in Chapter 3,I,B (Scheme 82) takes place at

120–130°C only in the presence of an excess of sulfur; no reaction has been observed in the absence of an oxidant [70KG1384].

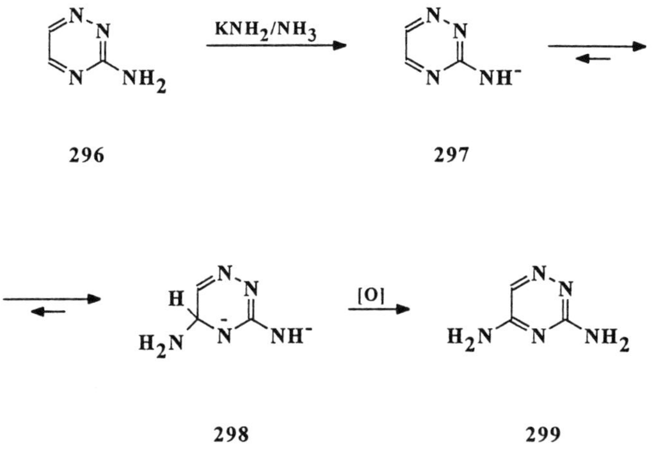

Scheme 169

The advantages of oxidative aromatization are the following: (1) the S_N^H reactions occur without replacement of other functional groups in the ring, even when these groups have a good leaving ability; and (2) air oxygen can be used as oxidant and affords water which is rather attractive from both technological and ecological points of view.

The disadvantages of oxidative aromatization are: (1) in "spontaneous" oxidations a considerable part of the aromatic substrate used is spoiled due to the formation of by-products; and (2) an added oxidant may oxidize the nucleophilic reagent. Although, due to some of these disadvantages, oxidative processes in aromatization cannot universally be applied [89RCR747], there is no reason to consider the oxidative version of the S_N^H reaction as a nonperspective one. Just the opposite, in many cases addition of oxidant provides high or nearly quantitative yields of the S_N^H products. In particular, successful amination of a variety of azines by liquid ammonia or potassium amide in the presence of potassium permanganate is a brilliant illustration for this approach [87KG1011; 93ACSA95]. According to a recent publication [92USP5117063] the oxidative amination of nitrobenzene may find industrial application.

In *auto*-aromatization reactions the abstraction of the sp^3-hydrogen is assisted by departure of the readily leaving auxiliary group A. The departing group takes two electrons while hydrogen atom is eliminated as proton (see Chapter 1). A great many examples of *auto*-aromatization reactions are provided by vicarious nucleophilic substitutions, as well as *tele*-substituton reactions, which have been discussed in this book (see Chapters 2 and 3).

As far as vicarious S_N^H reactions are concerned, two plausible pathways for *auto*-aromatization of the intermediate σ^H-adducts are discussed in the literature: (1) the intramolecular 1,2-shift of the hydride ion from the ring sp^3-carbon to C-α of the side-chain substituent accompanied by elimination of an auxiliary group A in a concerted manner; and (2) a concerted base-induced β-elimination of HA followed by protonation of the C-α [83JOC3860; 89RCR747; 92PJC3]. In order to choose between these two possibilities the influence of various concentrations of potassium *t*-butoxide on the reaction of *para*-fluoronitrobenzene with chloromethyl phenyl sulfone has been thoroughly investigated [83JOC3860]. The results of these studies revealed that yields of the S_N^H product [4-fluoro-2-(phenylsulfonyl)methyl-1-nitrobenzene, see Scheme 33 in Chapter 2,III,A, X= F] were in straight dependence on concentration of the base, thus indicating that aromatization of the intermediate σ^H-adduct occurs via β-elimination of HCl [83JOC3860]. The same conclusion has been reached by studying two concurrent reactions, that is, substitution of H-4 in nitrobenzene and the displacement of the fluoro atom in *para*-fluoronitrobenzene by action of the bulky tertiary carbanion of phenylphenoxyacetonitrile which is able to attack only the *para* position relative to the nitro group. It has been found that with an excess of base the displacement of hydrogen in nitrobenzene prevails over possible substitution of the fluoro atom in *para*-fluoronitrobenzene which remains completely unchanged; lack of base results in the conventional nucleophilic substitution of halogen [83JOC3860].

Kinetic isotope effect k_H/k_D measured for the displacement of hydrogen in nitrobenzene and *para*-D-nitrobenzene by action of chloromethyl phenyl sulfone proved to be very low (0.9), while k_H/k_D values obtained for other vicarious substitution reactions were in the range 3–7 [88JOC690; 89RCR747]. Since these data seem to be rather controversial, further measurements are necessary.

The discussion concerning the intricate mechanism of dehydroaromatization of σ-adducts launched more than 30 years ago is still proceeding since these reactions are considered to be a model of the important metabolism processes which occur with participation of coenzymes, such as NAD, NADH, and FAD [76MI4; 77BCJ1535; 83JA5886; 88JOC1646; 89CL1227; 89JCS(CC)941; 90BCJ2682; 90JOC3647; 93BCJ1191].

In order to investigate how the sp^3-hydrogen is eliminated from dihydroazines the hydrogen exchange in the solvents of different polarities has been studied [61JA712]. Other approaches are based on the analysis of the relationship between the rate constants for the hydrogen transfer from the donor to the acceptor and equilibrium constants for the nucleophilic addition of the cyanide ion to pyridinium salts [59LA149], the elucidation of the ESR spectra [72CL369; 90CL1275], measuring of the isotope effects [77CJC2741; 77JCS(CC)181; 90BCJ2682], and the electrochemical modeling of the reactions [76MI3; 79LA918; 80MI2; 91JOC6736; 92CL1583].

Elimination of hydrogen from σ-adducts is often initiated by single electron transfer (SET) [76MI3; 84KG1299; 88JOC1646; 89CL1227; 90BCJ2682; 91JOC6736]. In particular, electrochemical oxidation of 9-methyl and 9-aryl-substituted 10-methyl-9,10-dihydroacridines is realized as a three-step process involving an electron–proton–electron (EPE) transfer [76MI3]. The sequence of elementary steps in this multistep process can be different and is dependent on both the nature of the dihydro compound and the conditions used. For instance, 9-cyano-10-methyl-9,10-dihydroacridine is oxidized electrochemically via transfer of an electron–electron–proton (EEP) process, while in the presence of sodium nitrite the oxidative stage involves a proton–electron–electron sequence of elementary steps (PEE) [80MI2; 84KG1299]. The latter probably accounts for the acidic character of H-9 at the sp^3-carbon due to the presence of the cyano group [80MI2].

When analyzing results of the recent studies on the mechanism of dehydrogenation reactions, one can see that the data on the stepwise mechanism of the hydride ion elimination dominate in the literature [76MI3; 84KG1299; 88JOC1646; 88T1; 89CL1227; 90BCJ2682; 91JOC6736], although there are also reports on the one-stage transfer of the hydride ion [77CJC2741; 84KG1011; 84KG1299; 88T1].

C. Kinetic Studies

As already mentioned, the literature contains a huge amount of data concening rate constants for the formation of σ-adducts from nitro-activated arenes and heteroarenes [82CRV77; 84MI1; 91MI2]. Since rate constants for the addition step may be very high, a special stopped-flow or temperature-jump technique is often required for their determination [91MI2].

The question of whether σ-adducts are the reaction coordinates or not can be answered by means of a kinetic study provided that the measuring of concentration changes for a minimum of three components can be performed. This condition is usually not fulfilled and, therefore, suitable models are necessary. It seems that heteroarenes provide better synthetic possibilities in choosing the appropriate models for kinetic studies. The reaction of the acridinium cation **300** with arylamines can serve as one of them [76KG1227], because (1) the acridinium cation has one electrophilic center, therefore no isomeric adducts can be formed, and (2) the annelation of the pyridine ring with two benzene rings enhances the stabilities of the σ^H-adducts, making it possible to measure them by spectroscopic methods. The reaction can proceed either under oxidizing conditions (air bubbling through the reaction solution) resulting in the formation of product **303** or, in case an oxidant is absent, cation **300** performs the dehydrogenation of **302** into **303** .by hydrogen transfer from position C-9 in **302** to C-9 in **300** under formation of dihydroacridine **304** (Scheme 170).

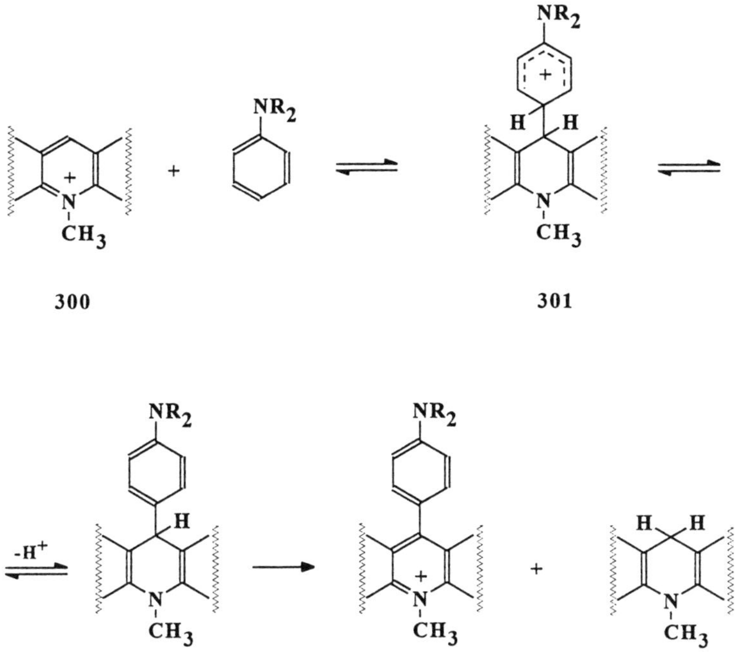

300 301

302 303 304

Scheme 170

Rate constants for the reaction of N-methylacridinium iodide with arylamines in the presence of air bubbling through the solution of these reagents in DMF were obtained [76KG1227]. The kinetic isotope effect measured for the reaction with 2,4,6-D$_3$-aniline k_H/k_D= 2.2 proved to be rather small; base catalysis was not found. ^{1}H NMR spectroscopy showed that the dihydroacridines **302** were always present in the reaction mixtures and, in a number of cases, these adducts could even be isolated as crystalline substances [78ZOR140]. The dihydroacridines **302** exhibit the typical properties of intermediates in a nucleophilic substitution. When treated with acetic acid in an inert atmosphere they undergo the dissociation reaction to yield the starting materials, that is, the cation **300** and the corresponding arylamine, while in the presence of an oxidant σ–adducts **302** are aromatized into the S_N^H products **303**.

These facts show that the reaction proceeds via a stepwise mechanism. The first step is the addition of arylamine to cation **300** to form σ^H-adduct **301**. The abstraction of a proton from the arylamine moiety in **301** leading to **302** does not occur simultaneously with the addition process; deprotonation of **301** takes place at the next stage which is not a rate-determining one. The low value of the kinetic isotope effect indicates a considerable reversibility of the addition step, that is, comparatively large value of k_{-1}. The fact that the presence of **302** could be proved indicates that the proton abstraction from adduct **301** does not occur simultaneously with the departure of the C-9 hydrogen, yielding **303**.

The problem whether the dihydroacridines **302** are intermediates in the formation of **303** or only side products of the kinetic systems has been solved by a ^{1}H NMR kinetic study of the reaction between N-methylacridinium iodide and *ortho*-toluidine in a DMSO-d$_6$ solution at 35°C in the absence of an added oxidant. Experimental curves obtained by measuring concentrations of the cation **300**, the σ-adduct **302**, and the product **303** have been compared with the data of theoretical calculations in which two alternative mechanistic schemes have been simulated by mathematical modeling of the reaction [79ZOR117]. The first one involves dihydroacridine **302** as the reaction intermediate (Scheme 171), while the second scheme corresponds to parallel reactions with dihydro compounds **302** as the side products (Scheme 172).

300 + arylamine ⇌ **302**

302 + **300** ⇌ **303** + **304**

Scheme 171

300 + arylamine ⇌ **302**

300 + arylamine + **300** ⇌ **303** + **304**

Scheme 172

A better correspondence between theoretical and experimental curves has been observed for the first variant (Figure 4), thus justifying the conclusion that substitution of hydrogen in acridinium cation proceeds via a two-step mechanism with the intermediacy of the adduct **302** [79ZOR117].

The same kinetic features have been established for the other examples of aminoarylation reactions [74KG675; 78ZOR134]. A series of detailed kinetic studies has also been performed on oxidative hydroxylation of a number of N-alkylazinium cations in a basic 20% acetonitrile–80% water solution in the range pH 11–14 at 25°C [78JOC1132; 84CJC729; 86JOC2060; 86JOC2068]. Rate parameters for the ferricyanide ion oxidation of a number of N-methylazinium cations were determined [86JOC2060; 86JOC2068]. The reaction was shown to be consistent with the generalized mechanism, as outlined in Scheme 173 for 2-methylisoquinolinium ion.

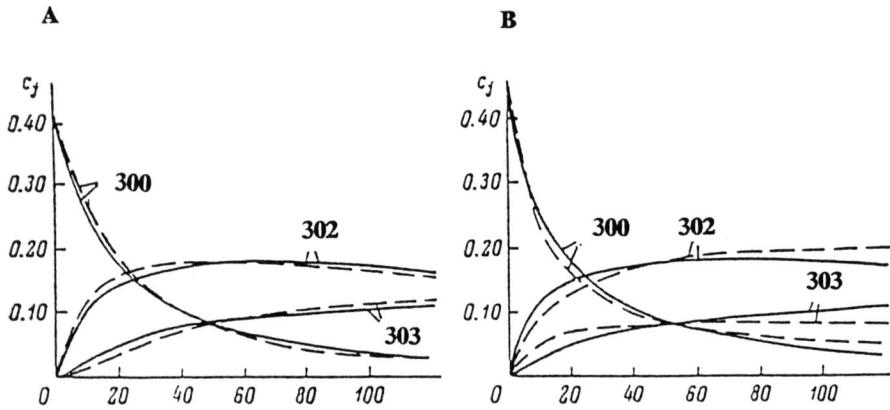

Figure 4. The experimental (solid lines) and theoretical (dotted lines) curves for the reaction of N-methylacridinium iodide (**300**) with *ortho*-toluidine leading to the formation of 10-methyl-9,10-dihydroacridine (**304**) and the S_N^H product **303** via intermediacy of the dihydroacridine **302** (A) or with the formation of **302** and **303** in two parallel reactions (B).

It involves an electron transfer from isoquinoline anhydrobase anions **306** to ferricyanide ion to give nitrogen radical cation species **307**. Substituent effects were shown to be in agreement with electron transfer from the endocyclic nitrogen atom rather than from the exocyclic oxyanion in **306**. The radical **307** undergoes a base-catalyzed deprotonation from C-1, as shown by measuring deuterium kinetic isotope effects, followed by electron transfer from **308** to ferricyanide ion, yielding 2-methyl-1-isoquinolone (Scheme 173) [86JOC2060].

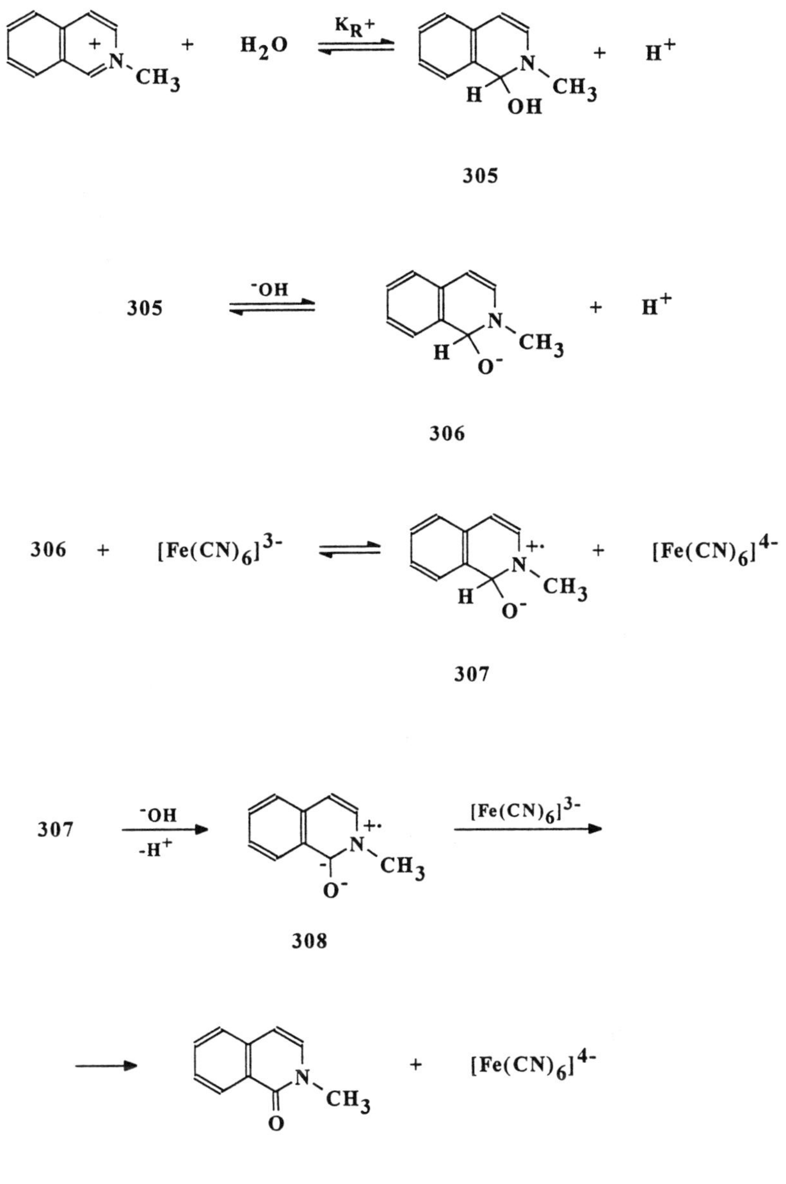

Scheme 173

II. THE ELIMINATION-ADDITION MECHANISM [$S_N^H(EA)$]

There is a group of nucleophilic aromatic substitution reactions in which not only the ring carbon, bearing leaving group, but also a neighboring carbon at the *ortho*-position is involved in the bond formation with a nucleophile. As a result, the group enters an *ortho*-position with respect to that vacated by the leaving group. The process is defined as a *cine*-substitution reaction, and it can formally be regarded as an S_N^H reaction facilitated by elimination of an auxiliary group.

Cine-substitutions in aromatics can be divided into two classes: (1) reactions which occur by an initial addition of the nucleophile at a C-H position followed by elimination of the leaving group from the neighboring position; that is, the $S_N^H(AE)$ substitution; and (2) reactions which occur by initial deprotonation and elimination of the leaving group, followed by the addition of the nucleophile, that is, $S_N^H(EA)$ substitution.

In this section we want to discuss shortly the $S_N^H(EA)$ *cine*-substitutions. Since this subject has been exhaustively reviewed in monographs [65AGE543; 65AGE731; 67MI2; 69MI2; 77AHC121] and textbooks on organic chemistry [92MI1; 92MI2] we do not dwell on *cine*-substitution reactions in this book.

$S_N(AE)$ *cine*-substitutions happen to occur, for example, in reactions of nonactivated halobenzenes with nucleophiles of strong basic character, such as potassium amide, potassium hydroxide and the like. A well-known example of these *cine*-substitutions is the reaction of chlorobenzene with potassium amide in liquid ammonia, yielding aniline (Scheme174) [67MI2].

In order to account for the fact that aniline obtained from the amino-dehalogenation reaction on ^{14}C-1-labeled chlorobenzene contains 50% of the label on the *ipso*-carbon and 50% on the carbon at the *ortho*-position, the "elimination–addition" mechanism $S_N(EA)$ was advanced [53JA3290; 56JA601; 58MI2]. It has been suggested to involve the formation of dehydrobenzene intermediate **310** due to elimination of a proton and the chloride anion, followed by addition of ammonia (Scheme 174).

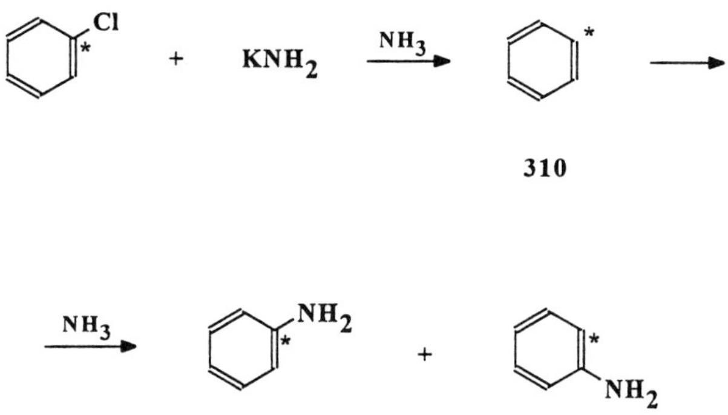

Scheme 174

Further studies have provided experimental evidence for the formation of didehydrobenzene [67MI2]. Also, it has been shown that this amino-dehalogenation reaction is greatly influenced by the nature of halogen. In the conversion of chlorobenzene into aniline, elimination of HCL is a two-step E1cB reaction, while bromobenzene eliminates HBr in a concerted manner. The reason is that the bromo substituents cannot provide effective stabilization for the phenyl anion formed on deprotonation of bromobenzene. Indeed, no proton/deuterium exchange could be detected in the reaction of 2-D-1-bromobenzene with ammonia (Scheme 175), and the amino-debromination reaction was found to exhibit a marked kinetic isotope effect ($k_H/k_D = 6$).

A great deal of $S_N(AE)$ *cine*-substituion reactions in aromatics [65AGE543; 65AGE731; 67MI2] as well as in heteroaromatic compounds [77AHC121; 78JCS(P1)1140] have been established to occur via the aryne mechanism. A well-established example of a hetaryne mechanism is found in the amination reaction of

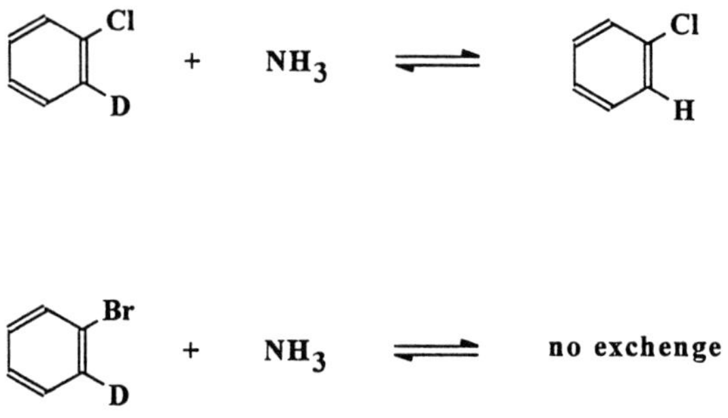

Scheme 175

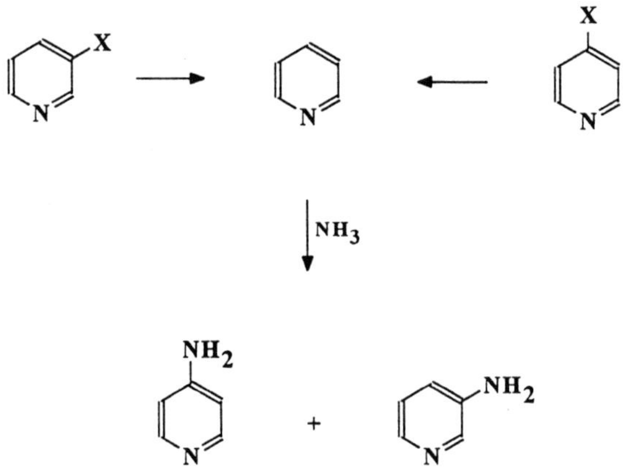

Scheme 176

X = Cl, Br, I

3- and 4-X-pyridines (X= Cl, Br, I) by potassium amide in liquid ammonia. From all six compounds a mixture of 3- and 4-aminopyridines is obtained in high yield; the ratio between these two amino-compounds (2:1) is in all reactions the same, independent of the position and nature of the halogen atom in the starting materials (Scheme 176) [77AHC121].

III. THE S_N^H(ANRORC) MECHANISM

It has been demonstrated that many nucleophilic substitutions in aza aromatic compounds, involving an amide anion, ammonia, or amine as nucleophile, occur via a so-called ANRORC mechanism [78ACR462]. This means that the nucleophilic substitution takes place via a series of steps: (1) addition of the nucleophile; (2) ring opening; and (3) ring closure. Convincing proof for this ring-opening mechanism is obtained by extensive ^{15}N-labeling work with a number of aza aromatics. So, for instance, amination of 4-bromo-6-phenylpyrimidine, being ^{15}N-labeled in the ring nitrogen atoms, with potassium amide in liquid ammonia, gives 4-amino-6-phenylpyrimidine in which the pyrimidine ring has partly lost its ^{15}N-label (Scheme 177) [71RTC1239].

Also in amino-dehydrogenation reactions the ANRORC mechanism is observed. A well-studied example is the Chichibabin amination of 4-phenylpyrimidine **311** [79JOC4677]. Treatment of 4-phenylpyrimidine with ^{15}N potassium amide in liquid ammonia for 70 hours and quenching the solution with ammonium chloride gave 15% of 6-amino-4-phenylpyrimidine **315*** and 60% of 2-amino-4-phenylpyrimidine **318***. Degradation studies and mass spectroscopic determination showed that in the 6-amino compound **315*** the ^{15}N-label is nearly all present in the amino group, but that in the 2-amino compound **318*** the ^{15}N-label is 92% (!) present in the nitrogen of the

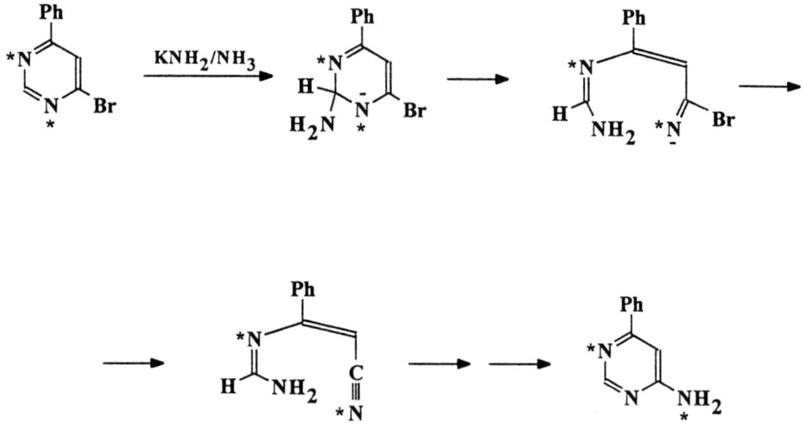

Scheme 177

pyrimidine ring. These results clearly show that the formation of the 2-amino compound **318*** almost exclusively occurs via a ring–opening–ring–closure mechanism and that on the contrary, the 6-amino compound **315*** appears to be formed without following this ANRORC process. ^{1}H and ^{13}C NMR spectroscopy of a solution of **311** in KNH$_2$/NH$_3$ showed the presence of two σ-adducts, that is, 6-amino-4-phenyl-dihydropyrimidinide **313** and the 2-amino-6-phenyldihydropyrimidinide **312**. The ratio **313/312** is 80:20 about 20 minutes after dissolving **311** in KNH$_2$/NH$_3$, but after a few hours σ-adduct **312** is completely absent. Apparently we deal with the kinetically favored formation of σ-adduct **312**, which slowly converts into the more stable adduct **313** via **311**.

Based on all these data the following S_N^H(ANRORC) mechanism can be proposed for the Chichibabin amination of 4-phenylpyrimidine (Scheme 178).

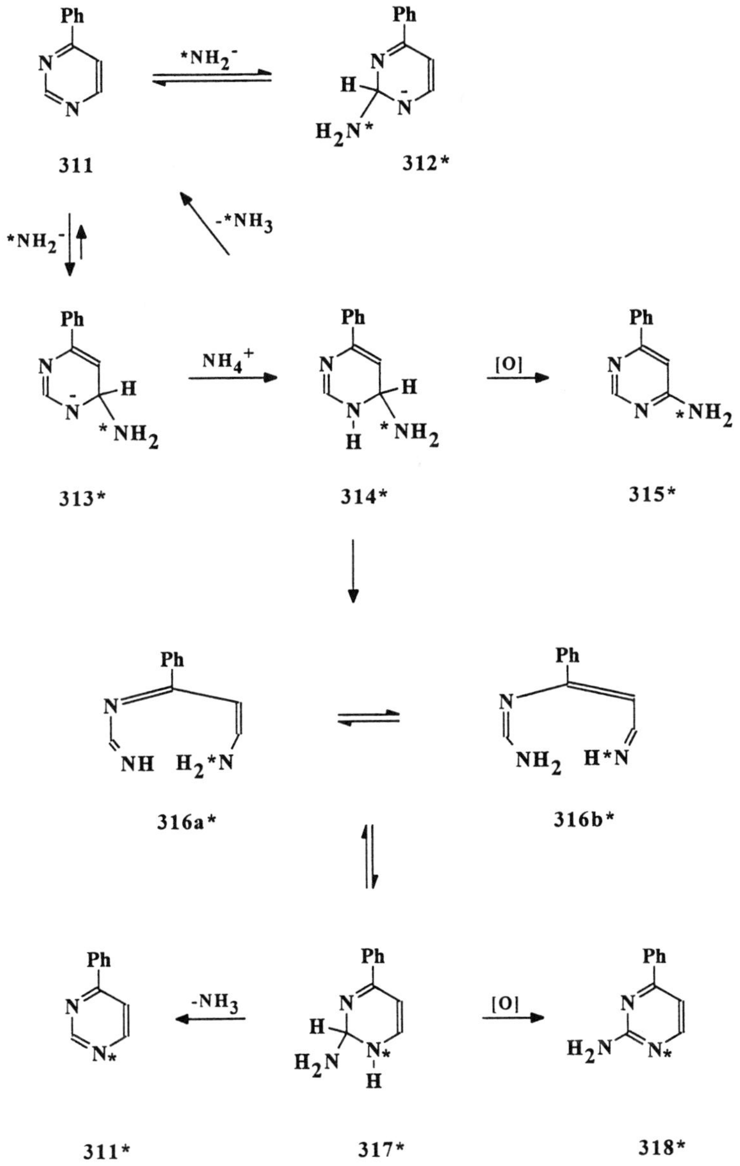

Scheme 178

Quenching of the solution of **313*** in KNH_2/NH_3 with ammonium chloride, being a strong acid in this medium, not only neutralizes the amide ion, but also protonates the ring nitrogen atom of **313*** into the uncharged 1,6-dihydropyrimidine **314*** from which the exocyclic labeled 6-aminopyrimidine **315*** is formed. Ring opening of **314*** into 6-amino-1,3-diazacyclohexa-1,3,5-triene **316a*** ⇌ **316b***, followed by ring closure gives the 2-aminodihydropyrimidine **317***, from which the ring-labeled 2-amino compound **318*** is formed. A further support for this mechanism is the experimental fact that recovered starting material **311*** is considerably ring labeled.

The S_N^H (ANRORC) mechanism has also been found in the amination of 5-phenylpyrimidine [82MI2] and 2-phenyl-1,3,5-triazine [76RTC125], however, not in the amination of 4-*t*-butylpyrimidine [82MI2] and 2-phenylpyrazine [82MI2].

References

1871CB21   V. von Richter, *Chem. Ber.* **4**, 21 (1871).

1871CB451   V. von Richter, *Chem. Ber.* **4**, 451 (1871).

1871CB553   V. von Richter, *Chem. Ber.* **4**, 553 (1871).

1882LA344   P. Hepp, *Liebigs Ann. Chem.* **215**, 344 (1882).

1883CB2152   W. Koenigs and G. Koerner, *Chem. Ber.* **16**, 2152 (1883).

1886CB2155   J.V. Janovsky and L. Erb, *Chem. Ber.* **19**, 2155 (1886).

1891CB971   J.V. Janovsky, *Chem. Ber.* **24**, 971 (1891).

1899CB3486   A. Wohl, *Chem. Ber.* **32**, 3486 (1899).

01CB2442   A. Wohl and W. Aue, *Chem. Ber.* **34**, 2442 (1901).

05CB1603   A. Reissert, *Chem. Ber.* **38**, 1603 (1905).

05CB1610   A. Reissert, *Chem. Ber.* **38**, 1610 (1905).

05CB3415   A. Reissert, *Chem. Ber.* **38**, 3415 (1905).

06CB2533   J. Meisenheimer and E. Patzig, *Chem. Ber.* **39**, 2533 (1906).

06CB3006   F. Sachs, *Chem. Ber.* **39**, 3006 (1906).

06CB3023   F. Sachs, *Chem. Ber.* **39**, 3023 (1906).

06CB3081   F. Sachs, *Chem. Ber.* **39**, 3081 (1906).

09CB3776	A. Kaufmann and A. Albertini, *Chem. Ber.* **42**, 3776 (1909).
14MI1	A.E. Chichibabin and O.A. Zeide, *Zh. Russ. Phys.-Khim. Obsh.* **46**, 1216 (1914) [*CA* **9**, 1901 (1915)].
30CB749	T. Reichstein, *Chem. Ber.* **63**, 749 (1930).
30LA80	E. Bergman, O. Blum-Bergman, and A.F. Christiaani, *Liebigs Ann. Chem.* **483**, 80 (1930).
30LA123	K. Ziegler and H. Wollschnitt, *Liebigs Ann. Chem.* **479**, 123 (1930).
30MI1	A.E. Chichibabin and A.P. Kursanova, *Zh. Russ. Phys.-Khim. Obsh.* **62**, 1211 (1930).
31LA174	K. Ziegler and H. Zeiser, *Liebigs Ann. Chem.* **485**, 174 (1931).
32JCS1254	W. Bradley and R. Robinson, *J. Chem. Soc.*, 1254 (1932).
37CR56	M. Sommelet, *C. R. Hebd. Seances Acad. Sci.* **205**, 56 (1937).
48JA2172	W.E. Knox and W.I. Grossman, *J. Am. Chem. Soc.* **70**, 2172 (1948).
49MI1	M. J. Dewar, "The Electronic Theory of Organic Chemistry," Oxford, London, 1949, p. 77.
50CB10	K. Bauer, *Chem. Ber.* **83**, 10 (1950).

50JA2181	H. Gilman and D.A. Shirley, *J. Am. Chem. Soc.* **72**, 2181 (1950).
50JOC481	J.F. Bunnet, J.F. Cormac, and F.C. McKay, *J. Org. Chem.* **15**, 481 (1950).
50YZ444	M. Ishidate and T. Sakagushi, *Yakugaku Zassi* **70**, 444 (1950).
51CRV273	J.F. Bunnet and R.E. Zahler, *Chem. Rev.* **49**, 273 (1951).
51JA3325	N.J. Leonard, H.A. de Walt, and G.W. Leubner, *J. Am. Chem. Soc.* **73**, 3325 (1951).
51JA4122	Ch.R. Hauser and S.W. Kantor, *J. Am. Chem. Soc.* **73**, 4122 (1951).
52MI1	V. Veijola, *Suom. Kemi* **B 25**, 1 (1952).
53JA3290	J.D. Roberts, H.E. Simmons, L.A. Carlsmith, and C.W. Vaughau, *J. Am. Chem. Soc.* **75**, 3290 (1953).
54MI1	M. Colonna and A. Risaliti, *Ann. Chim. (Rome)* **44**, 1029 (1954).
55JA5051	J.F. Bunnet and R. Morath, *J. Am. Chem. Soc.* **77**, 5051 (1955).
56JA601	J.D. Roberts, D.N. Semenov, H.E. Simmons, and L.A. Carlsmith, *J. Am. Chem. Soc.* **78**, 601 (1956).
56LA176	F. Krouhnke and K. Ellegast, *Liebigs Ann. Chem.* **600**, 176 (1956).
57CB2215	F. Krouhnke and H. Honig, *Chem. Ber.* **90**, 2215 (1957).

57JOC1370	N.B. Edoly, J.G. Murky, and E.L. May, *J. Org. Chem.* **22**, 1370 (1957).
57LA221	K. Dimroth, G. Arnoldy, S. Eicken, and G. Schiffler, *Liebigs Ann. Chem.* **604**, 221 (1957).
58JOC358	Ch.R. Hauser, J.K. Lindsey, and D. Lendicer, *J. Org. Chem.* **23**, 358 (1958).
58MI1	J.F. Bunnet, *Quart. Rev.* (London) **12**, 1 (1958)
58MI2	E.F. Jenny, M.C. Caserio, and J.D. Roberts, *Experientia* **14**, 349 (1958).
59LA149	K. Wallentels, *Liebigs Ann. Chem.* **621**, 149 (1959).
60AG294	J. Sauer and R. Huisgen, *Angew. Chem.* **72**, 294 (1960).
59AG310	H.V. Dobenek, H. Deuber, and F. Jerchel, *Angew. Chem.* **71**, 310 (1959).
60CB1579	W. Schneider and B. Muller, *Chem. Ber.* **93**, 1579 (1960).
60JA3797	M. Rosenblum, *J. Am. Chem. Soc.* **82**, 3797 (1960).
60JCS989	M.L.H. Green, L. Pratt, and G. Wilkenson, *J. Chem. Soc.*, 989 (1960).

60JOC1884	R.B. Davis and L.C. Pizzini, *J. Org. Chem.* **25**, 1884 (1960).
61JA712	R.H. Abelen, R.F. Hutton, and F.H. Westheimer, *J. Am. Chem. Soc.* **83**, 712 (1961).
61MI1	O.N. Kursanov and N.K. Barahetskaya, *Izv. Akad. Nauk. SSSR ser. khim.*, 1703 (1961) [*CA* **56**, 3447 (1962)].
62CB1484	H. von Dobeneck and W. Goltzsche, *Chem. Ber.* **95**, 1484 (1962).
62JA4153	G.A. Russel and E.G. Janzen, *J. Am. Chem. Soc.*, **84**, 4153 (1962).
62JCS367	T.J. King and C.E. Newall, *J. Chem. Soc.*, 367 (1962).
63CPB411	M. Hamana and M. Yamazaki, *Chem. Pharm. Bull.* **11**, 411 (1963).
63CPB1331	M. Hamana and H. Noda, *Chem. Pharm. Bull.* **11**, 1331 (1963).
63JA1356	J.A. Dixon and D.H. Fishman, *J. Am. Chem. Soc.* **85**, 1356 (1963).
64AHC285	G. Illuminati, *Adv. Heterocycl. Chem.* **4**, 285 (1964).
64CPB783	S. Takehashi and S. Kano, *Chem. Pharm. Bull.* **12**, 783 (1964).
64JA1807	G.A. Russel, E.G. Janzen, and E.T. Strom, *J. Am. Chem. Soc.* **86**, 1807 (1964)
64JCS2806	J.D. Loudon and D.M. Smith, *J. Chem. Soc.*, 2806 (1964).
64JOC3381	H.J. Richter and N.E. Rustad, *J. Org. Chem.* **29**, 3381 (1964).

64MI1	M. Hamana and O. Hoshino, *Yakugaku Zasshi*, 35 (1964).
64MI2	A.N. Kost, A.K. Sheinkman, and N.F. Kazarinova, *J. General Chem. USSR* **34**, 2044 (1964) [*CA* **61**, 8271 (1964)].
64TL613	J.A. Dixon, D.H. Fishman, and R.S. Dudinyak, *Tetrahedron Lett.*, 613 (1964).
64TL867	H. Metzger, H. Konig, and K. Seelert, *Tetrahedron Lett.*, 867 (1964).
64TL3295	D. Bryce-Smith and B.J. Wakefield, *Tetrahedron Lett.*, 3295 (1964).
65AGE543	T.H. Kauffmann, *Angew. Chem., Int. Ed. Eng.* **4**, 543 (1965).
65AGE731	T.G. Wittig, *Angew. Chem., Int. Ed. Eng.* **4**, 731 (1965).
65AHC145	R.G. Shepherd and J.L. Fedric, *Adv. Heterocycl. Chem.* **4**, 145 (1965).
65CPB912	M. Hamana and H. Noda, *Chem. Pharm. Bull.* **13**, 912 (1965).
65JOC910	T. Kato and H. Yamanaka, *J. Org. Chem.* **30**, 910 (1965).
65MI1	S. Kitaura, *Proc. Faculty Sci. Tokai Univ.* **1**, 63 (1965).
65TL4615	R. Lyle and G. Gauthier, *Tetrahedron Lett.*, 4615 (1965).

65ZC64	H. Meinert, *Z. Chem.* **5**, 64 (1965).
66CPB518	T. Okamoto, M. Hirobe, and A. Oksawa, *Chem. Pharm. Bull.* **14**, 518 (1966).
66CBP762	M. Hamana and H. Noda, *Chem. Pharm. Bull.* **14**, 762 (1966).
66HCA2046	R. Wizinger and H. Angliker, *Helv. Chim. Acta* **49**, 2046 (1966).
66JA3376	M. Saunders and E.N. Gold, *J. Am. Chem. Soc.* **88**, 3376 (1966).
66JOC243	V.J. Traynelis and J.V. McSweeney, *J. Org. Chem.* **31**, 243 (1966).
66JOC248	G.A. Russel and S.A. Weiner, *J. Org. Chem.* **31**, 248 (1966).
66TL1123	H. Nozaki, Y. Yamamoto, and R. Noyuri, *Tetrahedron Lett.*, 1123 (1966).
67CPB363	M. Hamana and I. Kumadaki, *Chem. Pharm. Bull.* **15**, 363 (1967).
67CPB474	M. Hamana and H. Noda, *Chem. Pharm. Bull.* **15**, 474 (1967).
67JA3744	A. Efraty and P.M. Maitlis, *J. Am. Chem. Soc.* **89**, 3744 (1967).
67KG248	A.K. Sheinkman, A.N. Kost and A.N. Prilepskaya, *Khim. Geterotsikl. Soedin.*, 248 (1967) [*CA* **68**, 12828 (1868)].
67MI1	E. Ochiai, "Aromatic Amine Oxides," Elsevier Publishing Co., Amsterdam, 1967.

67MI2	R.W. Hoffmann, "Dehydrobenzene and Cycloalkynes," Academic Press, New York, 1967.
67TL2087	H.C. van der Plas, M. Wozniak, and A. Veldhuizen, *Tetrahedron Lett.*, 2087 (1967).
67ZOR1617	S.S. Gitis, A.I. Glaz, V.V. Grigores, A.Ya. Kamiskii, A.S. Martymenko, and P.I. Saukov, *Zh. Org. Khim.* **3**, 1617 (1967) [*CA* **68**, 12104 (1968)].
68ACSA2338	C. Bjorklund and M. Nilsson, *Acta Chem. Scand.* **22**, 2338 (1968).
68ACSA2581	C. Bjorklund and M. Nilsson, *Acta Chem. Scand.* **22**, 2581 (1968).
68AHC1	F.D. Popp, *Adv. Heterocycl. Chem.* **9**, 1 (1968).
68JA1606	R.L. Eppley and J.A. Dixon, *J. Am. Chem. Soc.* **90**, 1606 (1968).
68JHC561	W.W. Paudler and T.J. Kress, *J. Heterocycl. Chem.* **5**, 561 (1968).
68JOC403	R.G. Landolt and H.R. Snyder, *J. Org. Chem.* **33**, 403 (1968).

68MI1	J. Miller, "Aromatic Nucleophilic Substitution," Elsevier, Amsterdam, 1968.
68MI2	S.V. Krivun, *Dokl. Akad. Nauk SSSR, ser. khim.* **180**, 615 (1968) [*CA* **69**, 106425 (1968)].
68MI3	S.V. Krivun, *Dokl. Akad. Nauk SSSR, ser. khim.* **182**, 347 (1968).
68TL4625	H. Nozaki, Y. Yamamoto, and T. Nisimara, *Tetrahedron Lett.*, 4625 (1968).
69AG136	G. Buck, *Angew. Chem.* **81**, 136 (1969).
69CHE645	A.F. Pozharskii, A.M. Simonov, E.A. Zvezdina, and V.A. Anisimova, *Chem. Heterocycl. Compd. (Engl. Transl.)*, 645 (1969) [*CA* **72**, 111370 (1970)].
69JA6381	T. Tokuhira and G. Fracukel, *J. Am. Chem. Soc.* **91**, 6381 (1969).
69JCS(C)2024	I.U. Khand, P.L. Pauson, and W.E. Watts, *J. Chem. Soc. (C)*, 2024 (1969).
69JOC655	F.M. Heshenson and L. Bauer, *J. Org. Chem.* **34**, 655 (1969).

69MI1	"The Chemistry of Nitro and Nitroso Groups," Interscience, New York, 1969 (H. Feuer, ed.), vol 1 and 2.
69MI2	F. Pietra, *Quart. Rev.* **23**, 504 (1969).
69TL433	J. Knabe and H. Holtje, *Tetrahedron Lett.*, 433 (1969).
70ACSA2757	V.D. Parker, *Acta Chem. Scand.* **24**, 2757 (1970).
70AG518	G. Frankel, J.W. Cooper, and C.M. Fink, *Angew. Chem.* **82**, 518 (1970).
70CHE987	A.F. Pozharskii, A.M. Simonov, V.M. Mar'janovskii, and R.P. Zinchenko, *Chem. Heterocycl. Compd. (Engl. Transl.)*, 987 (1970).
70CHE1075	I.S. Kashparov, A.F. Pozharskii, and A.M. Simonov, *Chem. Heterocycl. Compd. (Engl. Transl.)*, 1075 (1970) [*CA* **74**, 53648 (1971)].
70GEP2006205	Germany Patent 2006205 [*CA* **73**, 98593 (1970)].
70JA4682	C.F. Bernasconi, *J. Am. Chem. Soc.* **92**, 4682 (1970).
70JA5442	T. Agawa and S. Miller, *J. Am. Chem. Soc.* **92**, 5442 (1970).

70JCS(C)800   G.V. Boyd, A.W. Ellis, and M.D. Harms, *J. Chem. Soc. (C)*, 800 (1970).

70JHC1071   J. Bergman, *J. Heterocycl. Chem.* **7**, 1071 (1970).

70KG733   S.V. Krivun, G.N. Dorofeenko, and A.S. Kovalevskii, *Khim. Geterotsikl. Soedin.*, 733 (1970) [*CA* **73**, 98769 (1970)].

70KG1291   A.K. Sheinkman, A.P. Kucherenko, and S.N. Baranov, *Khim. Geterotsikl. Soedin.*, 1291 (1970).

70KG1292   A.K. Sheinkman, S.G. Potashnikova, and S.N. Baranov, *Khim. Geterotsikl. Soedin.*, 1292 (1970) [*CA* **74**, 125377 (1971)].

70KG1384   V.E. Possazhennikova, O.N. Chupakhin, and I.Ya. Postovskii, *Khim. Geterotsikl. Soedin.*, 1384 (1970) [*CA* **74**, 64191 (1971)].

70KG1454   S.V. Krivun and S.V. Dulskaya, *Khim. Geterotsikl. Soedin.*, 1454 (1970) [*CA* **74**, 53411 (1971)].

70KG1515   A.K. Sheinkman, A.N. Prilepskaya, and A.N. Kost, *Khim. Geterotsikl. Soedin.*, 1515 (1970) [*CA* **74**, 53477 (1971)].

70MI1   V.A. Trofimov, O.N. Chupakhin, and Z.V.Pushkareva, *Dokl. Akad. Nauk SSSR*, **191**, 1302 (1970) [*CA* **76**, 59417 (1972)].

70ZOR614   A.K. Sheinkman, S.G. Potashnikova, and S.N. Baranov, *Zh. Org. Khim.* **6**, 614 (1970) [*CA* **72**, 132479 (1970)].

71CHE624   A.F. Pozharskii, M.M. Medvedeva, E.A. Zvezdina, and A.M. Simonov, *Chem. Heterocycl. Compd. (Engl. Transl.)*, 624 (1971) [*CA* **76**, 126865 (1972)].

71CHE1036   A.F. Pozharskii, V.V. Kuz'menko, and A.M. Simonov, *Chem. Heterocycl. Compd. (Engl. Transl.)*, 1036 (1971) [*CA* **76**, 153676 (1972)].

71CPB1809   T. Okamoto and H. Takahashi, *Chem. Pharm. Bull.* **19**, 1809 (1971).

71JCS(B)131   R.A. Abramovitch and A.R. Vinutha, *J. Chem. Soc. (B)*, 131 (1971).

71JCS(B)2423 A. Albert and H. Mizuno, *J. Chem. Soc. (B)*, 2423 (1971).

71JCS(CC)1120 J.H. Gorvin, *J. Chem. Soc., Chem. Commun.*, 1120 (1971).

71JOC772   R.E. Lyle and E. White, *J. Org. Chem.* **36**, 772 (1971).

71JOC2907   E.T. McBee, E.P. Wesseler, T. Hodgins, *J. Org. Chem.* **36**, 2907 (1971).

71JPC3636   C.F. Bernasconi, *J. Phys. Chem.* **75**, 3636 (1971).

71KG82   A.K. Sheinkman, A.K. Tokarev, and S.N. Baranov,

            Khim Geterotsikl. Soedin., 82 (1971) [*CA* **75**, 35669 (1971)].

71KG112   V.A. Trofimov, O.N. Chupakhin, Z.V. Pushkareva, and

            V.L. Rusinov, *Khim Geterotsikl. Soedin.*, 112 (1971);

            Chem. Heterocycl. Compd. USSR (Eng. Transl.), **6**, 104 (1971)

            [*CA* **75**, 20164 (1971)].

71KG643   A.K. Sheinkman, A.K. Tokarev, S.G. Potashnikova, A.A. Deikalo,

            A.P. Kucherenko, and S.N. Baranov, *Khim. Geterotsikl. Soedin.*,

            643 (1971) [*CA* **76**, 59409 (1972)].

71KG648   A.K. Sheinkman, A.N. Kost, S.G. Potashnikova, A.O. Ginzburg,

            and S.N. Baranov, *Khim. Geterotsikl. Soedin.*, 648 (1971)

            [*CA* **76**, 72342 (1972)].

71KG1148   A.K. Sheinkman and A.N. Prilepskaya, *Khim. Geterotsikl. Soedin.*,

            1148 (1971) [*CA* **76**, 126742 (1972)].

71MI1   H. Higashino, *Proceedings of 3-rd Congress of Heterocycl. Chem.*,

            Sendai, 362 (1971).

71MI2   A.K. Sheinkman, G.V. Samoilenko, and S.N. Baranov, *Dokl. Akad. Nauk SSSR, ser. khim.* **196**, 1377 (1971) [*CA* **75**, 20514 (1971)].

71MI3   A.R. Katritzky and G.M. Lagowsky, "Chemistry of Heterocyclic N-Oxides," Academic Press, London-New York, 1971.

71MI4   A.F. Pozharski and A.M. Simonov, "Chichibabin Amination of Heterocycles," Izd. Rost. Univ., Rostov-on-Don, USSR, 1971.

71MI5   I.M. Sosonkin, A.Ya. Kaminsky, S.S. Gitis, and E.G. Kaminskaya, *Dokl. Acad. Nauk SSSR, ser. him.* **197**, 635 (1971) [*CA* **75**, 19499 (1971)].

71RTC1239   J. de Valk and H.C. van der Plas, *Recl. Trav. Chim. Pays-Bas* **90**, 1239 (1971).

71T3655   W. Buratti, G.P. Gardini, F. Minisci, F. Bertini, R. Galli, and M. Perchinunno, *Tetrahedron* **27**, 3655 (1971).

71ZOR2388   V.N. Drozd and O.I. Trifonova, *Zh. Org. Khim.* **7**, 2388 (1971) [*CA* **76**, 58671 (1972).

72ACSA2883 H. Wennerstroum and O. Wennerstroum, *Acta Chem. Scand.* **26**, 2883 91972).

72BSF4549   F. Terrier and A.P. Chatrousse, *Bull. Soc. Chim. Fr.*, 4549 (1972).

72CHE731   T.P. Filipskikh, A.F. Pozharskii, V.N. Koroleva, A.M. Simonov, and E.A. Zvezdina, *Chem. Heterocycl. Compd. (Engl. Transl.)*, 731 (1972).

72CJC129   E. Buncell and J.G.. Webb, *Can. J. Chem.* **50**, 129 (1972).

72CJC919   J.W. Bunting and M.G. Meathrel, *Can. J. Chem.* **50**, 919 (1972).

72CL369   A. Ohno and N. Kito, *Chem. Lett.*, 369 (1972).

72JA682   J.A. Zoltewicz and L.S. Helmick, *J. Am. Chem. Soc.* **94**, 682 (1972).

72JCS(P1)2918   R.M. Acheson, N.D. Wright, and P.A. Tasker, *J. Chem. Soc., Perkin Trans. 1*, 2918 (1972).

72JCS(P2)1141   M.J. Foreman, G.R. Knox, P.L. Pauson, K.H. Todd, and W.E. Watts, *J. Chem. Soc., Perkin Trans. 2*, 1141 (1972).

72JOC3588   Y. Kobayashi, J. Kumadaki, and H. Sato, *J. Org. Chem.* **37**, 3588 (1972).

72KG216   V.L. Rusinov, O.N. Chupakhin, V.A. Trofimov, M.I. Kollegova, and I.Ya. Postovskii, *Khim. Geterotsikl. Soedin.*, 216 (1972); *Chem. Heterocycl. Compd. USSR (Engl. Transl.)* 7, 193 (1972) [*CA* 76, 126744 (1972)].

72KG529   A.K. Sheinkman, A.K. Tokarev, and A.N. Prilepskaya, *Khim. Geterotsikl. Soedin.*, 529 (1972) [*CA* 77, 48169 (1972)].

72KG1313   S.N. Baranov, M.A. Dumbai, and S.V. Krivun, *Khim. Geterotsikl. Soedin.*, 1313 (1972) [*CA* 78, 29544 (1973)].

72KG1425   A.Ya. Ilchenko and V.M. Rudenko, *Khim. Geterotsikl. Soedin.*, 1425 (1972) [*CA* 78, 43420 (1973)].

72KG1673   A.F. Pozharskii and A.A. Konstantinchenko, *Khim. Geterotsikl. Soedin.*, 1673 (1972) [*CA* 78, 71876 (1973)].

72MI1   E.E. Golteuzen, Z.V. Todres, and A.Ya. Kaminskii, *Izv. Akad. Nauk. SSSR ser. khim.*, 1083 (1972) [*CA* 77, 87205 (1972)].

72TL497   P.G. Gassman, G. Gruetzmacher, and R.H. Smith, *Tetrahedron Lett.*, 497 (1972).

72TL501	C.R. Johnson, C.C. Bacon, and W.D. Kingsbury, *Tetrhedron Lett.*, 501 (1972).
72TL3587	S. Julia, C. Huynh, and D. Michelot, *Tetrahedron Lett.*, 3587 (1972).
73CHE88	A.D. Garnovskii, A.M. Simonov, and V.I. Minkin, *Chem. Heterocycl. Compd. (Engl. Transl.)*, 88 (1973).
73ICA621	P.J.C. Walker and R.J. Mawby, *Inorg. Chim. Acta* 7, 621 (1973).
73JCS(D)622	P.J.C. Walker and R.J. Mawby, *J. Chem. Soc., Dalton Trans.*, 622 (1973).
73JCS(P1)2588	J.R. Brooks, D.N. Harcourt, and R.D. Weigh, *J. Chem. Soc., Perkin Trans. 1*, 2588 (1973).
73JOC1947	J.A. Zoltewicz, L.S. Helmic, T.M. Oestreich, R.W. King, and P.E. Kandetski, *J. Org. Chem.* **38**, 1947 (1973).
73JOC1949	J.A. Zoltewicz, T.M. Oestreich, J.K. O'Halloran, and L.S. Helmick, *J. Org. Chem.* **38**, 1949 (1973).
73MI1	R.A. Abramovitch and E.M. Smith, Pyridine-1-Oxides, *in* "Pyridine and Its Derivatives," Wiley and Sons, New York, 1973 (R.A. Abramovitch, ed.), Supplement Part II.

73MI2	I.Ya. Postovskii, O.N. Chupakhin, T.L. Pilicheva, and Yu. Yu. Popelis, *Dokl. Akad. Nauk SSSR, ser. khim.* **212**, 1125 (1973) [*CA* **80**, 27193 (1974)].
73MI3	H.C. van der Plas, "Ring Transformations of Heterocycles," Academic Press, New York, 1973, vol. I and II..
73MI4	L.U. Bryukhovetskaya, L.A. Mironova, and S.M. Shein, *Izv. Akad. Nauk. SSSR ser. khim.*, 1594 (1973).
73RTC708	P.J. Lont, H.C. van der Plas, and A. van Veldhuizen, *Recl. Trav. Chim. Pays Bas* **92**, 708 (1973).
73RTC1232	J.P. Geerts, H.C. van der Plas, and A. van Veldhuizen, *Recl. Trav. Chim. Pays-Bas* **92**, 1232 (1973).
73S1	F. Minisci, *Synthesis*, 1 (1973).
73TL1123	C. Paulmier, M.P. Simonin, A.P. Chatrousse, and F. Terrier, *Tetrahedron Lett.*, 1123 (1973)
73UK1416	A.K. Sheinkman, S.I. Suminov, and A.N. Kost, *Usp. Khim.* **8**, 1416 (1973) [*CA* **79**, 146302 (1973)].
73YZ1019	M. Hasegava and T. Okamoto, *Yakugaku Zasshi* **93**, 1019 (1973).

73ZOR156	V.N. Drozd, O.I. Trifonova, and V.V. Sergeichuk, *Zh. Org. Khim.* **9**, 156 (1973) [*CA* **78**, 110285 (1973)].
73ZOR2354	M.I. Kalin'kin, Z.N. Parnes, V.E. Puzanova, A.D. Khmelinskaya, S.M. Shein, and D.N. Kursanov, *Zh. Org. Khim.* **9**, 2354 (1973) [*CA* **80**, 36773 (1974)].
74H167	M. Hamana, H. Noda, and M. Aoyama, *Heterocycles* **2**, 167 (1974).
74JA7091	M.F. Semmelhack and H.T. Hall, *J. Am. Chem. Soc.* **96**, 7091 (1974).
74JA7092	M.F. Semmelhack and H.T. Hall, *J. Am. Chem. Soc.* **96**, 7092 (1974).
74JCS(CC)308	J. Clark and B. Parvizi, *J. Chem. Soc., Chem. Commun.*, 308 (1974).
74JOC1836	Y. Kobayashi, J. Kumadaki, Y. Hirose, and Y. Hanzawa, *J. Org. Chem.* **39**, 1836 (1974).
74KG3	A.K. Sheinkman, *Khim. Geterotsikl. Soedin.*, 3 (1974) [*CA* **80**, 95781 (1974)].
74KG675	V.E. Kirichenko and O.N. Chupakhin, *Khim. Geterotsikl. Soedin.*, 675 (1974) [*CA* **81**, 77237 (1974)].

74KG993	O.N. Chupakhin, E.O. Sidorov, and I.Ya. Postovskii, *Khim. Geterotsikl. Soedin.*, 993 (1974) [*CA* **81**, 120578 (1974)].
74KG1097	A.K. Sheinkman and M.M. Mestechkin, *Khim. Geterotsikl. Soedin.*, 1097 (1974).
74MI1	Z.V. Todress, *Zh. Vsesouznogo Khim. Obchsh.* **19**, 285 (1974) [*CA* **81**, 625410 (1974)].
74MI2	O.Yu. Okhlobystin, "Electron Transfer in Organic Reactions," Izd. Rostov University, Rostov-on-Don, 1974.
74MI3	G.V. Putyrskaya and J. Matusz, *Acta Chim. Acad. Hung.* **82**, 167 (1974).
74RTC114	E.A. Oostveen, H.C. van der Plas, and H. Jongejan, *Recl. Trav. Chim. Pays-Bas* **93**, 114 (1974).
74TL1361	K.K. Prasad, *Tetrahedron Lett.*, 1361 (1974).
74ZOR133	I.M. Sosonkin and G.L. Kalb, *Zh. Org. Khim.* **10**, 133 (1974).
75CIL520	L.S. Levitt and B.W. Levitt, *Chem. Ind. (London)*, 520 (1975).
75JA1247	M.F. Semmelhack, H.T. Hall, M. Joshifuji, and G. Clarke, *J. Am. Chem. Soc.* **97**, 1247 (1975).

75JA5531   J.A. Zoltewicz and J.K. O'Halloran, *J. Am. Chem. Soc.* **97**, 5531 (1975).

75JCS(CC)262 Y.-H. So, J.Y. Becker, and L.L. Miller, *J. Chem. Soc., Chem. Commun.*, 362 (1975).

75JCS(D)1683 P.L. Pauson and J.A. Segal, *J. Chem. Soc. (D),* 1683 (1975).

75JCS(P2)1568 J.H. Fendler, W.L. Hinze, and L.J. Liu, *J. Chem. Soc., Perkin Trans. 2*, 1568 (1975).

75JOC3746   R.B. Chapas, R.D. Knudsen, R.F. Nystrom, and H.R. Snyder, *J. Org. Chem.* **40**, 3746 (1975).

75JOM365   A.N. Nesmeyanov, N.A. Volkenau, L.S. Shilovtseva, and V.A. Petrakova, *J. Organomet. Chem.* **85**, 365 (1975) [*CA* **83**, 97528].

75KG387   O.N. Chupakhin, I.Ya. Postovskii, V.L. Rusinov, and V.N. Charushin, *Khim. Geterotsikl. Soedin.*, 387 (1975) [*CA* **83**, 28073 (1975)].

75KG1433   O.N. Chupakhin, E.O.Sidorov, and I.Ya. Postovskii, *Khim Geterotsikl. Soedin.*, 1433 (1975) [*CA* **84**, 43979 (1976)].

75M473	A. Pollak, B. Stanovnik, M. Tisler, and J. Venetic-Fortuna, *Monatsh. Chem.* **106**, 473 (1975).
75MI1	J.A. Zoltewicz, *Top. Curr. Chem.*, 33 (1975).
75MI3	E.O. Sidorov, "Nucleophilic Substitution of Hydrogen in Quinoxalines and Related Systems," Ph. D. thesis, Ekaterinburg, 1975.
75RTC45	A. Nagel, H.C. van der Plas, and A. van Veldhiuzen, *Recl. Trav. Chim. Pays-Bas* **94**, 45 (1976).
75T969	I.I. Bilkis and S.M. Shein, *Tetrahedron* **31**, 969 (1975).
76BCJ3227	T. Abe and Y.J. Kagani, *Bull. Chem. Soc. Jpn.* **49**, 3227 (1976).
76CL525	H. Minato, E. Yamazaki, and M. Kobayashi, *Chem. Lett.* 525, (1976).
76CPB1813	I. Takeuchi and Y. Hamada, *Chem. Pharm. Bull.* **24**, 1813 (1976).
76H987	S. Skonieczny, *Heterocycles* **6**, 987 (1976).
76JA5234	J.H. Pitman, *J. Am. Chem. Soc.* **98**, 5234 (1976).
76JCS(P2)662	L. Testafetti, M. Tiecco, P. Spagnalo, P. Zanitaro and G. Martelli, *J. Chem. Soc., Perkin Trans. 2*, 662 (1976).

76JOC1303	J.A. Zoltewicz, L.S. Helmick, and J.K. O'Halloran, *J. Org. Chem.* **41**, 1303 (1976).
76JOC2153	G. Baldini, G. Doddi, G. Illuminati, and F. Stegel, *J. Org. Chem.* **41**, 2153 (1976).
76JOC3250	R.E. Lyle and D.L. Comins, *J. Org. Chem.* **41**, 3250 (1976).
76JOC3448	G. Olah and H. Mayr, *J. Org. Chem.* **41**, 3448 (1976).
76KG266	O.N. Chupakhin, V.I. Shilov, I.Ya. Postovskii, and V.A. Trofimov, *Khim. Geterotsikl. Soedin.*, 266 (1976) [*CA* **85**, 94203 (1976)].
76KG976	P.B. Terentjev, V.G. Kartsev, and A.N. Kost, *Khim. Geterotsikl. Soedin.*, 976 (1976) [*CA* **85**, 192644 (1976)].
76KG1146	V.N. Charushin, O.N. Chupakhin, and I.Ya. Postovskii, *Khim. Geterotsikl. Soedin.*, 1146 (1976) [*CA* **86**, 5408 (1977)].
76KG1227	O.N. Chupakhin and V.L. Rusinov, *Khim. Geterotsikl. Soedin.*, 1227 (1976) [*CA* **86**, 54645 (1977)].
76MI1	M.F. Semmelhack, *in* "New Applications of Organometallic Reagents in Synthesis," American Elsevier, New York, 1976 (D. Seyferth, ed.), vol 1, pp. 361–395.

76MI2   S.M. Shein, *Zh. Vsesouznogo Khim. Obsh.* **21**, 256 (1976).

76MI3   I.M. Sosonkin, V.A.Subbotin, V.N.Charushin, and O.N. Chupakhin, *Dokl. Acad. Nauk SSSR, ser. khim.* **229**, 888 (1976) [*CA* **86**, 4589 (1977)].

76MI4   C. Williams, The Redox Chemistry of 1,4-Dihydronicotinic Acid Derivatives, *in* "Enzimes," Academic Press, New York, 1976 (F.De Boyer, ed.).

76OMR607   J.P. Geerts, A. Nagel, and H.C. van der Plas, *Org. Magn. Reson.* **8**, 607 (1976).

76RCR454   O.N. Chupakhin and I.Ya. Postovskii, *Usp. Khim.* **45**, 908 (1976); *Russ. Chem. Rev. (Engl. Transl.)*, 454 (1976) [*CA* **85**, 176367 (1976)].

76RTC113   Gy. Simig, H.C. van der Plas, and C.A. Landheer, *Recl. Trav. Chim. Pays-Bas* **95**, 113 (1976).

76RTC125   Gy.Simig and H.C. van der Plas, *Recl. Trav. Chim. Pays-Bas* **95**, 125 (1976).

76TL3537   W.P.K. Girke, *Tetrahedron Lett.*, 3537 (1976).

76ZOR2464	O.N. Chupakhin, E.O. Sidorov, S.M. Shein, and J.J. Bilkis, *Zh. Org. Khim.* **12**, 2464 (1976) [*CA* **86**, 106523 (1977)].
77AHC121	H.J. den Hertog and H.C. van der Plas, *Adv. Heterocycl. Chem.* **4**, 121 (1977).
77BCJ1535	A. Ohno, T. Kimura, H. Yamamoto, S.G. Kim, S. Oka, and Y. Ohnishi, *Bull. Chem. Soc. Jpn.* **50**, 1535 (1977).
77CB3561	J. Wellmann and E. Steckham, *Chem. Ber.* **110**, 3561 (1977).
77CJC2741	A.K. Colter, G. Saito, and F.J. Sharom, *Can. J. Chem.* **55**, 2741 (1977).
77JA4429	E. Buncell, J.G.K. Webb, and J.F. Wiltshire, *J. Am. Chem. Soc.* **99**, 4429 (1977).
77JCS(CC)181	T. Okamoto, A. Ohno, and S. Oka, *J. Chem. Soc., Chem. Commun.*, 181 (1977).
77KG684	O.N. Chupakhin, E.O. Sidorov, A.L. Kozerchuk, and Yu.I. Beilis, *Khim. Geterotsikl. Soedin.*, 684 (1977) [*CA* **87**, 84946 (1977)].
77KG690	O.N. Chupakhin, V.N. Charushin, I.M. Sosonkin, E.G. Kovalev, and I.Ya. Postovskii, *Khim. Geterotsikl. Soedin.*, 690 (1977).

77KG1554	M.F. Budyka, P.B. Terentjev, and A.N. Kost, *Khim. Geterotsikl. Soedin.*, 1554 (1977) [*CA* **89**, 129466 (1978)].
77MI1	O.N. Chupakhin, "Nucleophilic Substitution of Hydrogen in Azines," doctoral dissertation, Ekaterinburg, 1977.
77MI2	M.F. Semmelhack, *Ann. N.Y. Acad. Sci.* **295**, 36 (1977).
77MI3	Yu.G. Gololobov, P.P. Onysko, and V.P. Prokopenko, *Dokl. Akad. Nauk SSSR, ser. khim.* **237**, 105 (1977) [*CA* **88**, 50974 (1978)].
77MI4	A.V. Koblik, *Proceedings of the Polish-Soviet Symposium on the Chemistry of Oxygen-Containing Heterocycles*, p. 29, Torun, 1977.
77RTC151	H.C. van der Plas, M. Wozniak, and A. Veldhuizen, *Recl. Trav. Chim. Pays-Bas* **96**, 151 (1977).
77S862	K. Akiba, K. Ishikawa, and N. Inamoto, *Synthesis*, 862 (1977).
77ZOB2480	P.P. Onysko and Yu.G. Gololobov, *Zh. Obshch. Khim.* **47**, 2480 (1977) [*CA* **88**, 61752 (1978)].
77ZOR204	O.N. Chupakhin, E.O. Sidorov, and E.G. Kovalev, *Zh. Org. Khim.* **13**, 204 (1977) [*CA* **86**, 155027 (1977)].
78ACR147	C.F. Bernasconi, *Acc. Chem. Res.* **11**, 147 (1978).

78ACR462	H.C. van der Plas, *Acc. Chem. Res.* **11**, 462 (1978).
78AHC71	J.A. Zoltewicz and L.W. Deady, *Adv. Heterocycl. Chem.* **22**, 71 (1978).
78AJC2463	P.J. Newcombe and R.K. Norris, *Aust. J. Chem.* **31**, 2463 (1978).
78BCJ196	T. Abe and Y.J. Kagani, *Bull. Chem. Soc. Jpn.* **51**, 196 (1978).
78CPB3504	K. Funakoshi, H. Sonoda, Y. Sonoda, and M. Hamana, *Chem. Pharm. Bull.* **26**, 3504 (1978).
78H371	M. Hamana, F. Sato, Y. Kimura, M. Nishikawa, and H. Noda, *Heterocycles* **11**, 371 (1978).
78HCA449	F. Kienzle, *Helv. Chim. Acta* **61**, 449 (1978).
78JA7600	P.G. Gassman and R.H. Drewes, *J. Am. Chem. Soc.* **100**, 7600 (1978).
78JA7611	P.G. Gassman and D.R. Amick, *J. Am. Chem. Soc.* **100**, 7611 (1978).
78JCS(P1)1140	M. Novi, G. Guanti, F. Sancassan, and C. Dell'Erba, *J. Chem. Soc., Perkin Trans. 1*, 1140 (1978).

78JOC1132	J.W. Bunting, P.A. Lee-Young, and D.J. Norris, *J. Org. Chem.* **43**, 1132 (1978).
78JOC3662	J.W. Bunting and S. Kabir, *J. Org. Chem.* **43**, 3662 (1978).
78JOC4303	G. Doddi, F. Stegel, and M.T. Tanasi, *J. Org. Chem.* **43**, 4303 (1978).
78LA98	G. Scherowsky and H. Matloubi, *Leibigs Ann. Chem.*, 98 (1978).
78MI1	G. Jaonen, "Transition Metal Organometallics in Organic Synthesis," Academic Press, New York, 1978 (H.N.Y. Alper, ed.), vol. I and II.
78MI2	M. Wozniak, *Wiadomosci Chemiszne* **32**, 283 (1978).
78RCR260	Z.V. Todres, *Russ. Chem. Rev. (Engl. Transl.)*, **47**, 260 (11978).
78RCR1042	A.F. Pozharskii, A.F. Simonov, and V.N. Doron'kin, *Russ. Chem. Rev. Engl. Transl.)* **47**, 1042 (1978) [*CA* **90**, 71248f (1979)].
78RCR1061	S.S. Gitis and A.Ya. Kaminsii, *Russ. Chem. Rev. (Engl. Transl.)* **47**, 1061 (1978).
78RTC159	H.C. van der Plas and A. Koudijs, *Recl. Trav. Chim. Pays-Bas* **97**, 159 (1978).
78RTC273	A. Rykowsky, H.C. van der Plas and A. Veldhuizen, *Recl. Trav. Chim. Pays-Bas* **97**, 273 (1978).

78RTC288	C.A.H. Rasmussen and H.C. van der Plas, *Recl. Trav. Chim. Pays-Bas* **97**, 288 (1978).
78S206	B.C. Uff and R.C. Budhram, *Synthesis,* 206 (1978).
78T363	C. Mortelmans and G. van Binst, *Tetrahedron* **34**, 363 (1978).
78TL3495	J. Golin'ski and M. Makosza, *Tetrahedron Lett.,* 3495 (1978).
78ZOB342	P.P. Onysko, L.F. Kashukhin, and Yu.G. Gololobov, *Zh. Obshch. Khim.* **48**, 342 (1978) [*CA* **88**, 189536 (1978)].
78ZOR134	E.O. Sidorov and O.N. Chupakhin, *Zh. Org. Khim.* **14**, 134 (1978) [*CA* **88**, 189520a (1978)].
78ZOR140	V.N. Charushin, O.N. Chupakhin, E.O. Sidorov, and Yu.I. Beilis, *Zh. Org. Khim.* **14**, 140 (1978) [*CA* **88**, 169271 (1978)].
78ZOR431	O.N. Chupakhin, V.N. Charushin, I.Ya. Postovskii, N.A. Kluev, and E.N. Istratov, *Zh. Org. Khim.* **14**, 431 (1978) [*CA* **88**, 152561 (1978)].
79AHC1	J.W. Bunting, *Adv. Heterocycl. Chem.* **25**, 1 (1979).
79AHC187	F.D. Popp, *Adv. Heterocycl. Chem.* **24**, 187 (1979).
79AJC1949	R.K. Norris and R.D. Smith-King, *Aust. J. Chem.* **32**, 1949 (1979).

79AX(B)733   R. Destro, T. Pilati, and M. Simonetti, *Acta Crystallogr. Sect.* **B 35**, 733 (1979).

79CB1348   W.P.K. Girke, *Chem. Ber.* **112**, 1348 (1979).

79IS154   C.A.L. Mahaffy and P.L. Pauson, *Inorg. Synth.* **19**, 154 (1979).

79JA768   M.F. Semmelhack, J. Bisaha, and M. Czarny, *J. Am. Chem. Soc.* **101**, 768 (1979).

79JA3535   M.F. Semmelhack, H.T. Hall, R. Farina, M. Joshifuji, G. Clarke, T. Bargar, K. Hirotsu, and J. Clardy, *J. Am. Chem. Soc.* **101**, 3535 (1979).

79JA4268   K.Yoshida and S. Nagase, *J. Am. Chem. Soc.* **101**, 4268 (1979).

79JCS(CC)296 J. Cornforth, A.F. Sierakowski, and T.W. Wallace, *J. Chem. Soc., Chem. Commun.*, 296 (1979).

79JCS(P1)418 A.R. Katritzky, J.B. Bapat, R.J. Blade, B.P. Leddy, P.L. Nie, C.A. Ramsden, and S.S. Thind, *J. Chem. Soc., Perkin Trans. I*, 418 (1979).

79JCS(P1)1698 C.W.F. Leung, M.P. Sammes, and A.R. Katritzky, *J. Chem. Soc., Perkin Trans. I*, 1698 (1979).

79JCS(P2)469	L. Testaferri, M. Tiecco, and M. Tingoli, *J. Chem. Soc., Perkin Trans. 2*, 469 (1979).
79JHC301	A. Nagel, H.C. van der Plas, G. Geurtsen, and A. van Veldhuizen, *J. Heterocycl. Chem.* **16**, 301 (1979).
79JHC1209	M.J. Dimsdale, *J. Heterocycl. Chem.* **16**, 1209 (1979).
79JHC1409	A. Tanaka and T. Usui, *J. Heterocycl. Chem.* **16**, 1409 (1979).
79JOC2087	G. Bartoli, M. Bosco, A. Melandri, and A.C. Boicelli, *J. Org. Chem.* **44**, 2087 (1979).
79JOC4677	J. Breuker and H.C. van der Plas, *J. Org. Chem.* **44**, 4677 (1979).
79JOC4705	S.S. Gandhi, M.S. Gibson, M.L. Kaldas, and S.M. Vines, *J. Org. Chem.* **44**, 4705 (1979).
79KG1736	O.N. Chupakhin, V.I. Shilov, and V.F. Gryazev, *Khim. Geterotsikl. Soedin.*, 1736 (1979) [*CA* **91**, 210584 (1979)].
79LA918	W.L. Blaedel and R.G. Haas, *Liebigs Ann. Chem.* **42**, 918 (1979).

79MI1	M. Hamana and S.Saeki, "Some Advances in the Nucleophilic Reactions of Aromatic N-Oxides," *in* "The Chemistry of Heterocycles," Nankodo, 1979 (T. Kametani, T. Mukai, and S. Takano, eds.), Part IV.
79MI2	P. Tomasik and A. Wosznyk, *Pol. Pat.* 95820 [*CA* **90**, P 151997 (1979)].
79MI3	A.N. Kost, S.I. Suminov, and A.K. Sheinkman, "N-Acylpyridinium Salts," *in* "Iminium Salts in Organic Synthesis," Pergamon Press New York, 1979, Part II.
79ZOB39	P.P. Onysko and Yu.G. Gololobov, *Zh. Obshch. Khim.* **49**, 39 (1979) [*CA* **90**, 187044 (1979)].
79ZOR117	O.N. Chupakhin, E.O. Sidorov, and V.N. Charushin, *Zh. Org. Khim.* **15**, 117 (1979) [*CA* **90**, 18977 (1979)].
79ZOR206	O.N. Chupakhin, V.N. Charushin, E.O. Sidorov, and G.L. Rusinov, *Zh. Org. Khim.* **15**, 206 (1979) [*CA* **90**, 203839 (1979)].
80CC394	A. Treston, R. Blakely, and B. Zerner, *Chem. Comm.*, 394 (1980).
80H1033	F.D. Popp, *Heterocycles* **14**, 1033 (1980).

80H2015 J. Bunting, *Heterocycles* **14**, 2015 (1980).

80IC3092 T.A. Albright and B.K. Carpenter, *Inorg. Chem.* **19**, 3092 (1980).

80JA6430 A.J. Birch, A.L. Hide, and L. Radom, *J. Am. Chem. Soc.* **102**, 6430 (1980).

80JCS(CC)1147 O. Piepers and R.M. Kellogg, *J. Chem. Soc., Chem. Commun.*, 1147 (1980).

80JCS(P2)1331 P. Cogolli, F. Maiolo, L. Tastaferri, and M. Tingoli, *J. Chem. Soc., Perkin Trans. 2*, 1331 (1980).

80JHC293 M. Colonna, L. Greci, and M. Poloni, *J. Heterocycl. Chem.* **17**, 293 (1980).

80JHC305 M.M. Yousif, S. Saeki, and M. Hamana, *J. Heterocyclic Chem.* **17**, 305 (1980).

80JHC1029 M.M. Yousif, S. Saeki, and M. Hamana, *J. Heterocyclic Chem.* **17**, 1029 (1980).

80JHC1211 D. Bhattacharjee and F.D. Popp, *J. Heterocycl. Chem.* **17**, 1211 (1980).

80JOC1534 M. Makosza and J. Winiarski, *J. Org. Chem.* **45**, 1534 (1980).

80JOC2555   R.J. Card and W.S. Trayhanovsky, *J. Org. Chem.* **45**, 2555 (1980).

80JOC2560   R.J. Card and W.S. Trayhanovsky, *J. Org. Chem.* **45**, 2560 (1980).

80JOM207   P.L. Pauson, *J. Organomet. Chem.* **200**, 207 (1980).

80MI1   O.N. Chupakhin, *Izv. Sibirskogo Otd. Akad. Nauk SSSR, ser. khim.* **2**, 46 (1980) [*CA* **93**, 70192 (1980)].

80MI2   I.M. Sosonkin, O.N. Chupakhin, A.I. Matern, and G.N. Strogov, *Dokl. Akad. Nauk SSSR* **250**, 875 (1980).

80OMR426   A. Solladie-Cavallo and J. Suffert, *Org. Magn. Reson.* **14**, 426 (1980).

80ZOR2390   N.P. Shusherina, V.S. Pilipenko, O.K. Kireeva, B.I. Gellar and A.I. Stepanyants, *Zh. Org. Khim.* **16**, 2390 (1980) [*CA* **95**, 24739 (1981)].

81CCC503   O. Kocian and M. Ferles, *Collect. Czech. Chem. Commun.* **46**, 503 (1981).

81H1083   M.M. Yousif, S. Saeki, and M. Hamana, *Heterocycles* **15**, 1083 (1981).

81JCS(P1)503 M. Begtrup and J. Holm, *J. Chem. Soc., Perkin Trans. 1*, 503 (1981).

81JHC1349	H.J.W. van den Haak, H.C. van der Plas, and A. van Veldhuizen, *J. Heterocycl. Chem.* **18**, 1349 (1981).
81JOC2134	H.J.W. van den Haak, H.C. van der Plas, and A. van Veldhuizen, *J. Org. Chem.* **46**, 2134 (1981).
81JOC3805	A. Counotte-Potman, H.C. van der Plas, and A. van Veldhuizen, *J. Org. Chem.* **46**, 3805 (1981).
81JOM147	E. Rose, J.C. Boutonnet, L. Mordenti, O. Le Martret, and G. Precigoux, *J. Organomet. Chem.* **221**, 147 (1981).
81PAC2379	M.F. Semmelhack, *Pure Appl. Chem.* **53**, 2379 (1981).
81T3957	M.F. Semmelhack, G.R. Clarke, J.L. Garcia, J.J. Harrison, Y. Thebtaranonth, W. Wulff, and A. Yamashita, *Tetrahedron* **37**, 3957 (1981).
81TL4093	K. Akiba, H. Matsuoka, and M. Wada, *Tetrahedron Lett.*, 4093 (1981).
81ZOR418	E.O. Sidorov, A.I. Matern, and O.N. Chupakhin, *Zh. Org. Khim.* **17**, 418 (1981) [*CA* **95**, 6035 (1981)].

81ZOR880   A.V. Koblik, K.F. Suzdalev, G.N. Dorofeenko, and A.A. Loktionov, *Zh. Org. Khim.* **17**, 880 (1981) [*CA* **95**, 150343 (1981)].

81ZOR2402   E.V. Malykhin and V.D. Shteingarts, *Zh. Org. Khim.* **17**, 2402 (1981) [*CA* **96**, 68096 (1982)].

82CB3766   H. Werner and R. Werner, *Chem. Ber.* **115**, 3766 (1982).

82CHE1221   A.M. Simonov, *Chem. Heterocycl. Compd. (Engl. Transl.)*, 1221 (1982) [*CA* **98**, 125914 (1983)].

82CPB1680   M.M. Yousif, S. Saeki, and M. Hamana, *Chem. Pharm. Bull.* **30**, 1680 (1982).

82CPB1974   M.M. Yousif, S. Saeki, and M. Hamana, *Chem. Pharm. Bull.* **30**, 1974 (1982).

82CPB2326   M.M. Yousif, S. Saeki, and M. Hamana, *Chem. Pharm. Bull.* **30**, 2326 (1982).

82CRV77   F. Terrier, *Chem. Rev.* **82**, 77 (1982).

82CRV223   D.M. Stout and A.I. Meyers, *Chem. Rev.* **82**, 223 (1982).

82CRV427   G.A. Artamkina, M.P. Egorov, and I.P. Beletskaya, *Chem. Rev.* **82**, 427 (1982).

82H13	A.I. Meyers and N.R. Natale, *Heterocycles* **18**, 13 (1982).
82H177	M. Hamana, G. Iwasaki, and S. Saeki, *Heterocycles* **17**, 177 (1982).
82JCR(S)258	L. Forlani and V. Tortelli, *J. Chem. Res.* **62**, 258 (1982).
82JCS(P1)2299	J. Cornforth, A.F. Sierakowski, and T.W. Wallace, *J. Chem. Soc., Perkin Trans. 1*, 2299 (1982).
82JCS(P1)2749	M. Begtrup, *J. Chem. Soc., Perkin Trans. 1*, 2749 (1982).
82JHC1285	H. Hara and H.C. van der Plas, *J. Heterocycl. Chem.* **19**, 1285 (1982).
82JHC1527	H. Hara and H.C. van der Plas, *J. Heterocycl. Chem.* **19**, 1527 (1982).
82JOC2633	A.I. Meyers and R.A. Gabel, *J. Org. Chem.* **47**, 2633 (1982).
82JOC4315	D.L. Comins and A.H. Abdullah, *J. Org. Chem.* **47**, 4315 (1982).
82JOC5227	G. Bartoli, M. Bosco, R. Pal Pozzo, and F. Ciminali, *J. Org. Chem.* **47**, 5227 (1982).
82KG361	I.M. Sosonkin, A.K. Sheinkman, Z.M. Skorobogatova, G.N. Strogov, and T.P. Ikmer, *Khim. Geterotsikl. Soedin.*, 361 (1982)
82KG1278	Yu.V. Kurbatov and M.A. Solekhova, *Khim. Geterotsikl. Soedin.*, 1278 (1982) [*CA* **98**, 16552 (1983)].

82MI1	G. Jones and D.J. Baty, "Quinoline N-Oxides," *in* "Quinolines," J. Wiley and Sons, New York, 1982 (G. Jones, ed.), Part II.
82MI2	J. Breuker, "The Chichibabin Amination of Diazines," Ph.D. thesis, Wageningen, 1982.
82MI3	A.A. Grinfeld, G.A. Artamkina, and I.P. Beletskaya, *Izv. Akad. Nauk. SSSR ser. khim.*, 2635 (1982) [*CA* **98**, 106900 (1983)].
82MI4	L. Forlani, *Gazz. Chim. Ital.* **112**, 205 (1982).
82TL429	K. Akiba, Y. Iseki, and M. Wada, *Tetrahedron Lett.* **23**, 429 (1982).
82TL1709	K. Akiba, T. Kasai, and M. Wada, *Tetrahedron Lett.* **23**, 1709 (1982).
82TL3677	R.E. Ballard, J. Jones, and E. Sutherland, *Tetrahedron Lett.* **23**, 3677 (1982).
82ZOR1898	E.V. Malykhin, A.A. Shtark, and V.D. Shteingarts, *Zh. Org. Khim.* **18**, 1898 (1982) [*CA* **98**, 53352 (1983)].
83AHC95	H.C. van der Plas, M. Woznia, and H.J.W. van den Haak, *Adv. Heterocycl. Chem.* **33**, 95 (1983).
83AHC305	G. Illuminati and F. Stegel, *Adv. Heterocycl. Chem.* **34**, 305 (1983).

83JA2034	M.F. Semmelhack and A. Zask, *J. Am. Chem. Soc.* **105**, 2034 (1983).
83JA5886	K.P. Nambiar, D.M. Stauffer, P.A. Kolodziej, and S.A. Benner, *J. Am. Chem. Soc.* **105**, 5886 (1983).
83JA6962	E.P. Koundig, V. Desobry, and D.P. Simmons, *J. Am. Chem. Soc.* **105**, 6962 (1983).
83JCS(P1)3065	M. Ghavshon and D.A. Widdowson, *J. Chem. Soc., Perkin Trans. 1*, 3065 (1983).
83JHC9	M. Wozniak, H.C. van der Plas, and A. van Veldhuizen, *J. Heterocycl. Chem.* **20**, 9 (1983).
83JHC823	J.V. Cooney, *J. Heterocycl. Chem.* **20**, 823 (1983).
83JHC1239	D.L. Comins and N.B. Mantlo, *J. Heterocycl. Chem.* **20**, 1239 (1983).
83JOC1354	H.C. van der Plas, V.N. Charushin, and A. van Veldhuizen, *J. Org. Chem.* **48**, 1354 (1983).
83JOC1375	W.K. Fife, *J. Org. Chem.* **48**, 1375 (1983).
83JOC3860	M. Makosza and T. Glinka, *J. Org. Chem.* **48**, 3860 (1983).
83MI1	A.A. Gakh and V.M. Khutoretskii, *Izv. Akad. Nauk SSSR, ser. khim.*, 2655 (1983).

83MI2	M. Makosza, *in* "Current Trends in Organic Synthesis," Pergamon Press, Oxford, 1983 (H. Nozaki, ed.).
83MI3	O.A. Reutov, I.P. Beletskaya, and A.L. Kurts,"Ambident Anions," Consultants Bureau, New York, 1983.
83MI4	K. Yoshida, *in* "The Chemistry of Functional Groups," Wiley-Interscience, Chichester, 1983 (S. Patai and Z. Rappoport, eds.), Supplement C.
83RTC359	M. Wozniak, H.C. van der Plas, M. Tomula, and A. van Veldhuizen, *Recl. Trav. Chim. Pays-Bas* **102**, 359 (1983).
83RTC367	J. Breuker and H.C. van der Plas, *Recl. Trav. Chim. Pays-Bas* **102**, 367 (1983).
83RTC511	M. Wozniak, H.C. van der Plas, M. Tomula, and A. van Veldhuizen, *Recl. Trav. Chim. Pays-Bas* **102**, 511 (1983).
83S40	M. Makosza, J. Golinski, and J. Pankowski, *Synthesis,* 40 (1983).
83TL1801	R. Yamaguchi, Y. Nakazono, and M. Kawanisi, *Tetrahedron Lett.* **24**, 1801 (1983).
83TL3277	M. Makosza, J. Golinski, and A. Rykowski, *Tetrahedron Lett.* **24**, 3277 (1983).

83TL4735	J. Epsztajn, A. Bieniek, J.Z. Brzezinski, and A. Joz'wiak, *Tetrahedron Lett.* **24**, 4735 (1983).
83TL5269	K. Akiba, Y. Nishihara, and M. Wada, *Tetrahedron Lett.* **24**, 5269 (1983).
83USP4415578	G.Y. Lesher and B. Singh (Sterling Drug, Inc.), U.S. Pat. 4415578 (1983) [*CA* **100**, 85579 (1984)].
83ZOR663	Yu.V. Kurbatov and M.A.Solekhova, *Zh. Org. Khim.* **19**, 663 (1983) [*CA* **99**, 5487 (1984)].
84ACR109	G. Bartoli, *Acc. Chem. Res.* **17**, 109 (1984).
84ACSA(B)439	L. Eberson, *Acta Chem. Scand., Ser. B* **38**, 439 (1984).
84BAP57	M. Makosza and E. Kwast, *Bull. Acad. Pol. Sci., Ser. Chim.* **32**, 57 (1984).
84BAP69	M. Makosza and E. Slomka, *Bull. Acad. Pol. Sci., Ser. Chim.* **32**, 69 (1984).
84CB152	H. Werner, R. Werner, and C. Burschka, *Chem. Ber.* **117**, 152 (1984).
84CB161	R. Werner and H. Werner, *Chem. Ber.* **117**, 161 (1984).

REFERENCES

84CJC534 E. Buncel, S.K. Murarka, and A.R.Norris, *Can. J. Chem.* **62**, 534 (1984).

84CJC729 J.W. Bunting and G.M. Kauffman, *Can. J. Chem.* **62**, 729 (1984).

84H93 W.K. Fife, *Heterocycles* **22**, 93 (1984).

84H151 D.L. Comins, E.D. Stroud, and J.J. Herrick, *Heterocycles* **22**, 151 (1984).

84H339 D.L. Comins, R.K. Smith, and E.D. Stroud, *Heterocycles* **22**, 339 (1984).

84H795 B. Singh, *Heterocycles* **22**, 795 (1984).

84H1091 S. K. Dubey, E.E. Knaus, and C.S. Giam, *Heterocycles* **22**, 1091 (1984).

84H1811 G. Iwasaki, K. Wada, S. Saeki, and M. Hamana, *Heterocycles* **22**, 1811 (1984).

84H2375 W.K. Fife and E.F.V. Scriven, *Heterocycles* **22**, 2375 (1984).

84JA7146 F. Minisci, C. Giordano, E. Vismara, S. Levi, and V. Tortelli, *J. Am. Chem. Soc.* **106**, 7146 (1984).

84JCS(P1)2227 A.E. Hauck and C.S. Giam, *J. Chem. Soc., Perkin Trans. I*, 2227 (1984).

84JOC56	S.H. Rosenberg and H. Rapoport, *J. Org. Chem.* **49**, 56 (1984).
84JOC1488	M. Makosza and J. Golinski, *J. Org. Chem.* **49**, 1488 (1984).
84JOC1494	M. Makosza and J. Winiarski, *J. Org. Chem.* **49**, 1494 (1984).
84JOC4571	T.V. RajanBabu and T. Fukunaga, *J. Org. Chem.* **49**, 4571 (1984).
84KG1011	N.T. Berberova and O.Yu. Oklobystin, *Khim. Geterotsikl. Soedin.*, 1011 (1984) [*CA* **101**, 190665 (1984)].
84KG1299	I.Ya. Postovskii, O.N. Chupakhin, and A.I. Matern, *Khim. Geterotsikl. Soedin.*, 1299 (1984) [*CA* **102**, 5201 (1985)].
84KG1383	T.V. Stupnikova and V.V. Petrenko, *Khim. Geterotsikl. Soedin.*, 1383 (1984) [*CA* **102**, 149055 (1985)].
84LA8	M. Makosza, B. Chylinska, and B. Mudryk, *Liebigs Ann. Chem.* **660**, 8 (1984).
84MI1	E. Buncel, M.R. Crampton, M.J. Strauss, and F. Terrier, "Electron Deficient Aromatic and Heteroaromatic-Base Interactions," Elsevier, Amsterdam, 1984.
84MI2	A.V. Koblik, L.A. Muradjan, and B.A. Tertov, *Proceedings of the All-Union Conference on the Chemistry of Acetylenes,* Erevan, p. 132 (1984).

84MI3	K. Yoshida, "Electrooxidation in Organic Chemistry," Wiley-Interscience, New York, 1984.
84T433	S.A.G.F. Angelino, A. van Veldhuizen, D.J. Buurman, and H.C. van der Plas, *Tetrahedron* **40**, 433 (1984).
84T1843	M. Makosza, T. Glinka, and A. Kinowski, *Tetrahedron* **40**, 1843 (1984).
84TL803	R.A. Murhpy and M.P. Cava, *Tetrahedron Lett.* **25**, 803 (1984).
84TL3297	D.L. Comins and J.D. Brown, *Tetrahedron Lett.* **25**, 3297 (1984).
84TL3637	A. Dondoni, T. Dall'Occo, G. Galliani, and A. Medici, *Tetrahedron Lett.* **25**, 3637 (1984).
84TL3763	H.C. van der Plas and D.J. Buurman, *Tetrahedron Lett.* **25**, 3763 (1984).
84TL3897	G. Castaldi, F. Minisci, V. Torelli, and E. Vismara, *Tetrahedron Lett.* **25**, 3897 (1984).
84TL4791	M. Makosza and K. Wojciechowski, *Tetrahedron Lett.* **25**, 4791 (1984).
84TL4795	A. Rykowski and M. Makosza, *Tetrahedron Lett.* **25**, 4795 (1984).

84TL4867	D.L. Comins, A.H. Abdullah, and N.B. Mantlo, *Tetrahedron Lett.* **25**, 4867 (1984).
85AGE692	G. Heinisch and G. Lotsch, *Angew. Chem., Int. Ed. Engl.* **24**, 692 (1985).
85H1513	W. Sliwa and G. Matusiak, *Heterocycles* **23**, 1513 (1985).
85H2807	S. Konno, M. Sagi, Y. Yuki, and H. Yamanaka, *Heterocycles* **23**, 2807 (1985).
85JA5473	T.V. RajanBabu, G.S. Reddy, and T. Fukunaga, *J. Am. Chem. Soc,* **107**, 5473 (1985).
85JCA23	H.C. van der Plas, *Janssen Chim. Acta* **3**, 23 (1985).
85JCS(CC)417	A. Alemagna, C. Baldoli, P. Del Buttero, E. Licandro, and S. Maiorana, *J. Chem. Soc., Chem. Commun.,* 417 (1985).
85JCS(CC)1181	C. Baldoli, P. Del Buttero, S. Maiorana, and A. Papagni, *J. Chem. Soc. Chem. Commun.,* 1181 (1985).
85JHC353	H. Tondys, H.C. van der Plas, and M. Wozniak, *J. Heterocycl. Chem.* **22**, 353 (1985).
85JHC761	M. Wozniak, H.C. van der Plas, M. Tomula, and A. van Veldhuizen, *J. Heterocycl. Chem.* **22**, 761 (1985).

85JHC765   M. Wozniak, D.J. Buurman, and H.C. van der Plas, *J. Heterocycl. Chem.* **22**, 765 (1985).

85JHC1419 D.L. Comins and E.D. Stroud, *J. Heterocycl. Chem.* **22**, 1419 (1985).

85JOC287   R. Yamaguchi, M. Morijasu, M. Yoshika, and M. Kawanisi, *J. Org. Chem.* **50**, 287 (1985).

85JOC484   D.A. de Bie, B. Geurtsen, and H.C. van der Plas, *J. Org. Chem.* **50**, 484 (1985).

85JOC3091 G.B. Stahly, *J. Org. Chem.* **50**, 3091 (1985).

85JOC3435 M. Wozniak, H.C. van der Plas, and S. Harkema, *J. Org. Chem.* **50**, 3435 (1985).

85JOC4410 D.L. Comins and N.B. Mantlo, *J. Org. Chem.* **50**, 4410 (1985).

85JOC5660 N.R. Natale, J.J. McKenna, C. Niou, and M. Bort, *J. Org. Chem.* **50**, 5660 (1985).

85JOM183   E.P. Koundig, C. Perret, S. Spichiger, and G. Bernardinelli, *J. Organomet. Chem.* **286**, 183 (1985).

85KG669   V.N. Charushin, M.G. Ponizovskii, O.N. Chupakhin, E.O. Sidorov, and I.M. Sosonkin, *Khim. Geterotsikl. Soedin.*, 669 (1985) [*CA* **104**, 108771 (1986)].

85KG960	V.G. Baklykov, V.N. Charushin, O.N. Chupakhin, and N.N. Sorokin, *Khim. Geterotsikl. Soedin.*, 960 (1985) [*CA* **104**, 207229 (1986)].
85MI1	A.R. Katritzky, "Handbook of Heterocyclic Chemistry," Pergamon Press, Oxford, 1985.
85MI2	J. March, "Advanced Organic Chemistry," Wiley and Sons, NewYork, 1985, 3rd Edition, p. 588.
85MI3	A.J. Pearson, "Metallo-Organic Chemistry," Wiley-Interscience, 1985.
85MI4	A.F. Pozharskii, "Teoreticheskiye Osnovy Khimii Geterotsiklov," Khimiya, Moskva, 1985.
85S884	A. Rykowsky and H.C. van der Plas, *Synthesis*, 884 (1985).
85T237	H.C. van der Plas, *Tetrahedron* **41**, 237 (1985).
85T617	A. Citterio, A. Gentile, F. Minisci, M. Serravalle, and S. Ventura, *Tetrahedron* **41**, 617 (1985).
85T1199	G. Heinisch and G. Lotsch, *Tetrahedron* **41**, 1199 (1985).
85T4157	F. Minisci, A. Citterio, E. Vismara, and C. Giordano, *Tetrahedron* **41**, 4157 (1985).
85TL1027	G. Courtois, A. Al-Arnaut, and L. Miginiac, *Tetrahedron Lett.* **26**, 1027 (1985).

85TL3267   M. Wada, Y. Nishihara, and K. Akiba, *Tetrahedron Lett.* **26**, 3267 (1985).

85TL3989   J.C. Boutonnet, F. Rose-Munch, and E. Rose, *Tetrahedron Lett.* **26**, 3989 (1985).

85ZOR193   V.V. Zorin, Yu.B. Zelechonok, S.S. Zlotskii, and D.L. Rakhmankulov, *Zh. Org. Khim.* **21**, 193 (1985) [*CA* **103**, 141869 (1985)].

85ZOR1150 E.V. Malykhin, G.A. Kolesnichenko, and V.D. Shteingarts, *Zh. Org. Khim.* **21**, 1150 (1985) [*CA* **104**, 87872 (1986)].

86BCJ3905 A. Sugimori and T. Yamada, *Bull. Chem. Soc. Jpn.* **59**, 3905 (1986).

86BCJ3911 A. Sugimori, T. Yamada, *Bull. Chem. Soc. Jpn.* **59**, 3911 (1986).

86CCA33   H.C. van der Plas and M. Wozniak, *Croat. Chem. Acta* **59**, 33 (1986).

86CCA89   M. Hamana, *Croat. Chem. Acta* **59**, 89 (1986).

86H125   S.K. Dubey and E.E. Knaus, *Heterocycles* **24**, 125 (1986).

86H161   L. Bauer and S. Prachayasittikul, *Heterocycles* **24**, 161 (1986).

86H181   W. Sliwa, *Heterocycles* **24**, 181 (1986).

86H239   S. Konno, M. Sagi, M. Agata, Y. Yuuki, and H. Yamanaka, *Heterocycles* **24**, 239 (1986).

86H3199	D.L. Comins and E.D. Stroud, *Heterocycles* **24**, 3199 (1986).
86JA3409	M.L. Di Vona, G. Doddi, G. Ercolani, and G. Illuminati, *J. Am. Chem. Soc.* **108**, 3409 (1986).
86JHC477	H. Sladowska, J.W.G. de Meester, and H. C. van der Plas, *J. Heterocycl. Chem.* **23**, 477 (1986).
86JHC621	H. Tondys and H. C. van der Plas, *J. Heterocycl. Chem.* **23**, 621 (1986).
86JHC843	H. Sladowska, A. van Veldhuizen, and H. C. van der Plas, *J. Heterocycl. Chem.* **23**, 843 (1986).
86JHC1015	D.J. Buurman and H.C. van der Plas, *J. Heterocycl. Chem.* **23**, 1015 (1986).
86JOC1704	T.V. RajanBabu, B.L. Chenard, and M.A. Petti, *J. Org. Chem.* **51**, 1704 (1986).
86JOC2060	J.W. Bunting and D. Stefanidis, *J. Org. Chem.* **51**, 2060 (1986).
86JOC2068	J.W. Bunting and D. Stefanidis, *J. Org. Chem.* **51**, 2068 (1986).
86JOC4385	G. Doddi and G. Ercolani, *J. Org. Chem.* **51**, 4385 (1986).
86JOC4411	F. Minisci, E. Vismara, F. Fontana, G. Morini, M. Serravalle, and C. Giordano, *J. Org. Chem.* **51**, 4411 (1986).

86JOC5039   A.R. Katritzky and K.S. Laurenzo, *J. Org. Chem.* **51,** 5039 (1986).

86KG1380   V.N. Charushin, I.V. Kazantseva, M.G. Ponizovskii, L.G. Egorova, E.O.Sidorov, and O.N. Chupakhin, *Khim. Geterotsikl. Soedin.*, 1380 (1987).

86KG1544   T.L. Pilicheva, V.L. Rusinov, O.N. Chupakhin, N.A. Kluev, G.G. Alexandrov, and S.E. Esipov, *Khim. Geterotsikl. Soedin.*, 1544 (1986) [*CA* **107**, 77739 (1987)].

86MI1   F. Minisci, *in* "Substituent Effects in Radical Chemistry," D. Reidel Publ. Co., Dordrecht, 1986 (H. Viehe, ed.).

86MI2   K. Breuker, H.C. van der Plas, and A. van Veldhuizen, *Isr. J. Chem.* **27**, 67 (1986).

86MI3   B. Giese, "Radicals in Organic Synthesis: Formation of Carbon-Carbon Bonds," Pergamon Press, Oxford, New York, 1986.

86S50   M. Makosza and S. Ludwiczak, *Synthesis,* 50 (1986).

86T5973   G. Heinisch and G. Lotsch, *Tetrahedron* **42**, 5973 (1986).

86TL211   R. Yamaguchi, M. Moriyasu, and M. Kawanisi, *Tetrahedron Lett.* **27**, 211 (1986).

86TL3187	E. Vismara, M. Serravalle, and F. Minisci, *Tetrahedron Lett.* **27**, 3187 (1986).
86TL3271	T. Umemoto and K. Tomita, *Tetrahedron Lett.* **27**, 3271 (1986).
86ZOR806	E.V. Malykhin, G.A. Kolesnichenko, and V.D. Shteingarts, *Zh. Org. Khim.* **22**, 806 (1986) [*CA* **105**, 190580 (1986)].
87ACR282	M. Makosza and J. Winiarski, *Acc. Chem. Res.* **20**, 282 (1987).
87CPB1378	S. Konno, S. Ohba, M. Sagi, and H. Yamanaka, *Chem. Pharm. Bull.* **35**, 1378 (1987).
87CPB3628	T. Fujii, M. Ohba, and J. Sagaguchi, *Chem. Pharm. Bull.* **35**, 3628 (1987).
87H229	M. Hamana and Y. Fujimura, *Heterocycles* **25**, 229 (1987).
87H235	M. Hamana, Y. Fujimura, and Y. Nawata, *Heterocycles* **25**, 235 (1987).
87H731	G. Heinisch and G. Lotsch, *Heterocycles* **26**, 731 (1987).
87H2159	D.L. Comins and J.J. Herrick, *Heterocycles* **26**, 2159 (1987).
87JA3789	S. Rozen, D. Hebel, and D. Zamir, *J. Am. Chem. Soc.* **109**, 3789 (1987).

87JCS(P1)507   C. Glidewell, T. Shepherd, and D.M. Smith, *J. Chem. Soc., Perkin Trans. 1*, 507 (1987).

87JCS(P2)79   R.J. Catana, J.O. Singh, J.D. Anunziata, and J. Silber, *J. Chem. Soc., Perkin Trans. 2*, 79 (1987).

87JHC1377   D.J. Buurman and H.C. van der Plas, *J. Heterocycl. Chem.* **24**, 1377 (1987).

87JHC1657   A.C. Brouwer and H.C. van der Plas, *J. Heterocycl. Chem.* **24**, 1657 (1987).

87JOC488   E. Buncel, R.A. Renfrow, and M. Strauss, *J. Org. Chem.* **52**, 488 (1987).

87JOC730   F. Minisci, E. Vismara, F. Fontana, G. Morini, and M. Seravalle, *J. Org. Chem.* **52**, 730 (1987).

87JOC5643   M. Wozniak, A. Baranski, K. Nowak, and H.C. van der Plas, *J. Org. Chem.* **52**, 5643 (1987).

87JOM139   O.A. Artamkina, S.V. Kovalenko, I.P. Beletskaya, and O.A. Reutov, *J. Organomet. Chem.* **329**, 139 (1987).

87KG182   Yu.A. Zhdanov, E.A. Zvezdina, S.M. Statsenko, and A.N. Maximova, *Khim. Geterotsikl. Soedin.*, 182 (1987).

87KG309	Yu.A. Zhdanov, E.A. Zvezdina, S.M. Statsenko, V.A. Anisimova, A.N. Maximova, and V.V. Korsun, *Khim. Geterotsikl. Soedin.*, 309 (1987).
87KG1011	H.C. van der Plas, *Khim. Geterotsikl. Soedin.*, 1011 (1987).
87KG1118	L.M. Naumova, V.N. Charushin, and O.N. Chupakhin, *Khim. Geterotsikl. Soedin.*, 1118 (1987) [*CA* **108**, 154135 (1988)].
87KG1260	V.N. Charushin, L.M. Naumova, M.G. Ponizovskii, V.G. Baklykov, and O.N. Chupakhin, *Khim. Geterotsikl. Soedin.*, 1260 (1987).
87KG1385	O.N. Chupakhin, V.N. Charushin, I.V. Kazantseva, M.G. Ponizovskii, E.O. Sidorov, P.A. Torgashev, and A.V. Belik, *Khim. Geterotsikl. Soedin.*, 1385 (1987) [*CA* **109**, 128269 (1988)].
87MI1	V.N. Charushin, "Reactions of Azines with Bifunctional Nucleophiles," doctoral dissertation, Ekaterinburg, 1987.
87MI2	F. Minisci, E. Vismara, and F.Fontana, *Gazz. Chim. Ital.* **117**, 363 (1987).
87MI3	M. Makosza, *Proceedings of the 31-th IUPAC Congress, Section of Organic Chemistry*, p. 118, Sofia, 1987.

87MI4	E.P. Koundig, N.P. Do Thi, P. Paglia, D.P. Simmons, S. Spichiger, and E. Wenger, "Selective Reactions on Arenechromium Complexes," *in* "Organometallics in Organic Synthesis," Springer-Verlag, Berlin, Heidelberg, 1987 (A. de Meijere and H. tom Dieck, eds.), p. 265.
87MI5	C. Paradisi and G. Scorrano, *in* "Nucleophilisity, American Chemical Society, Washington D.C., 1987 (J.M. Harris and S.A. McManus, eds.),"ACS Chemistry Series."
87TL2705	T. Umemoto and G. Tomizawa, *Tetrahedron Lett.* **28**, 2705 (1987).
87ZOR1039	E.V. Malykhin and V.D. Shteingarts, *Zh. Org. Khim.* **23**, 1039 (1987).
88AHC2	C.K. McGill and A. Rappa, *Adv. Heterocycl. Chem.* **44**, 2 (1988).
88AHC199	D.L. Comins and S. O'Connor, *Adv. Heterocycl. Chem.* **44**, 199 (1988).
88AHC301	V.N. Charushin, O.N. Chupakhin, and H.C.van der Plas, *Adv. Heterocycl. Chem.* **44**, 301 (1988).
88H291	A. Sausins and G. Duburs, *Heterocycles* **27**, 291 (1988).
88H2659	H. Vorbrouggen and M. Maas, *Heterocycles* **27**, 2659 (1988).
88JHC831	A.T.M. Marcelis, H. Tondys, and H.C. van der Plas, *J. Heterocycl. Chem.* **25**, 831 (1988).

88JHC1769	M.C. Schroeder and J.C. Kiely, *J. Heterocycl. Chem.* **25**, 1769 (1988).
88JOC690	G.P. Stahly, B.C. Stahly, and J.R. Maloney, *J. Org. Chem.* **53**, 690 (1988).
88JOC1123	D. Hebel and S. Rozen, *J. Org. Chem.* **53**, 1123 (1988).
88JOC1646	D.D. Tanner and A. Kharrat, *J. Org. Chem.* **53**, 1646 (1988).
88JOC1729	G. Doddi and G. Ercolani, *J. Org. Chem.* **53**, 1729 (1988).
88JOC3978	A.R. Katritzky and K.S. Laurenzo, *J. Org. Chem.* **53**, 3978 (1988).
88JPR789	M. Makosza and S. Ostrowski, *J. Prakt. Chem.* **330**, 789 (1988).
88LA627	A. Rykowski and M. Makosza, *Liebigs Ann. Chem.*, 627 (1988).
88MI1	O.N. Chupakhin, V.N. Charushin, and A.I. Chernyshev, *Prog. NMR Spectroscopy* **20**, 95 (1988).
88MI2	D.J. Raber and N.K. Raber, "Organic Chemistry," West Publ. Co., St. Paul, New York, 1988.
88MI3	G.H. Frank and J.W. Stadlerhofer, "Industrial Aromatic Chemistry," Springer-Verlag, Berlin, 1988.
88OPP105	E. Vismara, F. Fontana, and F. Minisci, *Org. Prep. Proced. Int.* **20**, 105 (1988).

88OPPI591   D.J. Buurman and H.C. van der Plas, *Org. Prep. Proced. Int.* **20**, 591 (1988).

88S119   G. Heinisch and G. Lotsch, *Synthesis,* 119 (1988).

88SC1937   M.J. Shia, *Synth. Commun.* **18**, 1937 (1988).

88T1   O.N. Chupakhin, V.N. Charushin, and H.C. van der Plas, *Tetrahedron* **44**, 1 (1988).

88T1721   S. Ostrowski and M. Makosza, *Tetrahedron* **44**, 1721 (1988).

89ACSA995   F. Fontana, F. Minisci, M.C.N. Barbosa, and E. Vismara, *Acta Chem. Scand.* **43**, 995 (1989).

89AHC73   V.N. Charushin, S.G. Alexeev, O.N. Chupakhin, and H.C. van der Plas, *Adv. Heterocycl. Chem.* **46**, 73 (1989).

89CB493   R. Neidlein and G. Lautenschlager, *Chem. Ber.* **122**, 493 (1989).

89CL773   N. Nishiwaki, S. Minakata, M. Komatsu, and Y. Ohshiro, *Chem. Lett,* 773 (1989).

89CL1227   S. Fukuzumi, M. Ishikawa, and T. Tanaka, *Chem. Lett.*, 1227 (1989).

89H237   M. Alvarez, R. Lavilla, and J. Bosch, *Heterocycles* **29**, 237 (1989).

89H249   S. Rozen and D. Hebel, *Heterocycles* **28**, 249 (1989).

89H489   F. Minisci, E. Vismara, and F. Fontana, *Heterocycles* **28**, 489 (1989).

89JCS(CC)727 M. Ishikura and M. Terashima, *J. Chem. Soc., Chem. Commun.*, 727 (1989).

89JCS(CC)941 S. Fukuzumi, M. Chiba, and T. Tanaka, *J. Chem. Soc., Chem. Commun.*, 941 (1989).

89JHC1675 L.E. Lesheski, *J. Heterocycl. Chem.* **26**, 1675 (1989).

89JHC1751 G. Heinisch, G. Lotsch, S. Offenberger, B. Stanovnik, and M. Tisler, *J. Heterocycl. Chem.* **26**, 1751 (1989).

89JOC1726 T. Umemoto and G. Tomizawa, *J. Org. Chem.* **54**, 1726 (1989).

89JOC5224 F. Minisci, E. Vismara, and F. Fontana, *J. Org. Chem.* **54**, 5224 (1989).

89LA95 S. Ostrowski and M. Makosza, *Liebigs Ann. Chem.*, 95 (1989).

89LA545 M.K. Bernard, M. Makosza, B. Szafran, and U. Wrzeciono, *Liebigs Ann. Chem.*, 545 (1989).

89LA825 M. Makosza, J. Baran, D. Dziewonska-Baran, and J. Golinski, *Liebigs Ann. Chem.*, 825 (1989).

89MI1 G. Heinisch, *in* "Free Radicals in Synthesis and Biology," Kluwer Academic Publishers, 1989 (F. Minisci, ed.).

89MI2	E. Vismara, F. Minisci, M. C. Barbosa, and F. Fontana, *The Proceedings of the 12-th International Congress of Heterocyclic Chemistry*. Jerusalem, 1989.
89MI3	D.T. Hurst, *in* "Progress in Heterocyclic Chemistry," Pergamon Press, Oxford, 1989 (H. Suschitzky and E.F.V. Scriven, eds.), p. 228.
89MI4	G. Doddy, G. Ercolani, and P. Iaconianni, *Gazz. Chim. Ital.* **119**, 305 (1989).
89RCR747	M. Makosza, *Russ. Chem. Rev. (Engl. Transl.)* **58**, 747 (1989) [*CA* **111**, 213942 (1989)].
89SC317	R.B. Katz, J. Mistry, and M.B. Mitchell, *Synth. Commun.* **19**, 317 (1989).
89SC3523	C. Gesto, E. Delacuesta, and C. Avendano, *Synth. Commun.* **19**, 3523 (1989).
89TL4569	F. Minisci, E. Vismara, F. Fontana, and M.C.N. Barbosa, *Tetrahedron Lett.* **30**, 4569 (1989).
89ZOB1506	V.T. Abaev, L.I. Kisarova, S.E. Emanuilidi, A.A. Bumber, I.E. Mikhailov, I.B. Blank, A.I. Yanovskii, Yu.T. Struchkov, and O.Yu. Okhlobystin, *Zh. Obshch. Khim.* **59**, 1506 (1989).

90AHC117	H. Vorbrouggen, *Adv. Heterocycl. Chem.* **49**, 117 (1990).
90BCJ2682	M. Goto, Y. Mikata, and A. Ohno, *Bull. Chem. Soc. Jpn.* **63**, 2682 (1990).
90CL1275	S. Fukuzumi and T. Kitano, *Chem. Lett.*, 1275 (1990).
90H779	Y. Miura, Y. Fujimura, S. Takaku, and M. Hamana, *Heterocycles* **30**, 779 (1990).
90H1325	M. Hayashida, H. Honda, and M. Hamana, *Heterocycles* **31**, 1325 (1990).
90H2025	D. L. Comins and M.O. Killpack, *Heterocycles* **31**, 2025 (1990).
90JA8563	T. Umemoto, Sh. Fukami, G. Tomizawa, K. Harasawa, K. Wawada, and K. Tomita, *J. Am. Chem. Soc.* **112**, 8563 (1990).
90JHC79	F. Minisci, F. Fontana, and E. Vismara, *J. Heterocycl. Chem.* **27**, 79 (1990).
90JOC778	D.J. Buurman, A. van Veldhuizen, and H.C. van der Plas, *J. Org. Chem.* **55**, 778 (1990).
90JOC3647	J.F.J. Engbersen, A. Koudijs, and H.C. van der Plas, *J. Org. Chem.* **55**, 3647 (1990).
90JOC4979	M. Makosza and K. Sienkievicz, *J. Org. Chem.* **55**, 4979 (1990).

90KG579   Yu.B. Zelechonok, S.S. Zlotsky, V.V. Zorin, and D.L. Rakhmankulov, *Khim. Geterotsikl. Soedin.*, 579 (1990) [*CA* **113**, 231101 (1990)].

90KG1575   A.V. Gulevskaya, A.F. Pozharkii, and L.V. Lomachenkova, *Khim. Geterotsikl. Soedin.*, 1575 (1990) [*CA* **114**, 163853 (1991)].

90LA653   M. Wozniak, A. Baranski, K. Nowak, and H. Poradowska, *Liebigs Ann. Chem.*, 653 (1990).

90MI1   M. Wozniak, K. Nowak, and H. Poradowska, *Proceedings of the X-th Symposium on the Chemistry of Heterocyclic Compounds, Part II.* PO-148; Kosice, Chechoslovakia, August 13–17, 1990.

90MI2   M. Makosza and K. Sienkiewicz, *Abstracts of the 8 -th International IUPAC Conference on Organic Synthesis*, 240. Helsinki, July 23–27, 1990.

90MI3   O. Haglund and M. Nilsson, *Abstracts of the 8 -th International IUPAC Conference on Organic Synthesis*, 207. Helsinki, July 23–27, 1990.

90MI4   O.N. Chupakhin, *Izv. Sibirskogo Otd. Akad. Nauk SSSR, ser. khim.*, 110 (1990).

90MI5	A.A. Moroz and M.S. Shvartsenberg, *Izv. Sibirskogo Otd. Acad. Nauk SSSR, ser. khim.*, 115 (1990) [*CA* **114**, 184484].
90OPPI575	Z. Wrobel and M. Makosza, *Org. Prep. Proced. Int.* **22**, 575 (1990).
90OS129	T. Umemoto, K. Tomita, and K. Kawada, *Org. Synth.* **69**, 129 (1990).
90S942	O. Haglund, A.A.K.M. Hai, and M. Nilsson, *Synthesis,* 942 (1990).
90T2525	F. Fontana, F. Minisci, M.C. Nogueira Barbosa, and E. Vismara, *Tetrahedron* **46**, 2525 (1990).
90TL3217	N.R. Ayyangar, S.N. Naik, and K.V. Srinivasan, *Tetrahedron Lett.* **31**, 3217 (1990).
90TL6287	M.J. Shia, *Tetrahedron Lett.* **31**, 6287 (1990).
90TL7379	A.A. Gakh, A.S. Kiselev, and V.V. Semenov, *Tetrahedron Lett.* **31**, 7379 (1990).
90ZOR933	G.A. Artamkina, S.V. Kovalenko, I.P. Beletskaya, and O. A. Reutov, *Zh. Org. Khim.* **26**, 933 (1990) [*J. Org. Chem. USSR* **26**, 801 (1990)] [*CA* **113**, 211502 (1990)].
91BCJ1081	T. Umemoto, K. Harasawa, G. Nomizawa, K. Kawada, and K. Tomita, *Bull. Chem. Soc. Jpn.* **64**, 1081 (1991).

91CB577	M. Makosza, J. Golinski, S. Ostrowski, A. Rykowsky, and A.B. Sahasrabudhe, *Chem. Ber.* **124**, 577 (1991).
91CPB36	F. Sumija, N. Shirai, and Y. Sato, *Chem. Pharm. Bull.* **39**, 36 (1991).
91H1579	Y. Miura, S. Tekaku, Y. Nawata, and M. Hamana, *Heterocycles* **32**, 1579 (1991).
91JCS(P1)2877	N. Sato, Y. Shimomura, Y. Ohwaki, and R. Takeuchi, *J. Chem. Soc., Perkin Trans. 1*, 2877 (1991).
91JFC369	T. Umemoto, K. Harasawa, and G. Tomizawa, *J. Fluorine Chem.* **53**, 369 (1991).
91JFC392	A.S. Kiselev and A.A. Gakh, *J. Fluorine Chem.* **54**, 392 (1991).
91JHC1051	S. Prachayasittikul, G. Doss, and L. Bauer, *J. Heterocycl. Chem.* **28**, 1051 (1991).
91JHC1075	M. Grzegorek, M. Wozniak, A. Baranski, and H.C. van der Plas, *J. Heterocycl. Chem.* **28**, 1075 (1991).
91JOC2866	F. Fontana, F. Minisci, M. Barbosa, and E. Vismara, *J. Org. Chem.* **56**, 2866 (1991).
91JOC6298	D. Hebel and S. Rozen, *J. Org. Chem.* **56**, 6298 (1991).

91JOC6736	J. Klippenstein, P. Arya, and D.D.M. Wayner, *J. Org. Chem.* **56**, 6736 (1991).
91LA605	M. Makosza and J. Stalewski, *Liebigs Ann. Chem.*, 605 (1991).
91LA875	M. Wozniak, A. Baranski, and B. Szpakiewicz, *Liebigs Ann. Chem.*, 875 (1991).
91MC128	A.S. Kiselev, A.A. Gakh, N.D. Kagramanov, and V.V. Semenov, *Mendeleev Comm.*, 128 (1991).
91MI1	V.L. Rusinov and O.N. Chupakhin, "Nitroazines," Novosibirsk, Nauka, 1991.
91MI2	F. Terrier, "Nucleophilic Aromatic Displacement: The Influence of the Nitro Group," *in* "Organic Nitro Chemistry Series," VCH Publishers, Inc., New York, 1991 (H. Feuer, ed.).
91PJC323	M. Wozniak, K. Nowak, and H. Poradowska, *Pol. J. Chem.* **63**, 323 (1991).
91S103	M. Makosza, *Synthesis*, 103 (1991).
91S996	T. Kitano, N. Shirai, and Y. Sato, *Synthesis*, 996 (1991).
91T8573	N. Haider, G. Heinish, and J. Moshuber, *Tetrahedron* **47**, 8573 (1991).

91TL1311	J. Nasielski and C. Rypens, *Tetrahedron Lett.* **32**, 1311 (1991).
92BCJ55	S. Tamagaki, M. Ueno, and W. Tagaki, *Bull. Chem. Soc. Jpn.* **65**, 55 (1992).
92BSB579	G. Heinisch, *Bull. Soc. Chim. Belg.* **101**, 579 (1992).
92CB1369	M. Wieser, K. Sunkel, C. Robl, and W. Beck, *Chem. Ber.* **125**, 1369 (1992).
92CL1295	F. Coppa, F. Fontana, E. Lazzarini, F. Minisci, G. Pianese, and L. Zhao, *Chem. Lett.,* 1295 (1992).
92CL1583	S. Itoh, H. Fukushima, M. Komatsu, and Y. Ohshiro, *Chem. Lett.,* 1583 (1992).
92H1	H. Yamanaka, *Heterocycles* **33**, 1 (1992).
92H931	O.N. Chupakhin, B.V. Rudakov, S.G. Alexeev, and V.N. Charushin, *Heterocycles* **33**, 931 (1992).
92H1055	Y. Miura, S. Takaku, Y. Fujimura, and M. Hamana, *Heterocycles* **34**, 1055 (1992).
92H1129	M. Begtrup, *Heterocycles* **33**, 1129 (1992).
92HCA825	R. Neidlein and G. Schrouder, *Helv. Chim. Acta* **75**, 825 (1992).

92JA7708   R. Bacaloglu, A. Blasko, C. Barton, F. Ortega, and C. Zucco,

              J. Am. Chem. Soc. **114**, 7708 (1992).

92JA9237   M.K. Stern, F.D. Hileman, and J.K. Bashkin, *J. Am. Chem. Soc.*

              114, 9237 (1992).

92JCS(P1)333 K. Yoshida and K. Miyoshi, *J. Chem. Soc., Perkin Trans. I*, 333

              (1992).

92JCS(P1)2851 T. Kitano, N. Shirai, M. Motoi, and Y. Sato, *J. Chem. Soc.,*

              Perkin Trans. 1, 2851 (1992).

92JHC763   T. Kimachi, F. Yoneda, and T. Sasaki, *J. Heterocycl. Chem.* **29**,

              763 (1992).

92JOC4431   G. Doddy, G. Ercolani, and P. Mencarelli, *J. Org. Chem.* **57**, 4431

              (1992).

92JOC4784   M. Makosza and M. Bialecki, *J. Org. Chem.* **57**, 4784 (1992).

92JOC5034   T. Tanaka, N. Shirai, J. Sugimori, and Y. Sato, *J. Org. Chem.*

              57, 5034 (1992).

92KG107   S.V. Litvinenko, Yu. M. Volovenko, V.I. Savich, and F.S. Babichev,

              Khim. Geterotsikl. Soedin., 107 (1992) [*CA* **119**, 8774].

92KG1202	A.V. Gulevskaya, A.F. Pozharskii, A.I. Chernyshev, and V.V. Kuzmenko, *Khim. Geterotsikl. Soedin.*, 1202 (1992).
92LA19	G. Heinish and T. Huber, *Liebigs Ann. Chem.*, 19 (1992).
92LA813	H-G. Henning, G. Stemplinger, and K. Rothe, *Liebigs Ann. Chem.*, 813 (1992).
92LA899	M. Wozniak, A. Baranski, K. Nowak, and H. Poradowska, *Liebigs Ann. Chem.*, 899 (1992).
92MC151	O.N. Chupakhin, Yu. A. Azev, S.G. Alexeev, S.V. Shorshnev, E. Tsoi, and V.N. Charushin, *Mendeleev Commun.*, 151 (1992).
92MI1	J. McMurry, "Organic Chemistry," Brooks/Cole Publ. Co, Pacific Grove, California, 1992. 3rd Edition.
92MI2	J. March, "Advanced Organic Chemistry," Wiley-Interscience, New York, 1992. 4th Edition, p. 666.
92MI3	M.V. Gorelik and L.S. Efros, "Osnovy Khimii i Technologii Aromaticheskikh Soedinenii," Khimiya, Moskva, 1992.
92PJC3	M. Makosza, *Pol. J. Chem.* **66**, 3 (1992).
92PJC2005	Z. Wrobel and M. Makosza, *Pol. J. Chem.* **66**, 2005 (1992).
92S571	K. Wojciechowski and M. Makosza, *Synthesis*, 571 (1992).

92TL687	F. Coppa, F. Fontana, F. Minisci, G. Pianese, P. Torontelo, and L. Zhao, *Tetrahedron Lett.* **33**, 687 (1992).
92TL3057	F. Coppa, F. Fontana, E. Lazzarini, F. Minisci, G. Pianese, and L. Zhao, *Tetrahedron Lett.* **33**, 3057 (1992).
92TL4787	A. Ostrowicz, S. Baloniak, M. Makosza, and A. Rykowski, *Tetrahedron Lett.* **33**, 4787 (1992).
92USP5117063	M.K. Stern and K.J. Baskin, *US Patent* 5117063 (1992).
93ACSA95	M. Wozniak and H.C. van der Plas, *Acta Chem. Scand.* **47**, 95 (1993).
93BCJ1028	A. Yoshino, N. Nakamura, and K. Tanahashi, *Bull. Chem. Soc. Jpn.* **66**, 1028 (1993).
93BCJ1149	T. Sakakibara, Y. Ohwaki, and N. Sato, *Bull. Chem. Soc. Jpn.* **66**, 1149 (1993).
93BCJ1191	M. Okamura, Y. Mikata, N. Yamazaki, A. Tsutsumi, and A. Ohno, *Bull. Chem. Soc. Jpn.* **66**, 1191 (1993).
93JCS(P1)15	N. Sato, K. Kawahara, and N. Morii, *J. Chem. Soc., Perkin Trans. 1*, 15 (1993).

93JHC329   A.S. Kiselev, L. Strekowski, and V.V. Semenov, *J. Heterocycl. Chem.* **30**, 329 (1993).

93JHC691   N. Sato and E. Nagano, *J. Heterocycl. Chem.* **30**, 691 (1993).

93JOC959   E. Vismara, A. Donna, F. Minisci, A. Naggi, N. Pastori, and G. Torri, *J. Org. Chem.* **58**, 959 (1993).

93JOC4207   F. Minisci, F. Fontana, G. Pianese, Y.-M. Yan, *J. Org. Chem.* **58**, 4207 (1993).

93JOC6883   M.K. Stern and B.K. Cheng, *J. Org. Chem.* **58**, 6883 (1993).

93LA7   M. Wozniak, A. Baranski, and B. Szpakiewicz, *Leibigs Ann. Chem.*, 7 (1993).

93LA471   M. Wozniak and M. Tomula, *Liebigs Ann. Chem.*, 471 (1993).

93MI1   T. Kaija, K. Hasebe, K. Kohda, and Y. Kawazoe, *Abstracts of the 14-th International Congress on Heterocyclic Chemistry, PO3-230.* Antwerpen, 1993.

93MI2   S. Hibino, T. Choshi, and E. Sugino, *Abstracts of the 14-th International Congress on Heterocyclic Chemistry, PO3-235.* Antwerpen, 1993.

93MI3	O. Haglund, "Copper-Mediated Vicarious Nucleophilic Substitution," Ph.D. thesis, Chalmers University of Technology, Gouteborg, 1993.
93ZOR622	T.L. Pilicheva, V.L. Rusinov, A.V. Myasnikov, A.B. Denisova, G.G. Alexandrov, and O.N. Chupakhin, *Zh. Org. Khim.* **29**, 622 (1993).

Subject Index

4-Acetoxy-3-methoxyquinoline, 199

3-Acetyl-7,8-difluoro-4-ethoxycarbonyl-2-methyl-5-oxo-5,9b-dihydropyrazolo[5,1-*a*]quinoline, 245

5-Acetyl-2-methyl-2,3,4,5,6,7-hexahydro-1H-2,5-benzodiazonine, 88

Acridines, 108, 111, 117, 127-129, 131, 141

N-Acylazinium salts, reactions with

    - C-nucleophiles, 161

    - N,N-dimethylaniline, 163

    - indole, 164-166

    - organometallic reagents, 166

    - Grignard reagents, 167

    - trialkylphosphites, 170

N-Acyloxyazinium salts, 176, 187

Addition-elimination mechanism, 6, 109, 246-269

σ-Adducts 6-8, 11, 12, 37, 41, 49, 89, 114, 147, 152, 153, 177, 249-251, 255-268

Alizarin, 7

N-Alkoxyazinium salts, reactions with

    - alcohols, 188

    - arylamines, 193

    - cyanide ion, 190

    - enamines, 193

    - indoles, 193

    - thiols, 195, 196

SUBJECT INDEX

Alkylation of
- anthracene, 16
- benzene, 16, 17
- η^6-ethylbenzenetricarbonylchromium(0), 20
- naphthalene, 16
- nitroarenes, 68
- phenanthrene, 16

N-Alkylazinium salts, reactions with
- alkyl(aryl)thiolates, 158
- carbanions, 149
- Grignard reagents, 158
- hydroxide ion, 142
- liquid ammonia/potassium permanganate, 144-146
- organometallic compounds, 158
- trialkylphosphites, 158

Amination of
- acridine, 108
- N-alkylazinium salts, 144-146, 161
- 5-azacinnoline, 103
- 5-deazaflavins, 103
- 1,3-dimethyllumazine, 102
- imidazoles, 227
- naphthalene, 16
- naphthyridines, 10, 100, 125
- nitroarenes, 59-63, 71
- nitropyridines, 90
- nitroquinolines and isoquinolines, 96, 123, 124

- nitroquinoxalines, 99

- pyrazine, 110

- pyridazines, 109, 110

- pyrimidines, 90, 94, 109, 122

- 1,2,4-triazines, 90, 123

Aminoarylation of

- acridinium salts, 131, 150-153

- azines, 127

- isoquinoline, 173

- quinolinium salts, 127, 128

- quinoxalinium salts, 127, 128, 131

N-Aminoazinium salts, 216

4-Amino-3,6-dinitro-1,8-naphthyridine, 125

2-Amino-4-methoxy-5-nitropyrimidine, 95, 122

2-Amino-6-methoxy-3-nitropyrimidine, 95, 121

2-Amino-1,7-naphthyridine, 10

6-Amino-5-nitroisoquinoline, 124

5-Amino-4-nitro-6-phenylpyridazine, 123

4-Amino-3-nitroquinoline, 123

9-(4-Aminophenyl)acridine, 141

4-Aminopyrimidine, 122

2- and 4-Aminoquinolines, 110, 123, 124

5-Amino-1,2,4-triazine, 123

Anionic σ-adducts, 4, 31, 32, 109, 112, 263-268

ANRORC mechanism, 283-286

4-(9-Anthracenyl)quinazoline, 141

Arene-metal complexes, 18-31, 87

SUBJECT INDEX

Aromatization 8, 9, 11-13, 270-274

Arylation of

- diazanaphthalones, 104, 105

- quinazolinium trifluoroacetate, 131

Azine N-oxides, reactions with

- liquid ammonia/potassium permanganate, 180

- carbanions, 178, 179

- cyanide ion, 181

- Grignard reagents, 182

- vicarious C-nucleophiles, 183, 184

NH-Azinium salts, reactions with

- arylamines, 127, 128

- carbon-centered radicals, 132

(η^6-Benzene)iodo-*bis*(trimethylphosphan)osmium(II) hexafluorophosphate, 30

η^6-Benzenetricarbonylchromium, 20

2-Benzyl-10-methyl-4-phenyl-2,3,4,10-tetrahydro-1,2,4-triazino[5,6-*b*]quinoxalin-3-thione, 244

3-Bromo-4-phenylpyridine, 174

t-Butylation of

- benzene, 17

- η^6-benzenetricarbonylchromium, 20, 28, 29

- naphthalene, 17, 18

- pyridine, 119

t-Butylbenzene, 17, 20, 29

(η^6-*n*-Butylbenzene)iodo-*bis*(trimethylphosphan)osmium(II) hexafluorophosphate, 30

4-*n*-Butylpyridine, 240

3-Carbamoyl-1,6-dihydro-6-imino-1-methylpyridine, 161

Charge-transfer complexes, 246, 249

Chichibabin reaction, 3, 16, 59, 89, 90, 108, 109, 227-229, 252, 256

4-Chloro-2-ethylpyridine, 175

5-Chloro-2-nitrobenzyl cyanide, 69

2-Chloro-3-nitro-4-(phenylsulfonylmethyl)-1,8-naphthyridine, 125

Cine-substitution reactions, 9, 15, 177, 207, 209, 213, 280-282

Cyanation of

- N-acylazinium salts, 162

- N-acyloxyazinium salts, 190-192

- N-alkoxyazinium salts, 185

- N-alkylazinium salts, 148

- N-aminoazinium salts, 217

- arene-metal complexes, 26

- nitroarenes, 33

- N-oxides, 181

- pyrylium salts, 218

9-Cyano-10-anthranol, 34

9-Cyano-10-anthrone, 34

4-Cyano-2,6-diphenylpyrylium perchlorate, 226

4-Cyano-1-ethyl-2-quinoline, 149

Cyanomesitylene, 26

3-Cyano-2-methylbenzo[*b*]thiophen, 240

9-Cyano-10-nitroanthracene, 34

4-Cyano-2-phenyl-1,2,3-triazole, 235

2-Cyclohexenyl-4-methylquinoline, 140

9,9-Diacridanyl, 108

Diazanaphthalones, reactions with

    - arylamines, 104-107

    - N,N-dimethylaniline, 104-107, 253, 254

    - indole, 105, 106

4-(2,4-Dimethoxyphenyl)-2,6-diphenylpyrylium perchlorate, 227

4-(4-N,N-Dimethylaminophenyl)-2,6-diphenylpyrylium perchlorate, 226

1-(4-N,N-Dimethylaminophenyl)isoquinoline, 173

9-(4-N,N-Dimethylaminophenyl)-10-methylacridinium iodide, 160

4-(4-N,N-Dimethylaminophenyl)-2-phenylbenzo[*b*]pyrylium perchlorate, 227

N,N'-Dimethyldiacridanyl, 249

1,3-Dimethyl-2-imino-1,2-dihydropyridine, 146

1,3-Dinitrobenzene, reactions with

    - phenoxide ions, 34

    - acetone (Janovsky reaction), 247

1,4-Dinitrobenzene, reactions with methyl and phenyl radicals, 36

2,6-Dinitrobenzyl phenyl sulfone, 70

2',6'-Dinitrobiphenyl-4-ol, 70

2,4-Dinitrophenol, 248

2,4-Dinitrophenylacetone, 69

Electrochemical reactions, 72-75, 236-238

Electrochemical cyanation of

    - naphthalenes, 75

    - benzo[b]thiophens, 238

Elimination-addition mechanism, 280-283

2-Ethoxyquinoline, 198

Ethyl cyano(4-chloropyridyl-2)acetate, 198

2-Ethyl-4,6-dimethylpyrimidine, 138

Ethyl 2-pyrazinecarboxylate, 140

(9-Fluorenyl)tropylium trifluoroacetate, 80

N-Fluoroazinium salts, reactions with

- bases, 204

- carbanions, 210

- fluoride ion, 200

- methylene chloride, 208

- trialkylphosphins, 213

- trialkylphosphites, 213

2-Fluoro-4-*t*-butylpyridine, 215

Hydroxyarylation of azines, 111, 127

Hydroxylation of

- acridine, 110

- N-alkylazinium salts, 142

- benzimidazoles, 231

- naphthalene, 54-58

- pyridine, 110

- phenanthridine, 110

- quinoline and isoquinoline, 110

Hydroxymethylation of

- N-alkoxypyridinium salts, 185

- N-alkylazinium salts, 155

- NH-azinium salts, 139

SUBJECT INDEX

- pyridine N-oxides, 198

2-Hydroxymethyl-4-methylquinoline, 139

Indolo-dehydrogenation of

- N-acylazinium salts, 164-166
- N-alkylazinium salts, 153, 160

9-(Indolyl-3)-acridine, 173

9-(Indolyl-3)-10-methylacridinium iodide, 160

2-(Indolyl-3)-1-methylquinoxalinium iodide, 160

Intramolecular reactions, 241

Janovsky reaction, 247, 258

Kinetic data, 274-280

1,6-Methano[10]annulene, reactions with

- *t*-butylhydrazine, 79
- vicarious nucleophiles, 78, 81

1-Methoxy-1,2,3-triazolium salts, reactions with nucleophiles, 223

9-Methylacridinium salts, reactions with

- arylamines, 150
- N,N-dimethylaniline, 160
- N,N'-tetramethyl-*para*-phenylenediamine, 249
- indole, 160

Methylation of

- azines, 125, 126
- nitroarenes, 45
- phenanthrene, 17

4-(1-Methyl-3-indolyl)-2,6-diphenylpyrylium perchlorate, 226

1-Methylisoquinoline, 126

4-Methyl-5-methylamino-2-phenyl-1,2,3-triazole, 236

Methyl α-methyl-α-(4-nitrophenyl)propionate, 69

1-Methyl-4-nitro-5-(*para*-tosylmethyl)pyrazole, 236

1-Methyl-4-nitropyrazole, reactions with vicarious C-nucleophiles, 231

Methyl 6-phenylnicotinate, 176

9-Methylphenanthrene, 17, 18

Methylphenazinium salts, reactions with

    - carbanions, 156

    - morpholine, 156

6-Methyl-3-phenyl-1,2,4-triazine, 125

1-Methylpyridinium salts, reactions with

    - Grignard reagents, 158

    - nitromethane, 159

1-Methylquinolinium iodide, reaction with Grignard reagents, 158

1-Methylquinoxalinium iodide, reaction with indole, 160

MO Calculations, 93, 96, 97, 101, 251-255

4-Nitroaniline, 71, 72

Nitroarenes, reactions with

    - benzamide, 66, 71

    - butyl trifluoromethyl sulfone, 43

    - carbanions, 34

    - chloromethyl phenyl sulfone, 43

    - cyanide ion, 33, 34

    -Grignard reagnets, 37

- hydroxide ion, 1, 53, 54, 67

- trimethylsilyl ketene acetals, 38, 69

4-Nitrodiphenylamine, 63

Nitromethylation of N-alkylazinium salts, 159

Nitronaphthols, 54-57

Nitrophenols, 1, 7, 52, 53, 67, 255, 256

N-(4-Nitrophenyl)benzamide, 71

4-Nitrosodiphenylamine, 63

2-Nitrothiophen, reactions with

- chloromethyl phenyl sulfone, 231

- *t*-butyl hydroperoxide, 231

NMR studies of σ-adducts, 109, 259-260

2-Phenylethynylpyridine, 199

Phenyl isobutyronitrile, 30

Pyrazines, reactions with

- alkoxides, 117, 118

- carbon-centered radicals, 137, 140

Pyridines, reactions with

- aminating agents, 90, 91

- *n*-butyllithium, 119

- organometallic compounds, 119, 120

Pyridazines, reactions with carbon-centered radicals, 137

4-(2-Pyridyl)dimedone, 197

Pyrimidines, reactions with carbon-centered radicals, 138

Pyrylium salts, reactions with

- cyanide ion, 218

-carbon-centered radicals, 220

- N,N-dimethylaniline, 219

- indoles, 219

- organometallic compounds, 223

Quinazolinium salts, reactions with nucleophiles, 130, 131

NH-Quinolinium salts, reactions with

- arylamines, 127, 128

- carbon-centered radicals, 135, 140

Quinoxaline hydrochloride, reaction with N,N-dimethylaniline, 131

Rearrangements of

- von Richter, 84

- Smiles, 81

- Sommelet-Hauser, 82

SET Mechanism, 249, 251

Tele-substitutions, 10, 14, 117, 177

1,3,4-Thiadiazolium salts, reactions with

- acetone, 233

- alcohols, 233

- trialkylphosphites, 233

1,2,4-Triazines, reactions with

- Grignard reagents, 119, 121

- vicarious carbanions, 114-116

SUBJECT INDEX

1,2,4-Trinitrobenzene, reaction with carbon-centered radicals, 36, 37

1,3,5-Trinitrobenzene, reactions with

- carbon-centered radicals, 36

- dialkyl and trialkylphosphites, 66

- phenols, 52

- phenoxides, 34

- trimethylsilanes, 41

2,4,5-Trinitrodiphenyl, 36

2,4,6-Trinitrophenol, 52

Tropylium salts, 76

Vicarious nucleophilic substitutions, 4, 10, 12, 41, 47, 49, 59, 60, 78, 113-115

X-ray data for σ-adducts, 261-266

ISBN 0-12-174640-2